Y0-CBY-587

INTRODUCTION TO TRANSPORTATION ENGINEERING

INTRODUCTION TO TRANSPORTATION ENGINEERING

Everett C. Carter, Ph.D.

Professor of Civil Engineering and
Director, Transportation Studies Center
University of Maryland
College Park, Maryland

Wolfgang S. Homburger

Research Engineer and Lecturer
Institute of Transportation Studies
University of California
Berkeley, California

Institute of Transportation Engineers

RESTON PUBLISHING COMPANY, INC.
Reston, Virginia

A Prentice-Hall Company

Library of Congress Cataloging in Publication Data

Carter, Everett C
Introduction to transportation engineering.

Bibliography: p. 251
Includes index.
1. Transportation. I. Homburger, Wolfgang S., joint author. II. Title.
TA1145.C37 629.04 77-26913
ISBN 0-87909-388-9

A revision on *An Introduction to Highway Transportation Engineering,* 1968

10 9 8 7 6 5 4 3 2

Printed in the United States of America.

Contents

FOREWORD ix

ACKNOWLEDGMENTS xi

CHAPTER 1
THE CHALLENGE 3

CHAPTER 2
TRANSPORTATION CHARACTERISTICS 7

Broad Comparison of Transportation Systems, 7
National Transportation Characteristics, 9
National Highway System Characteristics, 10
Urban Transportation Characteristics, 11
References, 19

CHAPTER 3
VEHICULAR AND HUMAN CHARACTERISTICS 25

Motor Vehicle Characteristics, 25
Driver Characteristics, 31
Pedestrian Characteristics, 34
Bicycling Characteristics, 37
Traffic Accidents, 39
References, 43

CHAPTER 4
TRAFFIC STREAM CHARACTERISTICS 45

Speed and Travel Time, 45
Volumes, Headways and Gaps, 47
Density and Spacing, 52
Relationship Between Flow, Speed and Density, 52
Capacity and Level of Service Analysis, 54
Queueing Analysis, 60
References, 61

CHAPTER 5
TRAFFIC MEASUREMENTS 63

Principles of Traffic Measurements, 64
Types of Traffic Measurements, 65
Traffic Data Systems, 65
Inventories, 66
Traffic Observation Studies, 67
Interview Studies, 74
Accident Record Studies, 76
Statistical Studies, 76
Experimental Studies, 78
References, 78

CHAPTER 6
GEOMETRIC DESIGN OF STREETS AND HIGHWAYS 81

Design Controls and Criteria, 82
Elements of Design, 85
Cross Section Design, 98
Other Elements, 102
Lighting, 103
Summary, 104
References, 104

CHAPTER 7
GEOMETRIC DESIGN OF INTERSECTIONS AND INTERCHANGES 107

Intersectional Conflicts, 107
Types of Intersections, 108
Principles of Intersection Design, 108
Driveways, 110
Interchanges, 110
Intersectional Design Elements, 115
Tapers, 117
Pedestrians, 117
Channelization, 118
References, 118

CHAPTER 8
GEOMETRIC DESIGN OF PARKING AND LOADING FACILITIES 121

Parking Design for Automobiles, 122
Parking Garages, 125
Bus Terminals, 132
Truck Terminals, 132
References, 134

CHAPTER 9
REGULATION OF VEHICLES AND DRIVERS 137

Driver Controls, 139
Vehicle Controls, 144
References, 145

CHAPTER 10
TRAFFIC CONTROL DEVICES 147

Rules of the Road, 148
Traffic Signs, 149
Traffic Sign Design, 149
Pavement Markings and Markers, 157
Other Traffic Control Devices, 159
Traffic Signals, 161
References, 164

CHAPTER 11
TRAFFIC CONTROL TECHNIQUES 167

Intersection Control, 167
Traffic Signal Control, 168
Signal System Operation, 174
Control Systems, 175
Nonintersection Controls, 179
Transportation Systems Management, 183
References, 183

CHAPTER 12
MASS TRANSIT SYSTEMS 185

Transit Networks, 185
Local Transit Systems, 188
Semimetros, 193
Rapid Transit Systems, 193
Urban Railroads, 196
Paratransit, 196
Transit Systems in Major Activity Centers, 198
Operating Parameters, 198
Management Aspects Affecting System Performance, 203
References, 206

CHAPTER 13
URBAN TRANSPORTATION PLANNING 209

Organization for Urban Transportation Planning, 210
Goal and Policy Formation, 210
Inventory, 212
Travel Demand Model Formulation, 213
Land Use Allocation Models, 215
Economic, Employment, and Population Forecasts, 216
Formulation of Feasible Alternative Land Use and Transportation Plans, 216
Evaluation, 220
Program Adoption and Implementation, 221
Continuing Study, 221
References, 221

CHAPTER 14
SYSTEMS PLANNING ELEMENTS 223

Systems Planning Elements, 224
Classification, 226
Inventory and Needs Studies, 227
Forecasting, 229
Economic Analyses, 230
Evaluation, 231
Programming/Implementation, 232
References, 233

CHAPTER 15
TRANSPORTATION ENGINEERING MANAGEMENT AND ADMINISTRATION 237

Transportation Engineering Activities, 237
Traffic and Transportation Engineering Organizations, 238
Staffing, 240
Parking Administration and Finance, 244
Transportation Department Administration, 247
References, 250

APPENDIX
SELECTED REFERENCES IN: Air Transportation, Railroad Transportation, Water Transportation, Pipeline Transportation, Pedestrian Transportation 251

INDEX 253

Foreword

The Institute of Transportation Engineers is privileged to make this text available to those interested in an overview of the skills employed by engineers concerned with the planning, design and operation of transportation facilities.

The book reflects the challenge and excitement of their work in this significant area. Transportation is a major sector of our economy and adequate mobility is an ever-increasing necessity in our society. For these reasons, there is a continuing need for the infusion of capable and dedicated engineers into this realm of public service.

In 1968, the Institute published *An Introduction to Highway Transportation Engineering*. Recognizing that transit now plays a more essential role in the total transportation system, the Institute chose to revise that text and include increased emphasis on mass transportation. The new text was also undertaken to incorporate improvements in the techniques and practices of transportation engineering during the past decade. The Institute commissioned Dr. Everett C. Carter, Professor of Civil Engineering at the University of Maryland, and Wolfgang S. Homburger, Research Engineer and Lecturer at the University of California at Berkeley, to update the previous text to meet the above goals.

This text serves to provide a broad overview of transportation engineering. Additional information may be found in its companion publications: the *Transportation and Traffic Engineering Handbook* and the *Manual of Traffic Engineering Studies*.

William Marconi, President
Institute of Transportation Engineers
1978

Stadium-Armory
Stadium-Armory
Stadium-Armory

Acknowledgments

The authors wish to thank the many individuals who deserve credit for assisting in the preparation of this text. Three members of the Institute's Technical Council reviewed and commented on the first drafts. They were John J. DeShazo, Jr.; K.B. Johns; and Melvin B. Meyer. Thomas E. Mulinazzi reviewed selected chapters. Elizabeth R. Carter compiled part of the list of selected references. Jonathan Upchurch, the Institute's Director of Technical Affairs, provided able assistance in coordinating the efforts of the authors, the reviewers, and the publisher.

The contributing authors of the 1968 text, entitled *An Introduction to Highway Transportation Engineering*, from which this book was adapted were Donald G. Capelle; Samuel Cass; Donald E. Cleveland; Allen R. Cook; William E. Corgill; Donald R. Drew; Richard E. Futrell; Howard C. Hanna; Matthew J. Huber; Alan F. Huggins; Louis E. Keefer; Charles J. Keese; Harry Lampe; Burton W. Marsh; Woodrow W. Rankin; Edmund R. Ricker; and Gilbert T. Satterly, Jr. The editors of that text were Donald G. Capelle; Donald E. Cleveland; and Woodrow W. Rankin.

Everett C. Carter
Wolfgang S. Homburger

INTRODUCTION TO TRANSPORTATION ENGINEERING

BART
ba

1

THE CHALLENGE

Transportation has played a vital role in every aspect of ancient and modern civilization. The growth and decline of nations in history has been related to their ability to move on and protect their trade and military routes, harbors, and navigable rivers. The need to link the activities taking place in separate locations and to convey persons and goods over these links has increased as society has become more complex. Success in meeting this need has been a major contributor to increased standards of living around the world.

In the United States almost one-fifth of the gross national product represents transportation transactions. One of every eight dollars of personal expenditures is for transportation. More than half the population is licensed to drive, and over four-fifths of all households own one or more motor vehicles. Annual passenger travel in the U.S. averages almost 12,000 miles (19,000 km) for each inhabitant, and cargo shipments accumulate a similar distance per ton per capita annually.

The engineering profession is involved in all aspects of transportation. Civil engineers are responsible for the development, structural design, construction, and maintenance of the fixed facilities of transportation systems. Automotive, mechanical, and aeronautical engineers, as well as naval architects, design vehicles. Electrical and electronics engineers develop power, communications, and control systems. Human-factor specialists in industrial engineering study the complexities of human performance as vehicle operators and passengers. The transportation engineer operates to some extent in all these areas, while developing special skills in transportation planning, systems operations, and geometric design of facilities.

Today's engineering student finds it increasingly necessary to concentrate on only one of the many specialties in this profession. This book introduces the career of transportation engineering to acquaint the student with the fundamentals of this specialty and to provide a broader understanding of the role of transportation engineers in the operation of our nation's highways and transit systems.*

A broad approach to transportation is essential if this public service is to provide for the safe, efficient, and convenient movement of people and goods. Network considerations, terminal requirements, the relation of each system to other modes of transportation and to abutting land uses and environmental impacts must be considered by engineers in the planning, design, operation, and administration of the system.

These are responsibilities of the transportation engineer, although the title, in practice, may be traffic engineer, planning engineer, transportation director, or some other similar specialty.

Transporation engineers are total-system oriented. They recognize the need for balanced public expenditures for the various modes of public transportation, and they have a working knowledge of the many ways in which transportation serves society while influencing commmunity development.

In order to perform adequately, the engineer must have a well-developed concern for the public interest and a thorough knowledge of political processes. The engineer in transportation has the challenge of direct, meaningful public service.

The increasing complexities and rapid changes in the field of transportation require personnel qualified in the most modern analytic skills and technical subjects. For this reason and because of the many individual subject areas available for concentrated study, the engineering student desirous of pursuing a career in this field has unique opportunities for professional advancement with numerous governmental agencies, private organizations, and other employers. This introduction to transportation engineering will provide the incentive for some engineering students to give serious consideration to the opportunities and challenges offered by this field.

*No attempt is made, however, to describe transportation engineering in air transport, railroads, water transport, or pipelines. A selected list of references dealing with these modes is given in the appendix.

DYNAMIC
305
LAKELAND
RYDER
RENT
ONE
WAY
4296
CERAMI PONTIAC
North Jersey's Price Pacer
755 RT 17 PARAMUS 444-9200

2

TRANSPORTATION CHARACTERISTICS

Transportation is a service devised to serve society by linking the myriads of locations where activities take place. "No man is an island, entire unto itself; every man is a piece of the continent . . . ," wrote John Donne. Nor are the locations where civilization has found a use for land self-sufficient and insular, but part of an economy encompassing a region, a nation, and often much of the world.

The transportation mission is accomplished by the provision of links, means of moving persons and goods on these links, and terminals at which travel or shipment commences and ends. The links may be physical, as roads or railroad tracks, or navigational, as sea and air lanes. The means of movement may be discrete vehicles or continuous belts. The terminals range from sophisticated major complexes to nothing more than a space to load a cargo or pick up a passenger.

A variety of transportation modes have been used. Some have roots in antiquity; others are the by-products of the space age. Many are in competition with each other, and some decline and disappear as advancing technology replaces them with more efficient alternatives.

The world's major transportation activities take place on five major systems, each divided into two or more subsystems. In very general terms their principal missions are listed in Table 2-1. Although this book is limited to dealing with only the highway system and the rail transit portion of railroads, it is important to place these in the total context of a nation's transportation alternatives.

BROAD COMPARISON OF TRANSPORTATION SYSTEMS

A transportation system may be analyzed in terms of its *ubiquity, mobility,* and *efficiency.* The concept of ubiq-

TABLE 2-1
PRINCIPAL MISSIONS OF MAJOR TRANSPORTATION SYSTEMS

MAJOR SYSTEM	MODE	PASSENGER SERVICE	FREIGHT SERVICE
Highways	Trucks	None	Intercity and local; all commodities; generally small shipments; containers
	Buses	Intercity and local	Packages on intercity service
	Automobiles	Intercity and local	Personal items only
	Bicycles	Local; recreational	None
Rail transport	Railroads	Intercity-mostly <300 miles (500 km); commuters	Intercity; generally bulk and oversize shipments; containers
	Rail transit	Regional and intracity	None
Air transport	Air carriers	Intercity-mostly >300 miles (500 km); transocean	Shipments of high-value freight on long hauls only; containers
	General aviation	Intercity; recreational; business	Minor
Water transport	Transocean	Cruise traffic only	Bulk cargos, containers
	Coastal and inland	Ferry service	Mostly bulk cargos on ships and barges
Continuous flow transport	Pipelines	None	Oil and natural gas; long and short hauls
	Belts	Escalators and horizontal belts for short distances	Materials handling—mostly < 10 miles (15 km).
	Cables	Lifts and tows for short distances in rough terrain	Materials handling in rough terrain

uity includes the amount of accessibility to the system, directness of routing between these access points, and the capability to handle a variety of traffic.

Mobility can be defined by the quantity of traffic that can be handled, hence by the system's capacity and by the overall speed with which this traffic moves. Specific and indirect costs, environmental and energy impacts, reliability, and safety can be considered major indicators of efficiency.

The highway system is the most ubiquitous of the major transportation modes, thanks largely to the historic concept that every landowner should have direct access to a road. Thanks to the many routes available in the network, overall speeds are fairly good, but limited by the maximum speeds that human factors permit. Individual vehicles are relatively small, thus reducing capacity, especially in freight haulage. In many of the categories grouped under efficiency, the highway system does not score as well as some alternatives; however, the intensive use of the system testifies to the fact that users value the ubiquity attributes above the inefficiencies which they may face.

Rail transport is limited in ubiquity by the large investment that is required for each link and terminal, and the difficulty in finding capital funds for this purpose. Technologically, routes are less flexible than highways, being limited both in grade and curvature. However, railroads have penetrated some difficult terrain where construction and maintenance of roads have so far proved impracticable. Capacity can be quite high, thanks to train formation, and speeds twice those possible on highways are feasible. Many efficiency factors are high, but direct costs suffer from the labor-intensive nature of railroad operation.

The large cost of mitigating or solving environmental problems associated with airports reduces the number of access points to the air transport system. However, between any two of these direct routing is usually possible. Speeds are by far the highest of any transportation system, but capacity is relatively low. The high speed more than compensates for many inherent inefficiencies, especially those of higher costs.

The need for safe natural or artificial anchorages limits the accessibility to water transport, and ubiquity is further reduced by the geography of land masses and availability of navigable rivers. Canal construction can add to the water transport network, but only at high cost. Although the speed of ships is low, the carrying capacity is greater than that of any other vehicle, resulting in very high efficiency, especially for shipment of bulk cargos.

Pipeline ubiquity is similar to that of rail transport, in that each link requires a fixed investment. Unit costs per link are less than for highways and railroads, however, because land acquisition can be minimized by use of easements, and the laying of pipelines is a much simpler effort than road or railroad construction. Although essentially different in technology, belt systems have similar attributes of network extent, speed and capacity as pipelines.

NATIONAL TRANSPORTATION CHARACTERISTICS

All major transportation technologies play important roles in the United States (Table 2-2). Highways have the lion's share of domestic passenger transportation. However, air transportation performs a major function in carrying travelers over the longer distances. In domestic freight transportation the picture is quite different; the railroads carry one-third of all traffic (measured in weight-distance units), with highways, ships, and pipelines handling most of the remainder in roughly equal amounts. In international, transocean traffic, cargo moves almost entirely by ship and passengers by air.

The long-range trends in traffic carried by the various transportation systems have been generally upward. In the decade from 1964 to 1974 the largest rate of growth—over 10 percent annually—has been experienced by air transportation in both passenger and freight traffic. Pipeline shipments have increased at almost the same rate, while railroad and waterborne freight have grown at a rate between two and three percent per year. By contrast, passenger traffic on urban transit systems fell about two percent annually during the same decade. In long-distance rail passenger travel, there was a marked change after the creation of the National Passenger Railroad Corporation (AMTRAK): From 1964 to 1972 traffic declined at an average annual rate of about 10 percent, but each of the two succeeding years witnessed an increase of about 15 percent.

For the modes using the nation's highways, selected financial, inventory, and performance indicators are listed

TABLE 2-2

U.S. DOMESTIC TRANSPORTATION SYSTEM STATISTICS—1974

MODE	EXPENDITURES AND REVENUES % OF U.S. TOTAL	PASSENGER-MILES (OR km) % OF U.S. TOTAL	FREIGHT TON-MILES (OR m ton-km) % OF U.S. TOTAL	EXTENT OF NETWORK	
				miles X 10^3	km X 10^3
Highways	**85.5**	**93.0**	**20.2**	3816.0	6140.0
Passenger car[a]	41.9	89.7	...	...	...
Truck	41.9	...	20.2	...	...
Intercity bus	0.4	1.1	NA[b]	270.0	435.0
Local transit bus	0.5	0.8[c]	...	51.5[c]	82.9[c]
School bus	0.7	1.4	...	NA[b]	NA[b]
Rail	**6.8**	**0.9**	**35.1**	...	...
Intercity	6.6	0.4	35.1	200.0	322.0
Urban transit[d]	0.2	0.5[c]	...	0.7[c]	1.1[c]
Air	**6.3**	**5.9**	**0.2**	307.8[e]	495.2[e]
General aviation	1.6	0.4	...	...	...
Air carrier	4.6	5.5	0.2	...	...
Water[f]	**0.7**	**0.2**	**23.9**	25.5[g]	41.1[g]
Pipeline	**0.7**	...	**20.6**	1200.0	1930.0

[a] Includes motorcycles and taxis.
[b] NA-Data not available.
[c] Estimated from: U.S. Department of Transportation, *1974 National Transportation Survey. Urban Data Supplement.* Washington, D. C.: U.S. Government Printing Office, May 1976. Tables SC-5, SC-6, SD-23
[d] Includes streetcars.
[e] Airways.
[f] Excludes commercial fishing.
[g] Inland and coastal channels.

SOURCE: U.S. Department of Transportation. *Summary of National Transportation Statistics.* Washington, D. C.: U.S. Government Printing Office, June 1976 (except items marked[c]).

in Table 2-3. Most of the trends here are also upward. Because of inflation, the financial trend lines have risen more rapidly than those describing performance.

With global concern about energy resources increasing it is important to be aware of the extent of energy consumption by transportation systems. Almost all modes utilize petroleum derivatives, accounting for just over half of all these products used in the U.S. Relatively small proportions of the nation's use of natural gas, electricity, and coal are attributable to transportation. The generalized flow of energy in the U.S. is graphically shown in Fig. 2-1.

It is difficult to develop comparative energy efficiency measures for different transportation systems. First it must be realized that the construction of a transportation artery or terminal, including the manufacture and mining of the prerequisite materials, often consumes more energy than will be used for many years after the facility is in operation. Then the problem of relating energy use to transportation output, and accounting for energy losses in refining, generating, and transmission must be dealt with. Generalizations are, therefore, fraught with danger. The data in Table 2-4 give a very general indication based on calculations that are replete with assumptions about typical load factors, speeds, and trip lengths.

NATIONAL HIGHWAY SYSTEM CHARACTERISTICS

Further details about the use of motor vehicles in the United States are shown in Table 2-5. (It should be noted that this table reports the movement of vehicles rather than passenger and freight quantities, which were cited in Table 2-3.) Well over half of all travel takes place in urban areas on only one-sixth of the highway network. A more detailed breakdown, showing also average traffic volumes on the various rural and urban road systems, is shown in Fig. 4-6 in Chapter 4. The automobile accounts for three-quarters of all rural travel and for five-sixths of all urban travel, but each vehicle is used for less than 10,000 miles (16,000 km) annually. By contrast, truck-trailer combinations and commercial buses—intercity and urban transit—

TABLE 2-3
TRENDS IN SELECTED U.S. HIGHWAY STATISTICS

	ITEM	UNITS	1964	1974	AVG. ANNUAL CHANGE
FINANCIAL	Government expenditures on highway systems	$\$ \times 10^6$	12,985	24,506	+6.6%
	Private expenditures on automobiles	$\$ \times 10^6$	49,713	110,246	+8.3%
	Truck industry revenues	$\$ \times 10^6$	43,856	108,659	+9.5%
	Intercity bus revenues	$\$ \times 10^6$	595	1,062	+6.0%
	Urban bus revenues	$\$ \times 10^6$	1,010	1,377	+3.1%
INVENTORY	No. of automobiles	thousands	71,670	104,270	+3.8%
	No. of trucks	thousands	14,019	24,598	+5.8%
	No. of intercity buses	thousands	21	21	0.0
	No. of urban buses	thousands	49	49	−0.1%
	No. of school buses	thousands	222	354	+4.8%
	Rural highways	mi $\times 10^3$	3,153	3,178	+0.1%
		km $\times 10^3$	5,072	5,114	
	Municipal highways	mi $\times 10^3$	491	638	+2.6%
		km $\times 10^3$	791	1,026	
PERFORMANCE	Passenger travel-auto	*a*	1,506	2,190	+3.8%
		b	2,423	3,524	
	Passenger travel-intercity buses	*a*	23	28	+1.7%
		b	37	44	
	Passengers-urban buses	millions	4,729	3,998	−1.7%
	Freight traffic (intercity only)	*c*	347	495	+3.6%
		d	507	723	

[a] Billion passenger-miles
[b] Billion passenger-km
[c] Billion ton-miles
[d] Billion metric ton-km

SOURCE: U.S. Department of Transportation, *Summary of National Transportation Statistics*. Washington, D.C.: U.S. Government Printing Office, June 1976.

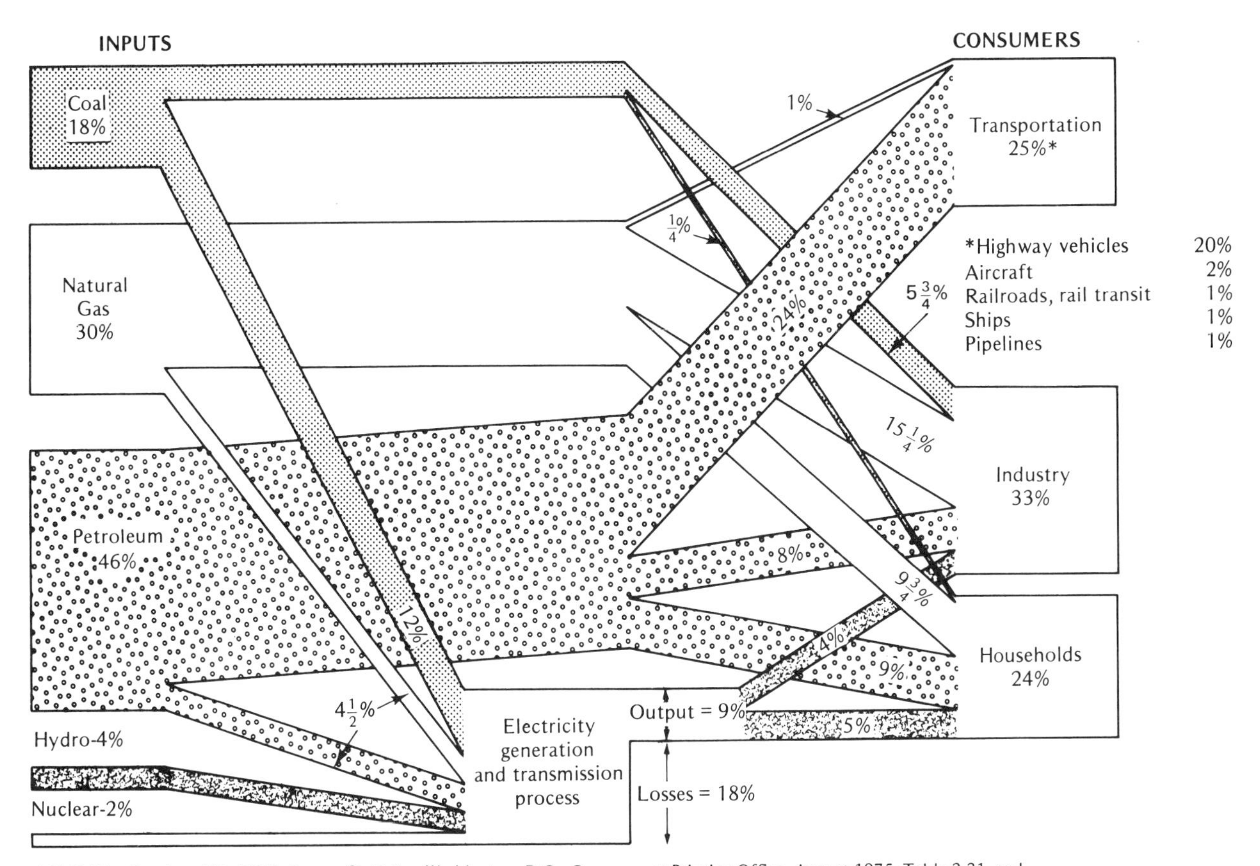

SOURCE: Based on U.S. DOT, *Energy Statistics*, Washington, D.C.: Government Printing Office, August 1975, Table 3-31, and U.S. DOT, *Summary of National Transportation Statistics*, Washington, D.C.: Government Printing Office, June 1976, Table 16, depicting data for 1974.

FIGURE 2-1
GROSS ENERGY CONSUMPTION PATTERN IN THE UNITED STATES

are utilized much more extensively. This pattern has remained fairly constant for more than twenty years.

The complex problems arising in urban areas because of the high density of land use and high concentration of traffic call for the special attention of the transportation engineer. In the remainder of this chapter, therefore, urban travel characteristics are explored in some detail.

URBAN TRANSPORTATION CHARACTERISTICS

In recent years a great amount of information about urban transportation characteristics has become available. In the United States, federal policy requires continuing urban transportation planning in urbanized areas of over 50,000 population. The types of studies conducted in such planning are summarized in Chapter 5, and the methods of data analysis in Chapter 13.

Additionally, the U.S. Department of Transportation conducted a national transportation study in 1974, in which special data were collected for 1972 with forward projections to 1980 and 1990. These data included capital and operating costs, physical facilities and equipment, demand, performance indicators, and social and environmental impacts for the highway and transit systems. From this study some basic urban travel characteristics have been summarized (Ref. 6).

Number of Urban Trips

Trip making is a manifestation of the types and amounts of activity taking place on urban land, as well as a reflection of the costs involved in making the trips. As shown in

TABLE 2-4
ENERGY EFFICIENCY OF TRANSPORT MODES

TRANSPORT MODE	ENERGY INPUTS	
Freight Transport	BTU/ton-mile	Joules/kg-km
Pipeline	450	330
Railroad	670	480
Waterway	680	490
Truck	2,800	2,020
Aircraft	42,000	30,400
Intercity Passenger Transport	BTU/psg-mile	10^3 Joules/psg-km
Bus	1,600	1,050
Railroad	2,900	1,900
Automobile	3,400	2,230
Aircraft	8,400	5,500
Urban Passenger Transport	BTU/psg-mile	10^3 Joules/psg-km
Bicycle	200	130
Walking	300	200
Mass transit	3,800	2,500
Automobile	8,100	5,300

SOURCE: Eric Hirst, *Energy Intensiveness of Passenger and Freight Transport Modes: 1950–1970.* Oak Ridge Tenn.: Oak Ridge National Laboratory. April 1973, Tables 11 and 12. [Bicycle and walking data from: Eric Hirst, "Energy-Intensiveness of Transportation," *Transportation Engineering Journal (ASCE)*, Vol. 99, no. TE 1 (Feb. 1973).]

TABLE 2-5
MOTOR VEHICLE USE IN THE UNITED STATES—1975

TYPE OF VEHICLE	RURAL TRAVEL		URBAN TRAVEL		AVERAGE ANNUAL TRAVEL PER VEHICLE	
	Veh-mi x 10^9	Veh-km x 10^9	Veh-mi x 10^9	Veh-km x 10^9	Miles	km
Automobiles[a]	440.9	709.6	609.6	981.0	9,406	15,138
Single-unit trucks	111.1	178.8	107.8	173.4	8,882	14,294
Truck combinations	45.7	73.9	9.9	15.9	49,125	79,059
Commercial buses	1.0	1.6	1.6	2.6	28,230	45,426
School buses	2.0	3.1	0.6	0.9	6,788	10,923
All vehicles	600.6	966.6	729.4	1173.9	9,644	15,521
	(45.2% of all travel)		(54.8% of all travel)			

[a]Includes taxis and motorcycles

SOURCE: U.S. Federal Highway Administration, *Highway Statistics 1975.* Washington, D.C.: U.S. Government Printing Office, Table VM-1, dated Jan. 1977.

Fig. 2-2, the total number of trips per capita is least in the largest metropolitan areas and greatest in the smaller urban centers. One primary reason is that travel in large cities tends to be more expensive both in travel time and cash costs, and this has an adverse effect on trip making. Hence, the range of daily trips per person is from 2.5 in regions exceeding two million population to more than four in small urban areas, with a national average of about three trips per person.

These trips are relatively short. One-half of all auto trips in the largest urban regions are less than about 6.4 miles (10 km), and three-fourths are less than 7.2 miles (11.6 km). As might be expected, trip lengths decrease in smaller regions. Bus transit trips are even shorter than automobile trips, but rail transit and commuter rail trips are longer (Fig. 2-3.) These data indicate that the need for very high speeds and for systems on which travelers are transported long distances are not necessary to serve intra-urban needs.

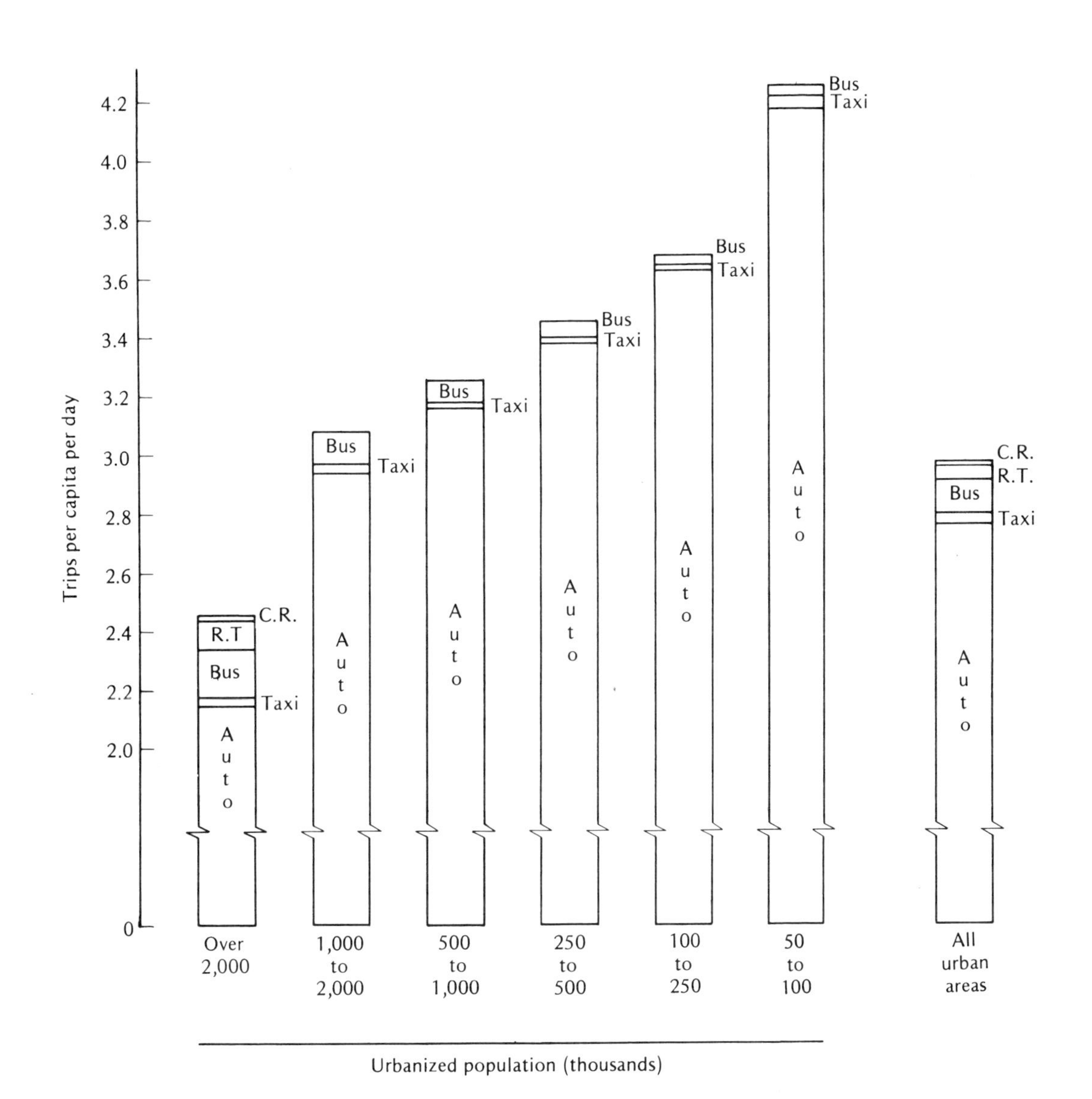

SOURCE: U.S. Department of Transportation, *1974 National Transportation Report, Urban Data Supplement.* Washington, D.C.: Government Printing Office, May 1976, Table SD-16.

FIGURE 2-2
AVERAGE DAILY PERSON TRIPS PER CAPITA BY MODE IN U.S. URBANIZED AREAS—1972

Urban Travel

The total motor vehicle travel occurring on urban highways, when expressed on a per capita basis, is relatively constant for all sizes of urbanized areas (Fig. 2-4). This is to be expected, since travel is the product of trips undertaken and their lengths; the first of these factors decreases with increased population per region, while the second factor increases. Roughly one-fifth of this travel takes place on interstate highways, somewhat more than one-

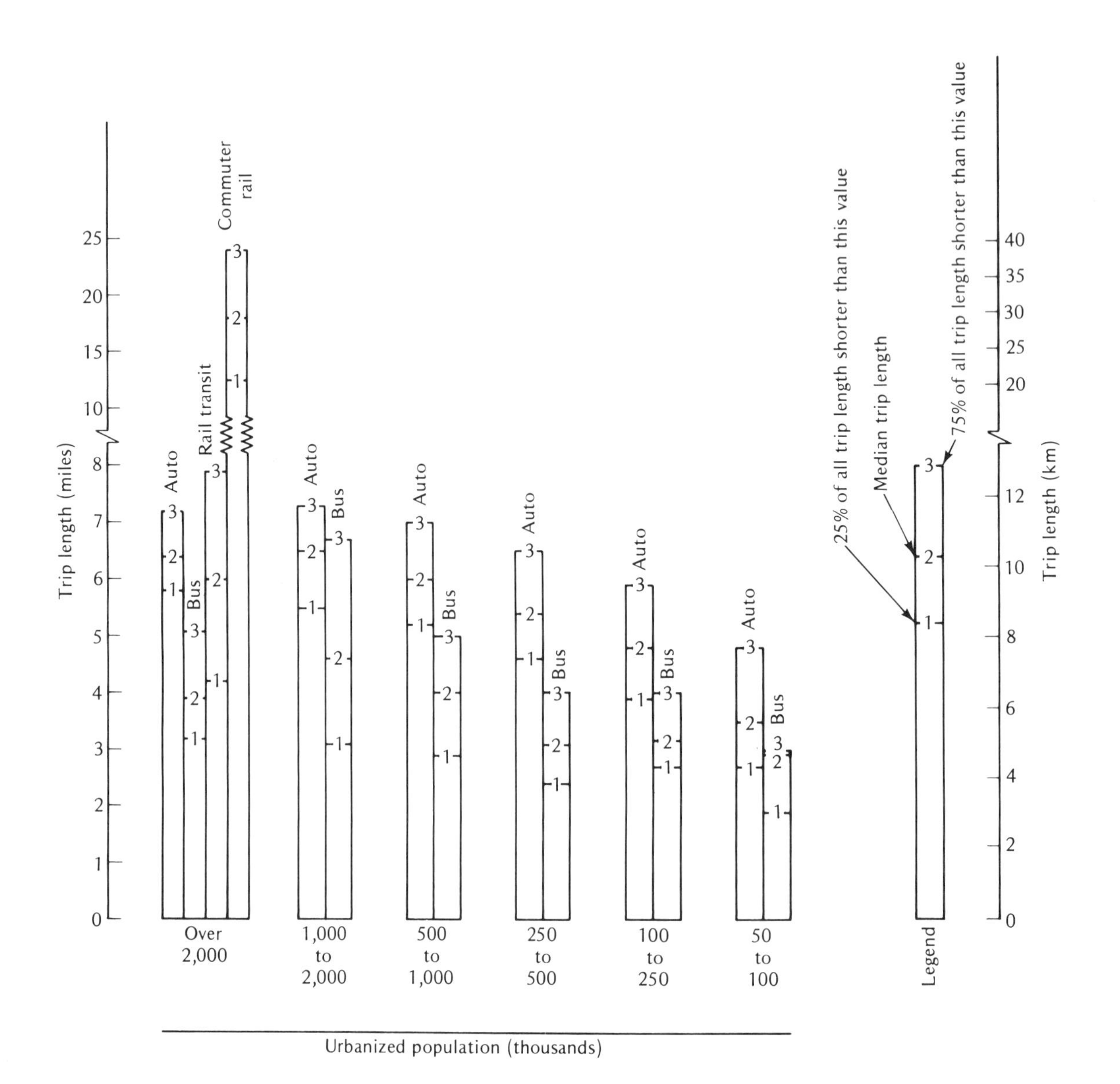

SOURCE: U.S. Department of Transportation, *1974 National Transportation Report, Urban Data Supplement.* Washington, D.C.: Government Printing Office, May 1976, Tables SE-3 and SE-8.

FIGURE 2-3

TRIP LENGTHS IN U.S. URBAN AREAS, BY TRAVEL MODE AND SIZE OF URBANIZED AREA—1972

half on other arterial streets, and the remainder on collector and local streets.

Figure 2-2 has already shown that transit systems carry fairly small proportions of all urban trips except in the largest metropolitan areas. These data are expressed in terms of travel units (passenger-miles or passenger-km) in Fig. 2-5. The very small number of commuter rail trips in areas exceeding two million population account for quite a substantial amount of travel because of the very large average trip length involved. Rail transit also makes a major contribution to the travel total in these large regions.

The various purposes for which urban trips are made

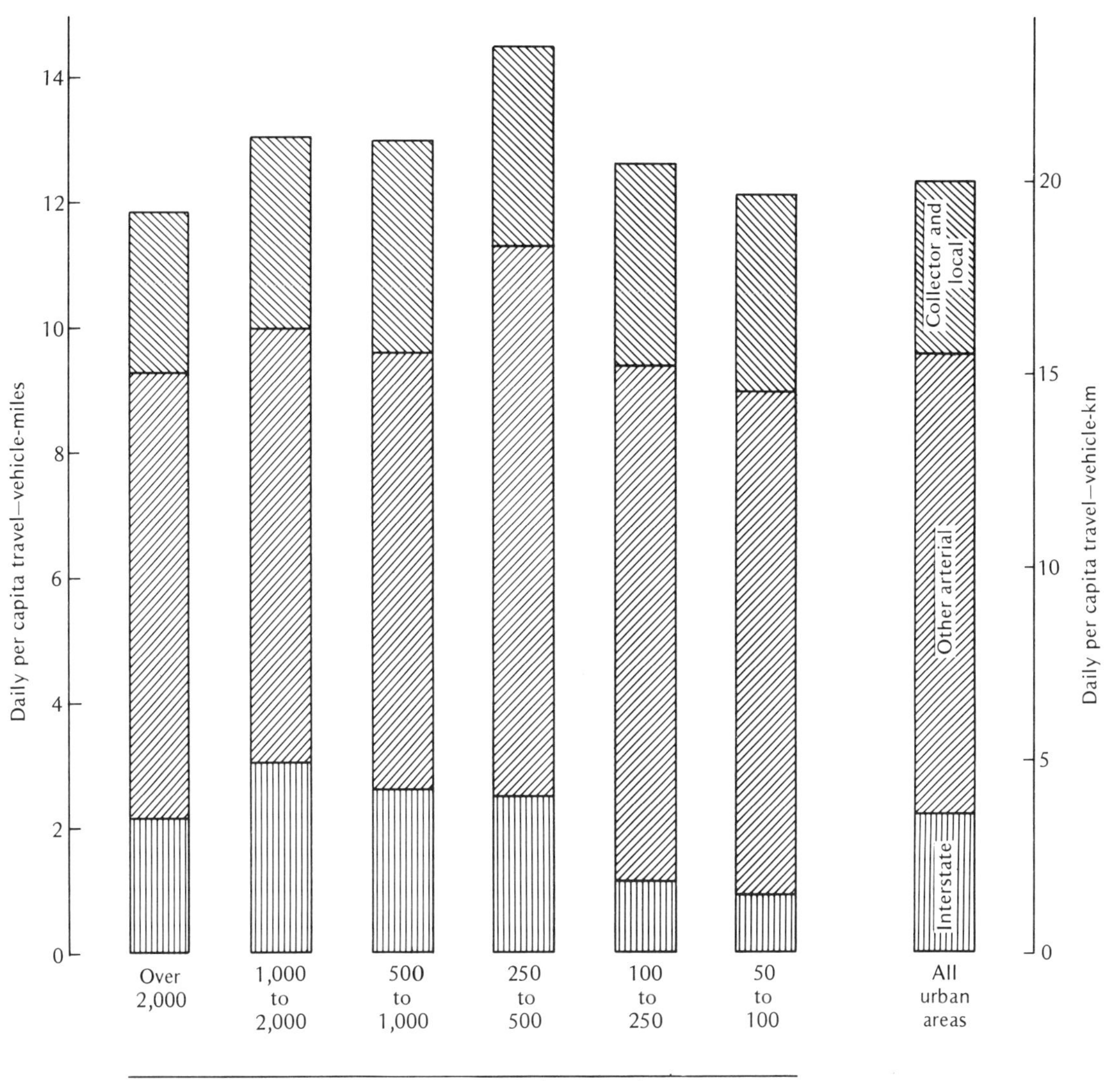

SOURCE: U.S. Department of Transportation, *1974 National Transportation Report, Urban Data Supplement.* Washington, D.C.: Government Printing Office, May 1976.

FIGURE 2-4
AVERAGE DAILY MOTOR VEHICLE TRAVEL PER CAPITA BY HIGHWAY SYSTEM IN U.S. URBANIZED AREAS—1972

are classified by mode in Table 2-6. In round numbers, 30 percent of these trips are made to earn a living, another 30 percent on family or personal business, 25 percent for social or recreational purposes, and 15 percent for educational, civic, or religious objectives. About 90 percent of these journeys occur in passenger cars or trucks, and 10 percent in public transit vehicles.

Peaking Patterns

A phenomenon in urban areas that gives rise to most of the transportation problems and defines most of the needs to be met is the commuting peak demand, commonly called the rush hours. Even though travel during these periods is much more time-consuming and uncomfortable

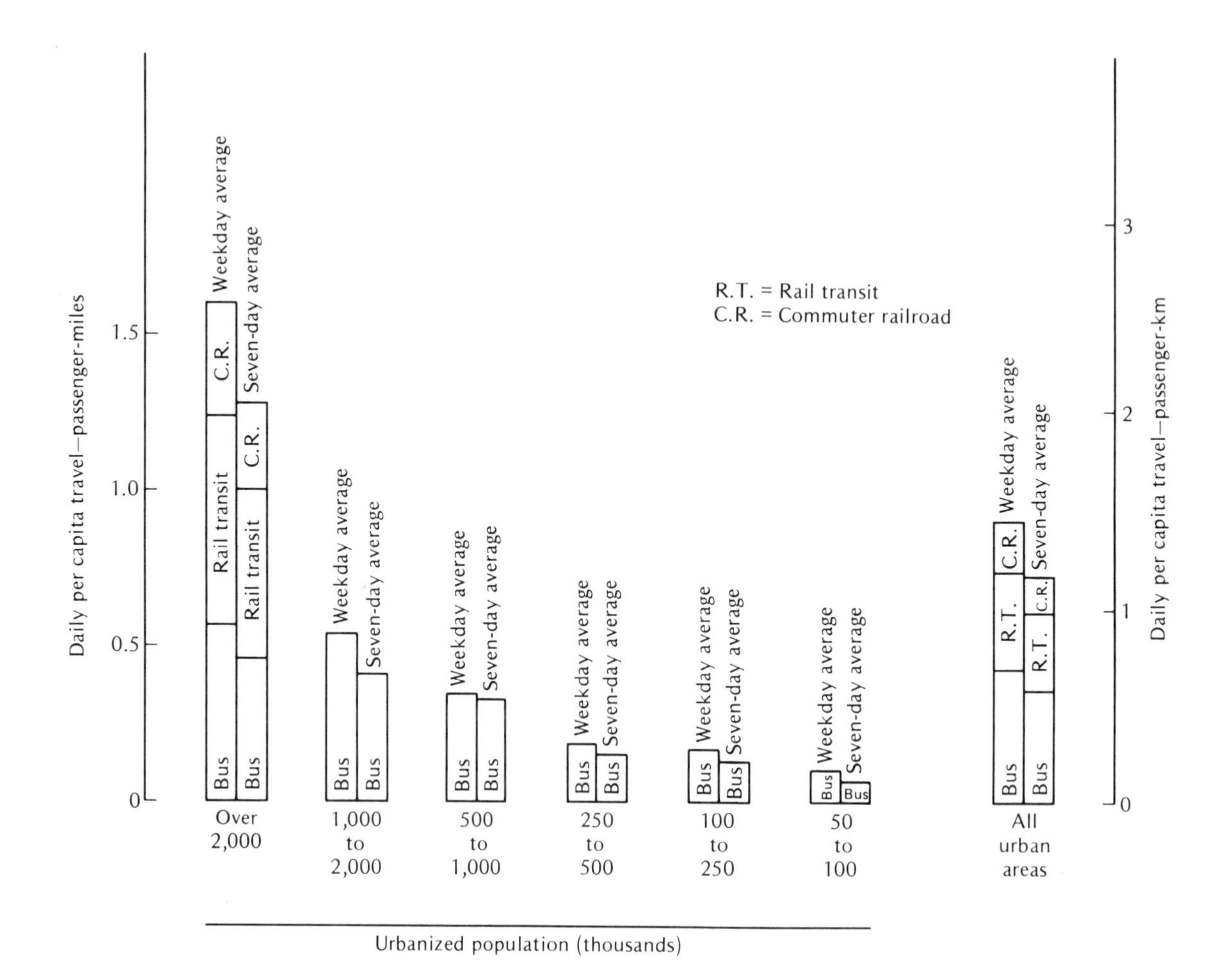

SOURCE: U.S. Department of Transportation, *1974 National Transportation Report, Urban Data Supplement.* Washington, D.C.: Government Printing Office, May 1976, Tables SD-22 and SD-24.

FIGURE 2-5
AVERAGE DAILY TRANSIT TRAVEL PER CAPITA IN U.S. URBANIZED AREAS—1972

than at other times of the day, the desire to commute during these peaks remains very strong. Small time shifts occur in response to especially overcrowded and congested situations as some employees participate in schemes of staggered work hours or flexible work schedules. However, the impact of such programs on total traffic volumes in the peaks is seldom noticeable.

In Fig. 2-6 this peaking condition is clearly illustrated. It will be noted that the two transit modes experience peaks which are even more pronounced than the peak for automobiles and trucks. Transit can compete with automobiles more favorably during commuting conditions than at other times, and therefore attracts a larger proportion of the total travel during these periods. Hourly variations in demand for highways are referred to again in Chapter 4: the consequences of this peak pattern to transit systems are discussed in Chapter 12.

The Central Business District

In each urbanized area there is at least one focal point where land use intensities are highest. This location is devoted to activities of major business concerns, retail shopping, tourist facilities, and specialized cultural or athletic

TABLE 2-6
DISTRIBUTION OF TRIPS BY PURPOSE AND MODE—1970 (IN U.S. STANDARD METROPOLITAN STATISTICAL AREAS)

TRIP PURPOSE	PERCENTAGE OF ALL TRIPS							
	Total for All Modes	Automobile		Truck	Rail Transit	Bus Transit	School Bus	Others[a]
		Driver	Psgr					
Earning a living								
To/from work	26.9	17.6	4.8	1.3	0.7	1.9	0.2	0.4
Related business	4.0	2.3	0.8	0.7	. . .	b	0.1	0.1
Total	**30.9**	**20.0**	**5.6**	**2.0**	**0.7**	**2.0**	**0.2**	**0.4**
Family Business								
Medical/dental	1.8	0.9	0.7	0.1	b	0.1	b	0.1
Shopping	15.9	8.9	6.1	0.3	0.1	0.4	b	0.1
Other	12.6	7.3	4.5	0.4	0.1	0.2	0.1	0.1
Total	**30.3**	**17.1**	**11.3**	**0.8**	**0.2**	**0.7**	**0.1**	**0.2**
Educational, civic, religious	**14.4**	**4.5**	**5.2**	**0.1**	**0.1**	**0.6**	**3.7**	**0.2**
Social/recreational	**23.4**	**10.8**	**11.3**	**0.5**	**0.1**	**0.5**	**0.2**	**0.1**
Other	**1.0**	**0.6**	**0.2**	**0.1**	b	b	b	b
Total for all purposes	**100.0**	53.0	33.6	3.5	1.0	3.8	4.2	0.9

Figures may not add due to rounding.
[a]"Other" includes motorcycle, taxi, railroad, and miscellaneous.
[b]Less than 0.05 per cent.

SOURCE: Computed from Reference (6), Tables IV-3 and IV-4.

centers, all of them depending on maximum accessibility to all parts of the urban region. This area is called the central business district (CBD).

Perhaps more than any other part of the region, the CBD poses major challenges to the transportation engineer. Much of the peak travel, mentioned above, has its origins and destinations here. The high density and high values of land use prevent transportation solutions that require much surface space, and solutions often involve the vertical as well as the horizontal dimension.

A generalized relationship between the number of trip destinations in a CBD and the size of the urbanized area is shown in Fig. 2-7. In the denser CBDs of large regions, more than 200,000 trips are destined for each square mile (80,000 to each square km) on a typical weekday. More than one-half of these arrive by transit in large CBDs, but the proportion on transit drops below this level in metropolitan areas of less than 2,000,000 population. The remaining trips are made by automobile, giving rise to the most intense parking problem facing transportation engineers. Some of its dimensions are described next.

Parking in Central Business Districts

Vehicle parking is an important element of the highway transportation system. On the basis of the average annual travel per automobile shown in Table 2-5, one can assume that the typical vehicle is used no more than 400 hours per year and is, therefore, parked about 95 percent of the time. It is particularly important to consider parking characteristics where parking demand is great and space availability is limited. This problem exists in the CBDs of most cities, and becomes more acute as city size increases. Figure 2-8 shows that the number of parking spaces in the CBD increases with the size of the urbanized area. When expressed in terms of spaces per capita, however, there is a decline with increasing size of urban areas.

The types of spaces available in the central area change as city size increases, as shown in Table 2-7. Curb parking spaces decline from 43 to 14 percent of the total as the intensity of land use in larger cities requires more parking space. The fraction of spaces in the more costly parking garages increases to almost one-third in centers with a population of over half a million.

The total number of parkers seeking these spaces also increases with city size (Fig. 2-9), although the proportion of the total population that drives to the CBD decreases because large regions can support outlying business and retail centers, which provide a competitive alternative to the CBD.

The length of time parked varies with city size, as shown in Table 2-8. Most parking durations are less than

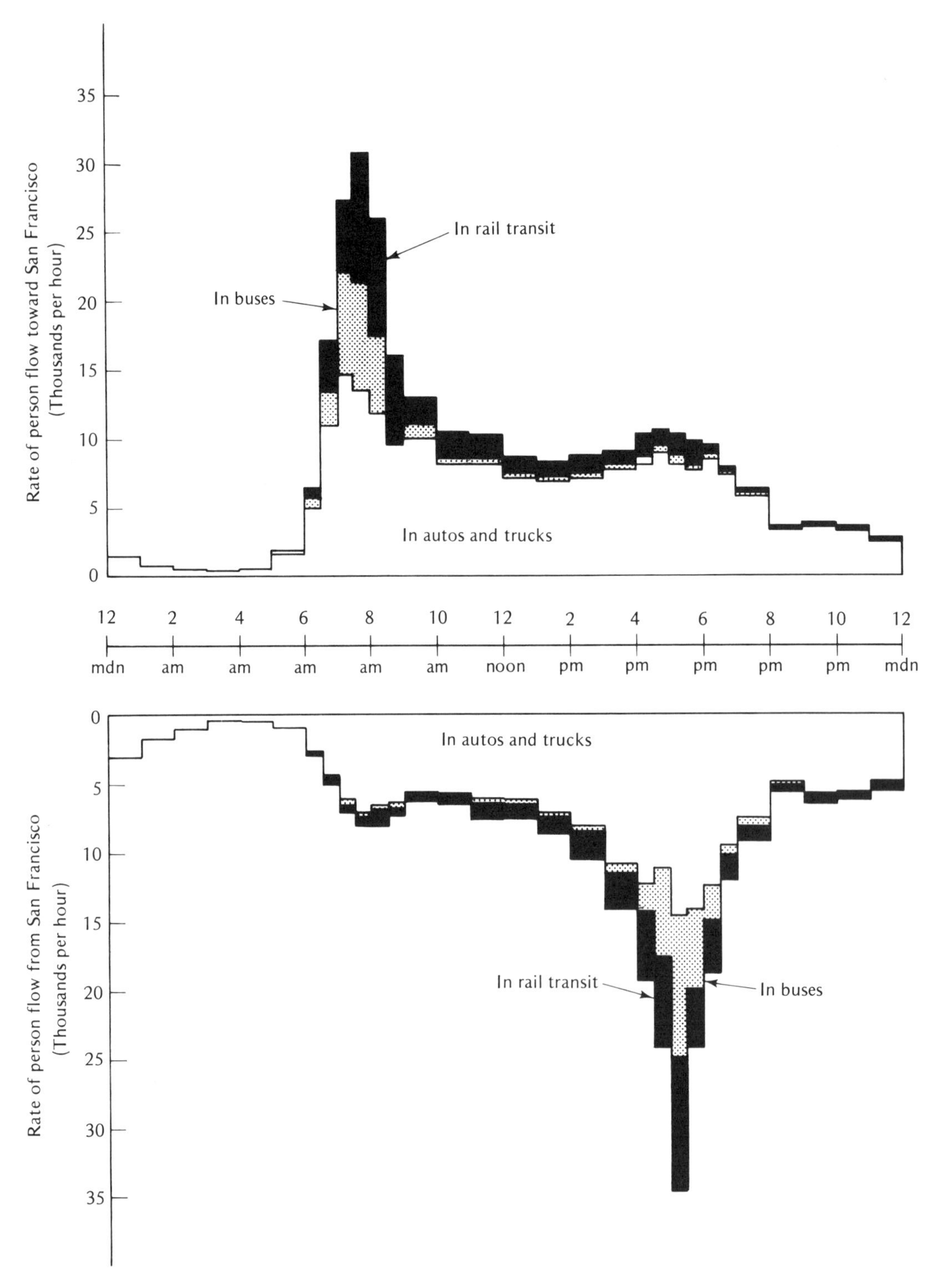

SOURCE: University of California Institute of Transportation Studies, *Traffic Survey Series A-47*. October 1976.

FIGURE 2-6

RATE OF PERSON FLOW ACROSS SAN FRANCISCO BAY SHOWING TYPICAL WEEKDAY PEAKING PATTERN

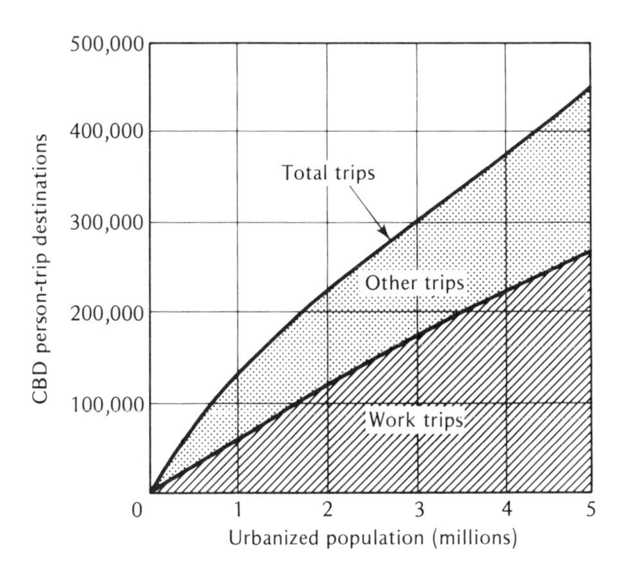

SOURCE: Wilbur Smith & Associates, *Urban Transportation Concepts*, Center City Transportation Project. September 1970, p. 21.

FIGURE 2-7
GENERALIZED CBD TRIP GENERATION

one hour in urban areas up to one-half-million population. This emphasizes the need for convenient parking facilities for short stops. Also notable is the high proportion of "errand-like" parking that requires less than 30 minutes per stop.

Table 2-9 shows walking distances from central area parking. As urban area size increases, so does the average walking distance. Since commuters traveling to work will park for the entire day, they are willing to walk the longest distances. Trips for shopping and for personal business involve relatively short parking durations, and parkers traveling for these purposes, therefore, seek spaces somewhat closer to their destinations.

Turnover, or the average numbers of vehicles using each space during the main parking demand period (usually 10 A.M. to 6 P.M.) is an important parking usage measurement. Table 2-10 shows that curb spaces, often subject to maximum parking time limits, have much higher turnover rates than do off-street spaces. Because parking durations increase with size of urban area, turnover rates are decreased in large cities.

Parking demand varies during the day. The maximum number of cars accumulated defines the maximum demand for spaces. This maximum accumulation occurs in most CBDs between 11 A.M. and 2 P.M. In the largest metropolitan areas slightly more than one-half of the total daily parkers are present at the time of maximum accumulation. This proportion drops with urban area size to 40 percent for regions of about one million population and 25 percent for those with 100,000 inhabitants.

REFERENCES

1. Baerwald, John E., *Transportation and Traffic Engineering Handbook.* Englewood Cliffs, N.J.: Prentice-Hall, 1976, Chapters 1, 4, 5.
2. Highway Research Board, *Parking Principles,* Special Report 125. Washington, D.C.: Transportation Research Board, 1971.
3. Motor Vehicle Manufacturers Association, *Motor Vehicle Facts and Figures.* Detroit, Mich.: Annual.
4. Smith (Wilbur) & Associates, *Parking in the City Center.* Detroit, Mich.: Automobile Manufacturers Association, May 1965.
5. United Nations, Department of Economic and Social Affairs, *Statistical Yearbook.* New York: Annual. (Contains tables of global statistics on railroads, motor vehicles, shipping, civil aviation, and tourist travel.)
6. U.S. Department of Transportation, *1974 National Transportation Report.* Washington, D.C.: U.S. Government Printing Office, July 1975. Also *Urban Data Supplement.* Washington, D.C.: U.S. Government Printing Office, May 1976.
7. ——————, *Summary of National Transportation Statistics.* Washington, D.C.: U.S. Government Printing Office, Annual. Also *Energy Statistics Supplement.* Washington, D.C.: U.S. Government Printing Office, annual.
8. U.S. Federal Highway Administration, *Highway Statistics.* Washington, D.C.: U.S. Government Printing Office, annual.

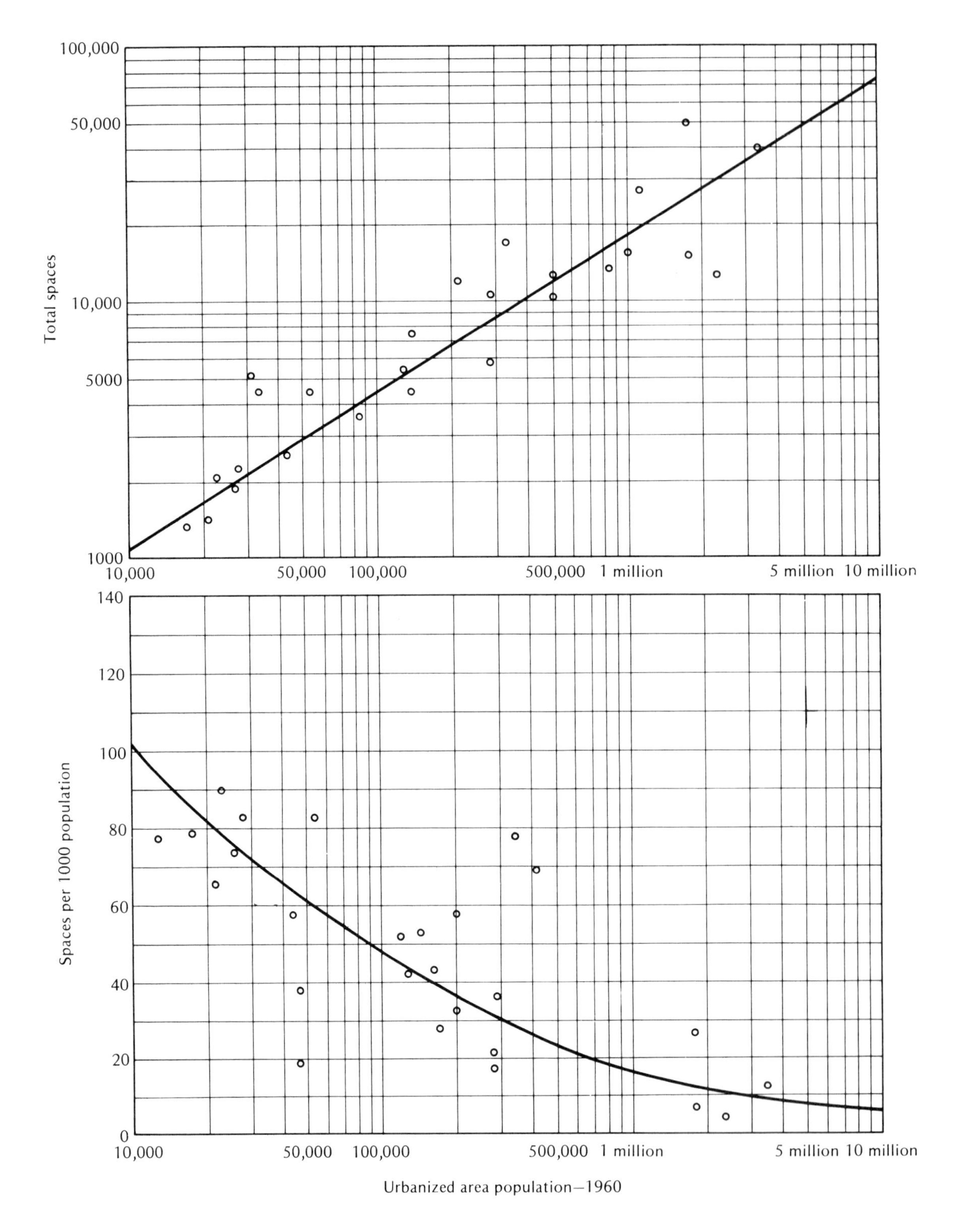

SOURCE: Wilbur Smith & Associates, *Parking in the City Center.* New Haven, Conn.: May 1965, p. 4.

FIGURE 2-8
CBD PARKING SPACES IN RELATION TO URBANIZED AREA POPULATION

TABLE 2-7
TYPE OF PARKING SPACES

URBANIZED AREA POPULATION (THOUSANDS)	PERCENTAGE OF SPACES At curb	In off-street lots	garages
10-25	43	57	0
25-50	38	59	3
50-100	35	60	5
100-250	27	62	11
250-500	20	64	16
500-1000	14	56	30
Over-1000	14	55	31

SOURCE: Reference (2), Table 2.3.

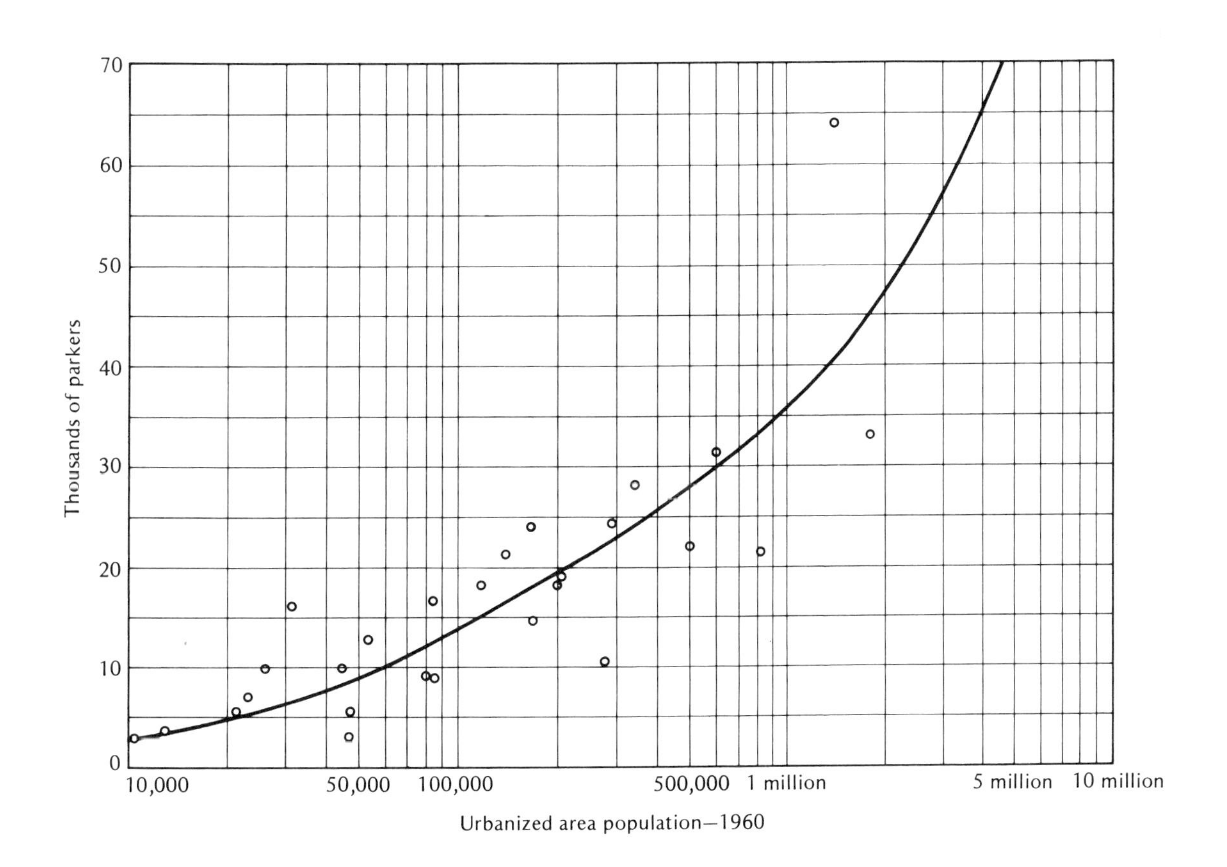

SOURCE: Wilbur Smith & Associates, *Parking in the City Center.* New Haven, Conn.: May 1965, p. 11.

FIGURE 2-9
DAILY CBD PARKERS IN RELATION TO URBANIZED AREA POPULATION

TABLE 2-8

PARKING DURATION

URBANIZED AREA POPULATION (THOUSANDS)	PERCENTAGE OF PARKERS STAYING LESS THAN			
	30 min.	1 hour	2 hours	5 hours
10-25	60	74	84	94
25-50	59	74	84	93
50-100	60	75	85	95
100-250	46	60	71	84
250-500	38	53	70	85
500-1000	24	36	49	67
Over-1000	16	28	48	60

SOURCE: Adapted from Reference (2), Table 2.10.

TABLE 2-9

AVERAGE WALKING DISTANCE BY TRIP PURPOSE

URBANIZED AREA POPULATION (THOUSANDS)	TRIP PURPOSE					
	Work		Shopping		Personal business	
	ft	m	ft	m	ft	m
10-25	270	90	200	65	200	65
25-50	400	130	280	90	240	80
50-100	410	135	350	115	290	95
100-250	500	165	470	155	390	130
250-500	670	220	570	185	450	150
500-1000	650	215	560	185	590	195

SOURCE: Reference (2), Table 2.11.

TABLE 2-10

EIGHT-HOUR (10 A.M.-6 P.M.) PARKING SPACE TURNOVER (PARKERS/SPACE)

URBANIZED AREA POPULATION (THOUSANDS)	CURB SPACES	OFF-STREET SPACES
10-25	6.7	1.8
25-50	6.4	1.5
50-100	6.1	1.6
100-250	5.7	1.5
250-500	5.2	1.4
500-1000	4.5	1.2
Over 1000	3.8	1.1

SOURCE: Reference (2), Table 2.14

3

VEHICULAR AND HUMAN CHARACTERISTICS

The transportation process may be considered to involve three principal elements: a system of links and terminals, fleets of vehicles (except in pedestrian, pipeline, or belt transportation), and human beings as operators, travelers, and shippers. The transportation infrastructure, its planning, design, and operation, are the subject of the remaining chapters of this book. In order to deal with those topics, however, some basic attributes of vehicles and human factors must be understood. In this chapter, information on vehicles is confined largely to highway motor vehicles, with some data on bicycles. Chapter 12 includes some attributes of mass transit vehicles. Human factors discussion here is confined to a few physical, physiological, and psychological aspects. The motivation for travel is included in the discussion of transportation planning in Chapter 13.

MOTOR VEHICLE CHARACTERISTICS

The vehicle is a significant element in many highway transportation engineering analyses. In this section some important vehicular characteristics are reviewed.

Dimensions and Weights

Motor vehicles in the traffic stream include passenger cars, buses, single-unit trucks, tractor trucks, and motorcycles, all varying widely in both size and weight. The design of these vehicles influences highway design both structurally and geometrically.

The semipermanency of highway geometry leads to a fixed limit on vehicle dimensions. Traffic lanes are limited to about 12 feet (3.65 m) in width, and the vertical clearance beneath structures is normally about 14.5 ft (4.40 m). It is likely that dimensions of motor vehicles will continue to be dictated by these highway dimensions.

The type and design of the pavement dictates permissible axle-loads or wheel-loads on the pavement. All states have maximum allowable loads. Current recommended practice on vehicle dimensions and weights is shown in Table 3-1.

Certain vehicle dimensions are of concern in specific design tasks. Figure 3-1 shows passenger car data required for layout of parking lots and garages. The turning radii shown are important in laying out aisles and ramps. When street intersections and channelization islands are designed, the turning ability of larger vehicles must be taken into account, each design being based on the largest vehicle expected to pass through the location on a regular basis. Further information about the geometry of vehicle paths is given in Chapter 6.

Acceleration and Deceleration

The accelerating and decelerating characteristics of motor vehicles are of particular concern to traffic engineers and highway designers. The rate at which drivers change their speed in normal circumstances is the basis for the design of transition lanes and tapers at expressway ramps, for location of warning signs, and for establishment of transition speed zones. The maximum acceleration capability comes into play when drivers execute overtaking maneuvers on two-lane roads and, therefore, defines the need for establishing passing prohibitions. Maximum deceleration rates and the minimum stopping distances related to them are used to analyze probable responses to emergency situations.

A basic law of motion states that the force required to move a vehicle is a function of the resistances to motion and of the rate of change of speed:

$$F = R + \frac{W \cdot a}{g} \qquad (3.1)$$

where

F = force required (in lb or kg)
R = resistances to motion (in lb or kg)
W = weight of the vehicle (in lb or kg)
a = Rate of change of speed (in ft/sec^2 or m/sec^2)
g = Acceleration due to gravity (32.2 ft/sec^2 or 9.81 m/sec^2)

Resistances to motion are usually classified as:

TABLE 3-1
RECOMMENDED MAXIMUM DIMENSIONS AND WEIGHTS OF MOTOR VEHICLES

Width	8.5 ft	(2.60 m)
Height	13.5 ft	(4.10 m)
Length		
Single-unit truck	40 ft	(12.2 m)
Bus	40 ft	(12.2 m)
Truck tractor and semitrailer	55 ft	(16.75 m)
Other combinations	65 ft	(19.8 m)
Axle Load[a]		
Single	10 s.t.[b]	(9.1 m.t.[c])
Tandem	17 s.t.	(15.4 m.t.)
Maximum Gross Weight[a]		
2-axle truck	15 s.t.	(13.6 m.t.)
3-axle truck	17-22 s.t.	(15.4-20.0 m.t.)
3-axle combination	21-25 s.t.	(19.0-22.7 m.t.)
4-axle combination	25-35 s.t.	(22.7-31.8 m.t.)
5-axle combination	35.8-42 s.t.	(32.4-38.1 m.t.)
6-axle combination	41.5-45 s.t.	(37.6-40.8 m.t.)

[a] Maximum permissible gross weight within ranges depends on axle spacing.
[b] s.t. = short tons. [c] m.t. = metric tons.

SOURCE: American Association of State Highway and Transportation Officials, *Recommended Policy on Maximum Dimensions and Weights of Motor Vehicles.* Washington, D.C.: American Association of State Highway and Transportation Officials, Rev. February 18, 1974. 20p.

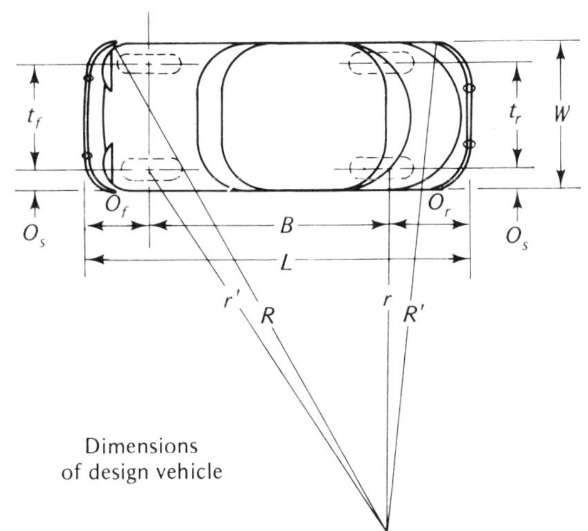

Symbol	Description	ft	m
L	Overall length	19.00	5.80
W	Overall width	6.67	2.05
B	Wheel base	10.33	3.15
O_f	Front overhang (front axle to bumper)	4.00	1.20
O_r	Rear overhang (rear axle to bumper)	5.25	1.60
O_s	Side overhang	0.65	0.20
t_r	Rear tread (center to center of tires)	5.35	1.65
t_f	Front tread (center to center of tires)	5.35	1.65
r	Min. turning radius (inside rear wheel)	13.75	4.20
r'	Min. turning radius (inside front wheel)	17.75	5.40
R	Min. turning radius (outside point, front bumper)	24.00	7.30
R'	Min. turning radius (outside point, front bumper)	20.00	6.10
	Overall height (sedans)	4.85	1.50
	Overall height (vans)	6.50	2.00
	Minimum running ground clearance	0.35	0.10

Note: These dimensions are exceeded only by a few luxury models which comprise less than 5% of all passenger cars in the United States.

SOURCE. Based on data published in *Automotive News Market Data Book 1976* and in *Parking Dimensions 1975 Model Cars*, published by the Motor Vehicle Manufacturers Association.

FIGURE 3-1
SELECTED PASSENGER CAR DIMENSIONS

1. *Rolling resistance.* A combination of roadway-tire resistances within the vehicle, and, if coasting or decelerating, resistance developed by engine compression.
2. *Air resistance.* A function of the frontal area of the vehicle, its shape, and the square of its velocity.
3. *Curve resistance.* The force acting through the front wheels to deflect the vehicle on a curved path.
4. *Grade resistance.* The force required to overcome the effect of gravity when a change of elevation of the vehicle takes place.

Vehicle weight is a major factor in all these resistances except that caused by air, and speed affects all except grade resistance.

In Eq. (3.1) positive values of a represent acceleration and negative values, deceleration. Resistances are positive, except that "grade resistance" on downhill alignments is actually a contribution to forward motion and is therefore given a negative sign. If the resulting F is positive, it indicates a tractive effort; if negative, a braking effort is involved.

Normal rates of acceleration of passenger cars are in

the range of 2.0 to 3.3 mph/sec (3.2 to 5.3 kph/sec). Maximum acceleration capability depends largely on the ratio of the available tractive effort to weight. This is seen if Eq. (3.1) is restated in the following form:

$$a = g\frac{F}{W} - \frac{R}{W} \qquad (3.2)$$

From a stop, maximum acceleration rates can range from 2.0 mph/sec (3.2 kph/sec) for large trucks to 10 mph/sec (16 kph/sec) for automobiles with 350-hp (260-kW) engines. At higher speeds the resistances come into play, and maximum acceleration rates drop. Figure 3-2 gives examples of the distance traveled by passenger cars and trucks under maximum acceleration conditions.

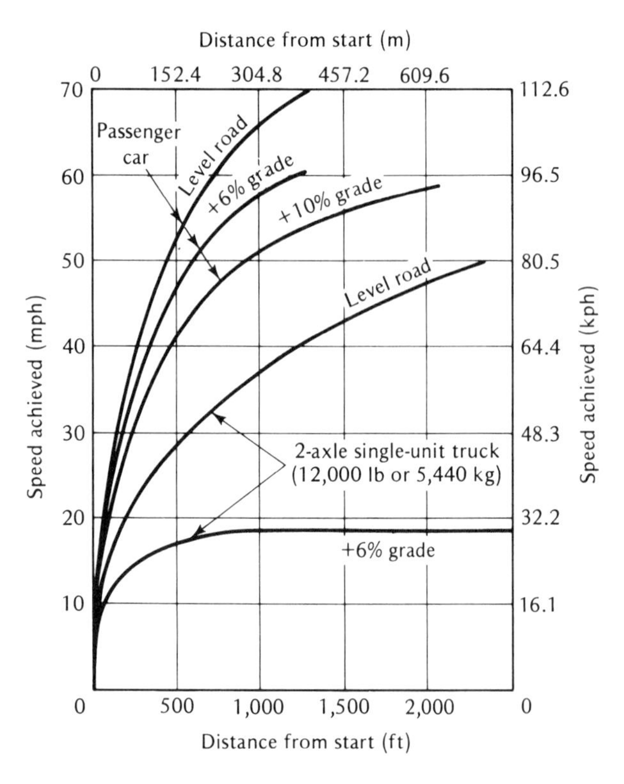

SOURCE: John E. Baerwald, ed., *Transportation and Traffic Engineering Handbook.* Arlington, Va.: Institute of Transportation Engineers, 1976, Chap. 2.

FIGURE 3-2
SPEED–DISTANCE RELATIONSHIPS OBSERVED DURING MAXIMUM RATE ACCELERATION

Maximum deceleration occurs when brakes are applied to a point just before wheels lock and skidding takes place. Under these conditions, well-designed brakes can produce a retarding force equal to or slightly more than vehicle weight. The resulting deceleration rate is intolerable except under emergency conditions with all vehicle occupants, it is to be hoped, using safety belts. However, in most emergencies, drivers will cause the wheels to lock. The retarding force then is no longer produced by the brakes, but by the friction forces generated between tires and road surface. This force is expressed as:

$$F = f \cdot W \qquad (3.3)$$

where f, the average friction coefficient, varies with the condition and type of the two materials which are in skidding contact. Also, since a skid causes smooth spots to develop in the tires, f varies inversely with the length of the skid. Figure 3-3 illustrates the wide range of values of this factor found in experiments.

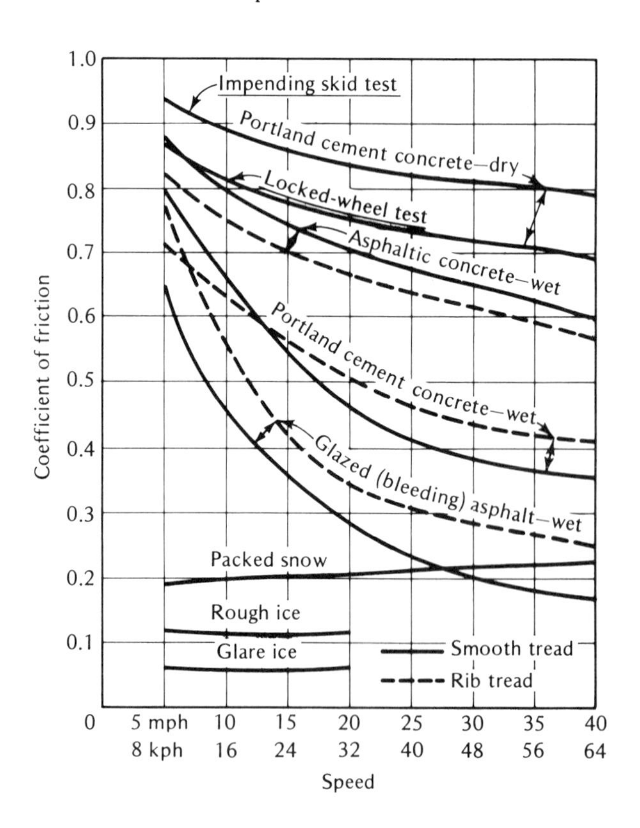

SOURCE: Ralph A. Moyer, "A Review of the Variables Affecting Pavement Slipperiness." *Proceedings, First International Skid Prevention Conference.* Charlottesville, Va.: the Conference, 1959.

FIGURE 3-3
EXAMPLES OF AVERAGE FRICTION FACTORS

Highway designers and traffic engineers are particularly concerned with the distances required by drivers to come to emergency stops. To be on the safe side in such calculations, skidding is assumed and the contribution of the resistances to deceleration, except for the effects of grade, are ignored. Allowance is made for the distances covered by the vehicle during the time in which the driver reacts to the situation. The resulting formula is

$$d = \frac{v^2}{2g(f + s)} + t_r \cdot v \qquad (3.4)$$

where

d = distance required to come to a stop (ft or m)
s = slope of the road (elevation change per unit of length)
t_r = driver reaction time (seconds)
v = original speed of vehicle (ft/sec or m/sec)
f and g are as defined earlier.

In application, it is easiest to measure speeds in mph or kph, and a driver reaction time of 1.0 second is often assumed. Equation (3.4) is then simplified to the following form:

$$d = \frac{v^2}{30(f + s)} + 1.5v \quad (d \text{ in feet, } v \text{ in mph}) \qquad (3.5a)$$

$$d = \frac{v^2}{254(f + s)} + 0.28v \quad (d \text{ in meters, } v \text{ in kph}) \qquad (3.5b)$$

Figure 3-4 illustrates the result of such calculations for some of the pavement and tire conditions shown in Fig. 3-3, as well as the stopping distances specified in the U.S. Federal Motor Vehicle Standard 105 when skidding does not take place. The distance corresponding to the driver reaction time is not included in these graphs.

Vehicle Passing

Vehicle passing maneuvers are related to the acceleration capabilities of vehicles already discussed. The characteristics of this maneuver and the importance to highway design are described in Chapter 6.

Driver Eye Height

The height of the drivers' eyes above the roadway has an effect on their ability to see and pass, particularly on crest vertical curves. In 1936 the median eye height was 4.75 ft (1.45 m). Because of changes in automobile styling, driver eye height has been lowered to current values of approximately 3.75 ft (1.15 m).

Vehicle Operating Costs

The costs of operating a motor vehicle may be divided into two groups. One group—out-of-pocket or operating costs—includes fuel, lubricating oil, tires, and maintenance. The other group is fixed costs, which are not a function of vehicle use. They include depreciation, insurance and licensing costs, and, in some states and countries, property tax levies on the vehicle. However, if one assumes an annual quantity of vehicle use, fixed costs can also be expressed in terms of unit distance traveled. Total costs vary with the unit costs of the items comprising them, such as fuel, wages, and taxes, as well as with the type of vehicle and type of use. Table 3-2 indicates typical costs per vehicle-mile for passenger cars driven 100,000 miles in a 10-year period and purchased in 1976. (Although the vehicle would operate through 1986, all prices used in the calculation for this table are those charged in 1976.)

Motor Vehicle Safety Standards

Since enactment of the Federal Motor Vehicle Safety Act of 1966, the U.S. Department of Transportation has promulgated a series of specifications (50 as of late 1976) dealing with safety standards of automobiles. They deal with such items as vehicle control layout in front of the driver, windshields, wipers and washers, brakes, lamps, tires and rims, theft protection devices, occupant protection (seat belts, door latches, padded dashboards, etc.), and fuel tank construction. Revisions and additions to these standards are under constant review, and are published in the *Code of Federal Regulations* [Ref. (7)].

Motor Vehicle Emission Standards

In the United States the federal government also establishes and enforces standards for maximum pollution emissions from motor vehicles. This program is part of the National Environmental Policy Act of 1969, and seeks to assure a healthy and pleasant level of air quality under all atmospheric and traffic conditions. Emissions are measured for a standard test run of engines under both cold and hot conditions. Cold engines generally emit more pollutants than warm ones, and the emission rate also varies

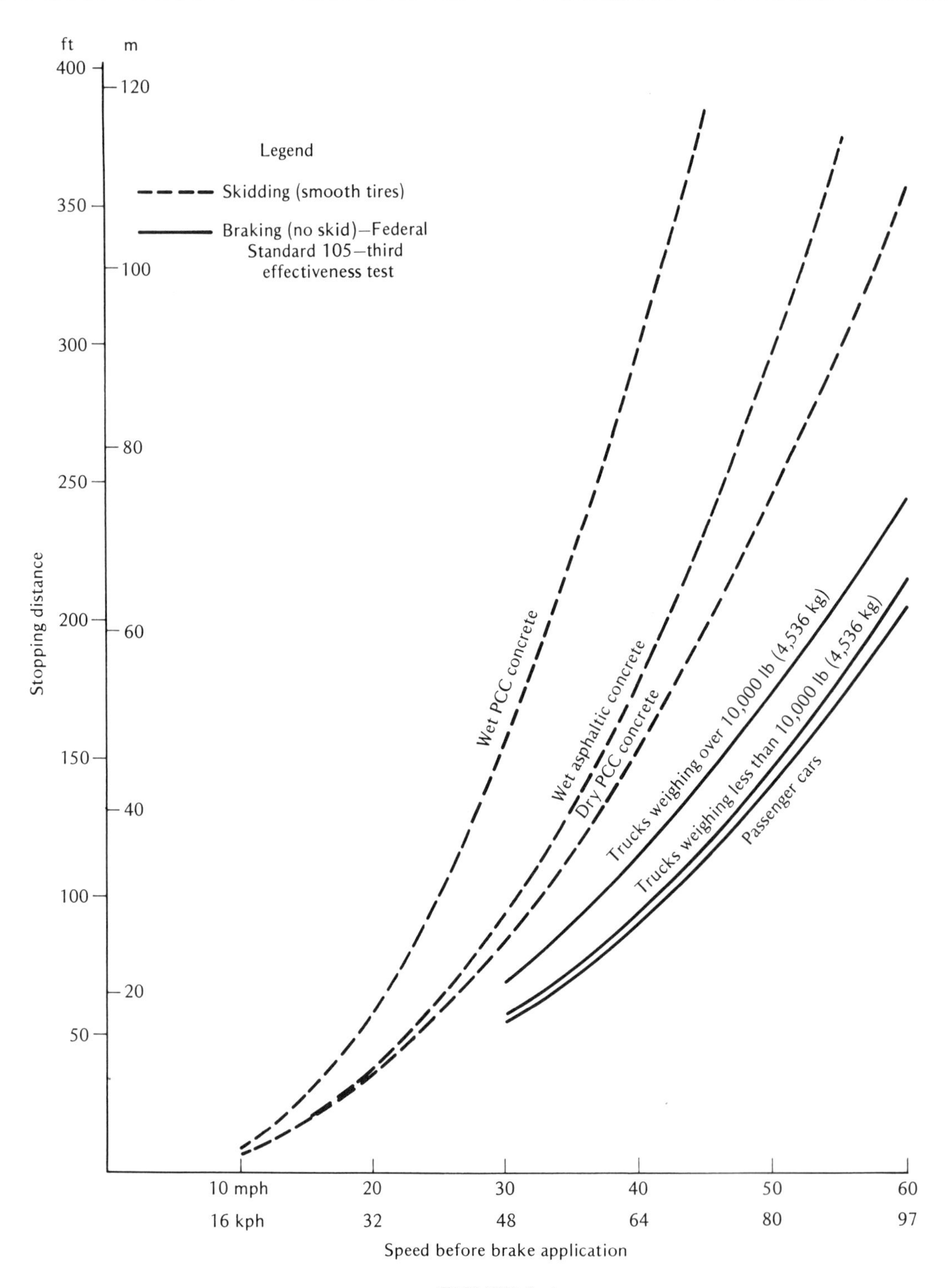

FIGURE 3-4

EMERGENCY STOPPING DISTANCES ON LEVEL ROADS UNDER SELECTED SKIDDING CONDITIONS AND FOR FEDERAL BRAKE STANDARDS (NO SKID) (*REACTION TIME NOT INCLUDED*)

TABLE 3-2
COST OF OWNING AND OPERATING AN AUTOMOBILE

Size	SUBURBAN-BASED OPERATION (TOTAL COSTS: CENTS PER MILE)						
	Original Vehicle Cost Depreciated	Maintenance, Accessories, Parts and Tires	Gas and Oil (Excluding Taxes)	Garage, Parking and Tolls	Insurance	State and Federal Taxes	Total Cost
Standard With standard equipment, weigh more than 4,000 lb. empty.	4.9	4.2	3.3	2.2	1.7	1.6	17.9
Compact Weigh more than 2700 lb but less than 3600 lb empty.	3.8	3.4	2.5	2.1	1.6	1.2	14.6
Sub Compact Weigh less than 2700 lb empty	3.2	3.1	1.8	2.1	1.5	0.9	12.6

Note: All costs are at 1976 levels.

SOURCE: Liston, L. L. and C. A. Aiken. *Cost of Owning and Operating an Automobile.* Washington, D. C.: U.S. Federal Highway Administration, 1976.

with the fuel-to-air ratio being fed into the engine. Therefore engine performance is measured with a standard test sequence simulating a typical mixture of cold and hot operation, acceleration, cruising, deceleration, and idling. Table 3-3 shows the standards applying to 1977 model year passenger cars and compares these to typical emission rates from vehicles without any emission controls and to the standards for the 1975 car models. The rapid rate at which standards have been raised is evident from this table.

TABLE 3-3
U.S. EXHAUST EMISSION STANDARDS FOR AUTOMOBILES

	MAXIMUM EMISSION (GRAMS PER MILE)		
	Carbon Monoxide	Hydrocarbons	Nitric Oxides
No exhaust controls	80	11	4
1975 models	15	1.5	3.1
1977 models	3.4	0.41	2.0

DRIVER CHARACTERISTICS

In designing and operating a highway transportation system the engineer must be continually aware of the requirements and capabilities of the highway user.

Driving is a complex task imposing varying demands on the motorist. Motorists, in turn, have varying abilities.

Drivers are faced with a continuing input of stimuli. Usually, they are capable of sorting out those which may be related to their driving. From these clues and observations they determine what action is required to continue on a safe path. Judgment, based on past experience or learning, supplements available information (e.g., the sight of a ball bouncing in the street indicates that unseen children may be present). Based on perceived events, analysis, and judgment, several alternative actions present themselves (to stop? slow down? swerve left?—right? how much? no action?). A comparison of these alternatives leads to a decision. This decision is translated into a set of "instructions" (e.g., remove foot from accelerator), which, in turn, involve certain muscular movements. There is a feedback of their responses (e.g., vehicle slows slightly) that will modify their information processing.

The above highly complex system in turn is further modified by the drivers' psychological and physiological state.

Driver Perception

Perception involves visual impressions in most driving situations. The driver sees the roadway, other vehicles, traffic controls, and obstacles to driving. Visual acuity, the ability to discern detail under average illumination, is most sensitive in an area subtended by a cone whose angle is three degrees about the center of the retina. A person with normal good vision (Snellen 20/20) can perceive an object that subtends one minute of angle within a 10-deg cone. At the same time a person with Snellen 20/40 vision would not be able to see an object unless it subtended two minutes of visual angle.

Beyond the cone of clear vision is the area of peripheral vision. This is a zone where the road user can see objects but without clear detail or color. In this area, which subtends an angle of from 120 deg to 160 deg for most drivers, a movement or bright light can alert drivers, but it requires that they move their eyes or head in order to bring the object into their cone of acute vision.

There is a time lapse in moving and fixing the eye on different objects. Drivers who have their eyes fixed on an object on the right of the roadway, then move their eyes to the left side to fix on an object, and return to view the right side of the roadway require from 0.5 to 1.3 seconds to accomplish this task. Similarly, a time interval of 0.5 to 1.5 seconds is required to shift fixation and discriminate detail at a distance after reading a speedometer.

The design and placement of traffic control devices are influenced by these elements of vision. Good design recognizes the need to place signs, barriers, and similar traffic devices where road users can see and comprehend them without moving their heads greatly.

Color is an important factor in perception. Under good illumination many colors can be differentiated, but as light decreases, the red and blue ends of the spectrum become less visible, while yellow remains relatively more visible. A substantial fraction of the population is color blind to some wavelengths.

Adaptability to light changes is also an important factor in visual acuity. Tunnels, headlights, and intermittent street lighting all present situations in which the eye can be subjected to rapid and sometimes severe changes in illumination. The eye may require several seconds to adapt to sudden decreases in illumination. A lesser time is required to adapt to an increase in light.

Other Senses

Tactual and *muscular* impressions may include the feeling of impending skidding, sway, pavement roughness, or deceleration rate. The sense of *hearing* may involve impressions of how the vehicle is operating, the presence of noise-producing rumble strips on the pavement, or of adjacent vehicles not in the line of sight. The sense of *smell* plays little direct role in the driving task, except to warn the driver of such problems as fire or noxious fumes in the environment.

Driver Reaction

The perception times discussed concern only the perception of objects in the driver's field of view. Driver reactions to stimuli include analytical operations, judgment, comparison, correlation, decision, and integration, in addition to perception. The total time for all these elements is called reaction time. Driving behavior studies indicate that drivers are aware of their own reaction time. In the case of cars following one another, for example, drivers who feel that they require little reaction time may follow more closely than persons who anticipate that they cannot make a decision rapidly.

Laboratory tests indicate that simple response time to touch, hearing, and sight stimuli is about 0.15 second. If the stimulus is related to muscular sensitivity or equilibrium, the reaction time ranges from 0.25 to 0.55 second. More complex tasks require greater reaction time. Braking responses to a brake light stimulus range from 0.4 second to above 1.0 second for some drivers.

Of even more concern is the likelihood for error in response as the driving situation becomes more complex. Figure 3–5 indicates the increase in average response time and percentage of error as more stimuli are introduced, thus complicating the problem. A consequence of an error in decision can be even more serious than an increase in reaction time, as, for example, a wrong-way entrance on a high-speed freeway ramp.

Effects of Age

All physical motor capabilities deteriorate with age. Table 3–4 shows that average visual acuity declines almost 0.5

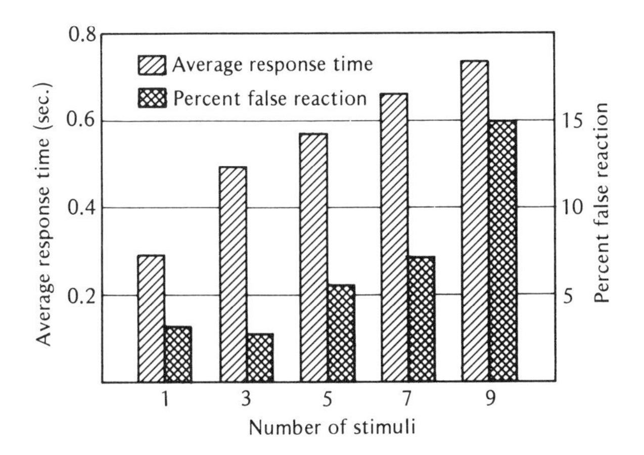

SOURCE: Adapted from T. W. Forbes, and M. S. Katy, *Summary of Human Engineering Research Data and Principles Related to Highway Design and Traffic Engineering Problems.* American Institute for Research, 1957, p. 16.

FIGURE 3-5

NUMBER OF STIMULI—INCREASE IN MEAN RESPONSE TIME AND ERRORS WITH INCREASING NUMBER OF STIMULI

TABLE 3-4

VISUAL ACUITY AND BRAKE REACTION TIME OF DRIVERS BY AGE

AGE GROUP	PERCENTAGE OF AVERAGE VISUAL ACUITY	AVERAGE REACTION TIME (SECONDS)
15-19	95	0.438
20-24	101	0.437
25-29	101	0.447
30-34	96	0.446
35-39	95	0.457
40-44	96	0.463
45-49	92	0.475
50-54	84	0.476
55-59	84	0.481
60-64	79	0.497
65-69	79	0.522
70-74	78	NA[a]
75-79	78	NA[a]

[a]NA not available.

SOURCE: John E. Baerwald, editor, *Traffic Engineering Handbook*, Institute of Traffic Engineers, 1965, pp. 82 and 84.

percent each year after 30 years of age, and that brake reaction time is seen to increase by 15 percent from ages 25 to 65. Older drivers also experience increasing difficulty in overcoming glare effects from vehicle headlights and other sources.

Driver Strategy

The perception-reaction cycle occurs throughout the driving process. Much of the time the driver notes that there is no reason to change the driving mode. When a condition requiring some change of vehicle motion does occur, the time required and available for the decision and reaction is of paramount concern. The use of a simple driving task model is useful in understanding the relationship between driving behavior and possible road and traffic hazards.

In Fig. 3-6 a vehicle, located near the left edge of the diagram, is moving at such a speed that it will traverse space "1" during the normal perception time, space "2" during the time required to reach a decision, and space "3" during minimum reaction time. If the decision is made to come to an emergency halt, space "4" within the arc S represents the minimum stopping distance (discussed earlier in this chapter). In most situations, however, the driver need take no emergency action.

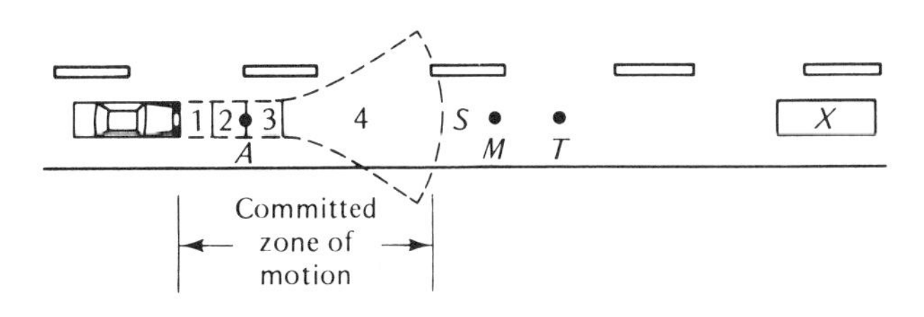

SOURCE: Adapted from Reference (1), p. 49.

FIGURE 3-6

CONCEPTUAL MODEL OF DRIVER PERCEPTION

Location X in Fig. 3-6 contains a condition that requires analysis and possible action. It may be another vehicle, a curve, a change of grade, a pedestrian, an obstruction. Point T is the *true point,* the last point at which physical and speed conditions require the decision to alter driving behavior (if any is required) to be implemented. The distance from T to X equals that from A to S. Point M is the *mental point,* the location where the driver believes the last chance for action to be. Point A is the *action point,* where the driver acts.

The distance AM is the margin of safety that drivers allow for their actions and judgment. MT is their perceptual error of the situation, a function of judgment and expertise. AT is the true margin of safety. A situation is safe so long as A remains upstream of T. If drivers misjudge a situation by placing M downstream of T, they usually compensate by a larger margin of safety. However, if A is downstream of T, an accident is unavoidable, because arc S is then downstream of X.

More detailed information about driver behavior can be found in Chapter 3 of Reference (1) and in Reference (2).

PEDESTRIAN CHARACTERISTICS

Virtually every passenger trip involves some walking, and everyone except infants and others physically unable to walk is a pedestrian. Transportation planners and engineers are, therefore, concerned with the provision of adequate pedestrian facilities as a basic part of all transportation systems. They need to understand the basic characteristics of walking. This knowledge leads to the proper design of sidewalks, crosswalks, passageways, stairs and escalators, and pedestrian paths. They must also know about the range of acceptable walking distances in order to determine feasible locations of parking facilities and transit stops. Special concern is caused by the possibilities of accidents, since pedestrians frequently come into propinquity to vehicular traffic.

Some general pedestrian characteristics are outlined here and, with regard to accidents, in the last section of this chapter. The relationship of walking speeds, flow rates, and density of pedestrian groups is covered in Chapter 4.

Pedestrian Dimensions

The relevant anthropomorphic data for planning pedestrian facilities are the width and depth of adults when walking in various postures. Figure 3-7 gives these data for selected situations. However, additional space is required not only for comfort but also for mobility. Figure 3-8 shows the floor area that should be available to a pedestrian without baggage under various conditions of queuing and standing. These values are reflected in the pedestrian speed-flow-density diagram in Fig. 4-10 in the next chapter.

Walking Speeds and Gap Acceptance

In analyzing pedestrian safety in streets and designing traffic signal timing programs, the transportation engineer takes into account the walking speeds of pedestrians and their willingness to walk through gaps of various sizes in the vehicular traffic stream.

Walking speeds vary with the physical condition and psychological attitude of each individual. Generally, about 90 percent of any group of pedestrians will walk at a speed of 4 ft/sec (1.2 m/sec) or greater, and this velocity is therefore used for many design purposes. However, when walkways are crowded, speeds drop (see Fig. 4-10).

The time gap between vehicles needed for pedestrian crossing depends on the type of operation (one-way or two-way) and width of street to be crossed. For a one-way street 44 feet wide, the average acceptable gap in traffic is about 5.7 seconds. However, there is much variability in pedestrian crossing gap acceptance. A British study summarized in Fig. 3-9 found that 50 percent of the sample who could cross when an approaching vehicle was about 4.5 seconds away did cross, that no one crossed when the approaching vehicle was less than 1.5 seconds away, and that everyone crossed when the approaching vehicle was more than 10.5 seconds away.

Pedestrian behavior is also strongly conditioned by such local conditions as enforcement and driver response to pedestrian behavior.

Trips and Walking Distances

Walking trips range from hiking all day to taking a few steps from the home to the garage or curb parking space. Unfortunately, in standard transportation studies, as described in Chapter 13, information about walking is seldom collected, and there is only fragmentary information about the motivation for walking entire trips and about the distances walked as a part of travel also involving a vehicular mode. One can speculate that the amount of walking increases as alternative forms of transportation become less available, less convenient, and more expensive. There is, therefore, more walking in large cities for automobile users because of inconvenient parking, whereas transit riders perhaps walk longer distances in smaller cities where transit routes are spaced more widely apart.

To estimate the quantity of pedestrians becomes a problem primarily in designing facilities, such as walkways, crosswalks, stairs, escalators, and waiting areas, in and adjacent to large transportation terminals. In these locations, "batches" of pedestrians arrive on buses, trains, planes, or ferries, creating sharp fluctuations in demand for space. This demand can be calculated from the traffic estimates of the systems serving the terminal. Similar problems arise, and similar analyses are made, for any location where crowds congregate, such as sports, cultural,

SKETCH	DEFINITION	DIMENSIONS		PROJECTED AREA
		Width ① ft (m)	Depth ② ft (m)	ft^2 (m^2)
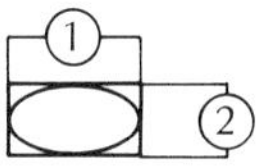	Basic area at thorax level	1.60–1.75 (0.49–0.53)	0.85–1.00 (0.26–0.31)	1.40–1.75 (0.13–0.16)
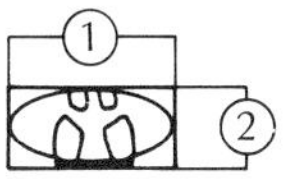	Maximum basic area projected on ground	2.00–2.15 (0.61–0.66)	0.95–1.00 (0.29–0.31)	1.85–2.25 (0.17–0.21)
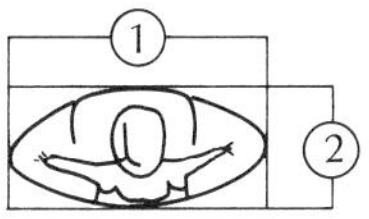	Stretched out elbows, arms horizontal, finger tips joined at chest	3.10 (0.95)	1.40 (0.43)	4.30 (0.40)
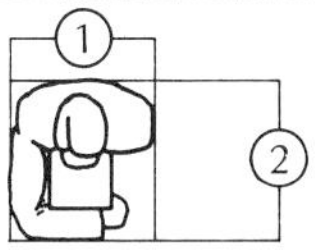	Parcel against chest held by bent arm	1.75 (0.53)	1.85 (0.57)	3.25 (0.30)
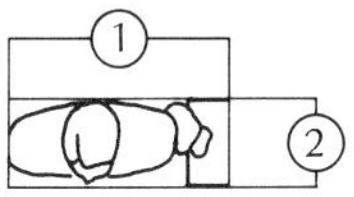	Hand baggage (one piece, small, at side)	2.70 (0.82)	1.00 (0.30)	2.70 (0.25)
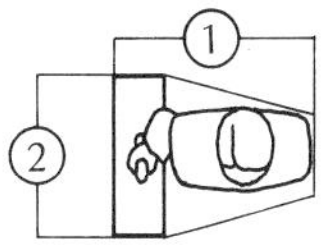	Hand baggage (one piece, large, at side)	2.40 (0.73)	1.95 (0.60)	3.90 (0.36)
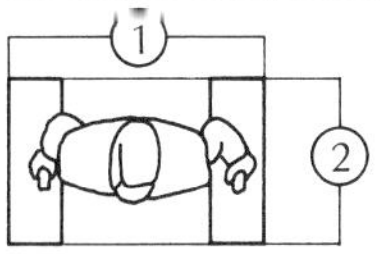	Hand baggage (two pieces, large, both sides)	3.05 (0.93)	1.95 (0.60)	6.05 (0.56)
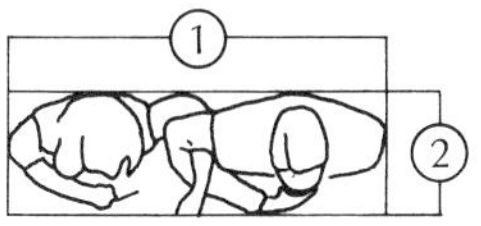	Couple, arm in arm	4.60 (1.40)	1.40 (0.43)	6.45 (0.60)

SOURCE: Batelle Research Center, *Synthesis of a Study on the Analysis, Evaluation and Selection of Urban Public Transport Systems.* Geneva: the Center, 1974.

Transportation Engineering Fig 3.7

FIGURE 3-7
WIDTH AND DEPTH DIMENSIONS OF PEDESTRIANS

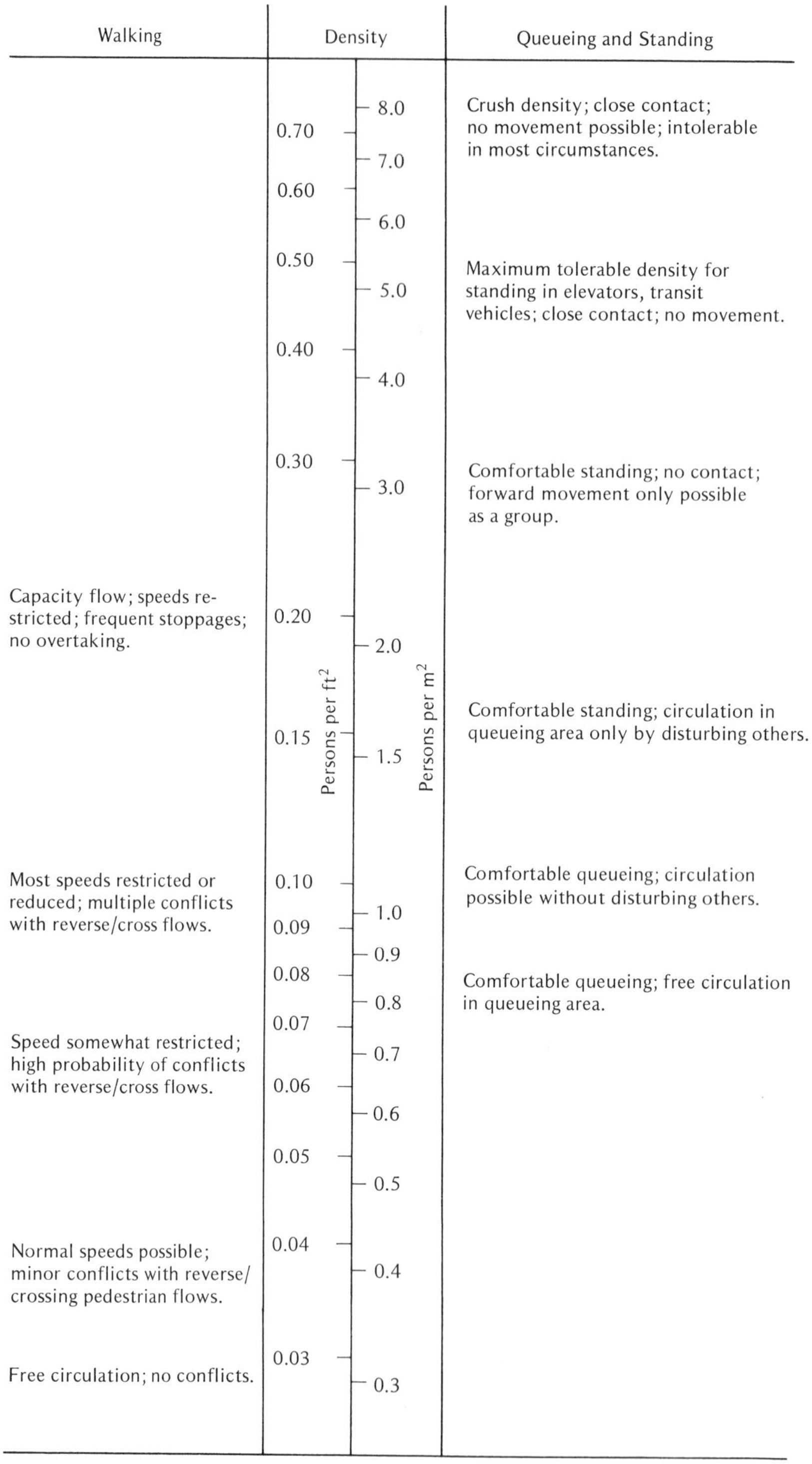

Walking	Density (Persons per ft^2)	Density (Persons per m^2)	Queueing and Standing
		8.0	Crush density; close contact; no movement possible; intolerable in most circumstances.
	0.70	7.0	
	0.60	6.0	
	0.50	5.0	Maximum tolerable density for standing in elevators, transit vehicles; close contact; no movement.
	0.40	4.0	
	0.30	3.0	Comfortable standing; no contact; forward movement only possible as a group.
Capacity flow; speeds restricted; frequent stoppages; no overtaking.	0.20	2.0	
	0.15	1.5	Comfortable standing; circulation in queueing area only by disturbing others.
Most speeds restricted or reduced; multiple conflicts with reverse/cross flows.	0.10	1.0	Comfortable queueing; circulation possible without disturbing others.
	0.09	0.9	
	0.08	0.8	Comfortable queueing; free circulation in queueing area.
Speed somewhat restricted; high probability of conflicts with reverse/cross flows.	0.07	0.7	
	0.06	0.6	
	0.05	0.5	
Normal speeds possible; minor conflicts with reverse/crossing pedestrian flows.	0.04	0.4	
Free circulation; no conflicts.	0.03	0.3	

SOURCE: Based on Reference (2) and Battelle, *op. cit.* in Fig. 3-7.

Transportation Engineering Fig 3.8

FIGURE 3-8
MINIMUM PEDESTRIAN SPACE REQUIREMENTS

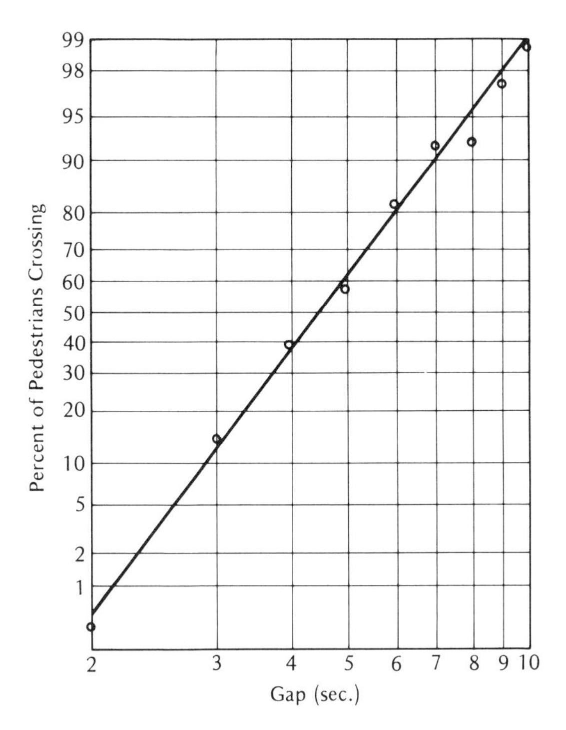

SOURCE: Adapted from J. Cohen, E. J. Dearnaley, and C. E. M. Hansel, "The Risk Taken in Crossing A Road", *Operational Research Quarterly*, Vol. 6, No. 3, 1955, p. 122.

FIGURE 3-9
PEDESTRIAN GAP ACCEPTANCE VARIABILITY

and recreational sites. In each situation the direction of various pedestrian flows is determined, and the tolerable time to handle peak demand or the maximum tolerable delay is assumed.

The maximum acceptable walking distances determine the commercial feasibility of parking facilities at downtown locations and in other major activity centers. They also indicate the adequacy of transit networks and stop locations. Figure 3-10 gives the result of various studies of pedestrians. It will be noted that in auto-oriented locations such as shopping centers and airports, walking distances are very short. One can often observe drivers cruising to find close-in parking spaces while rejecting the opportunity to park a few hundred feet further away. While the same hunting for convenient parking takes place in central business districts, the relatively greater scarcity of parking spaces forces many persons to walk longer distances.

In transit planning a "rule of thumb" suggests that service be offered within 1/4 mile (400 m) of origins and destinations in the densest parts of the city, and within 1/2 mile (800 m) elsewhere. In attitudinal studies it has been frequently observed that a trip-maker evaluates walking at from twice to three times in terms of trip disutility (effort, discomfort, etc.) as traveling in an automobile or transit vehicle.

BICYCLING CHARACTERISTICS

The bicycle preceded the automobile as the mode of individual mechanical transportation for large numbers of people in many countries. It was through the concerted efforts of bicyclists that road reconstruction and paving programs were inaugurated in the last quarter of the nineteenth century. In many countries, whose topography is flat and whose standard of living does not permit a high degree of automobile ownership, bicycling is still a major mode of transportation today.

In North America, however, the advent of mass-produced cars relegated the bicycle to a children's toy and transportation to school, and as a recreational outlet for a selected group of enthusiasts. Only since about 1970 has an increased transportation function for the bicycle in North America been recognized as harmonious with solving energy conservation and environmental pollution problems.

According to the Bicycle Manufacturers' Association of America, sales of bicycles in the United States exceeded those of automobiles from 1972 through 1974; during this three-year span, over 40 million bicycles were acquired, and by the end of it an estimated 75 million bicycles were in use.

Purposes of Bicycle Use

When analyzing bicycle use, it is important to distinguish between trips for which the bicycle is an end in itself and those in which it is a means of transportation to a "purposeful" destination. In the former category are bicycle riding by children in the neighborhoods of their homes and longer trips by cyclists of all ages for sightseeing, exercise, or racing. In a survey made in 1975 of a group of regular adult bicycle users [Ref. (3)] it was reported that about half the trips made by bicycle were in each of these two categories, but that the distance traveled for purposeful transportation was less than one-third of the total (Table 3-5). These statistics represent the habits of a special group of adults whose bicycle use is likely to be considerably above that of other bicycle owners.

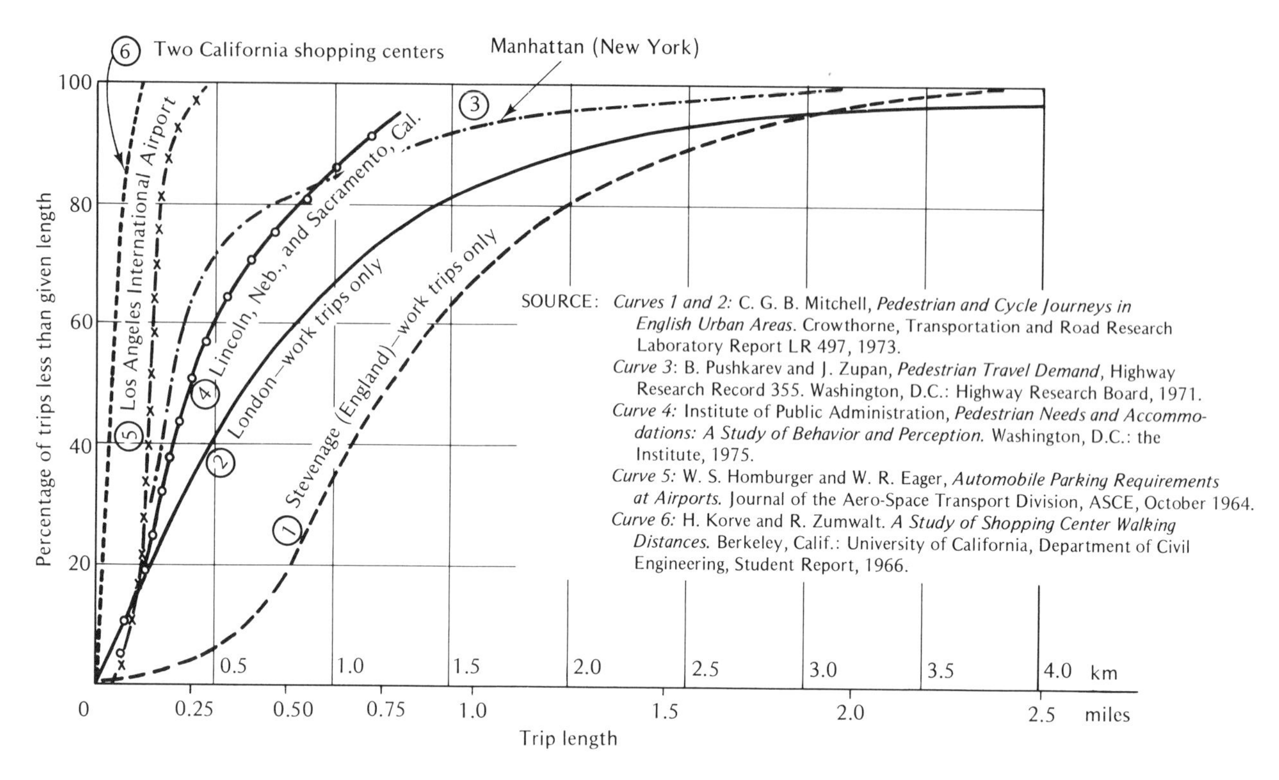

FIGURE 3-10
SELECTED CUMULATIVE WALKING DISTANCE DISTRIBUTIONS

TABLE 3-5
BICYCLE TRIP CHARACTERISTICS BY PURPOSE
(MEMBERS OF A BICYCLE LEAGUE)

TRIP PURPOSE	PERCENTAGE OF TOTAL TRIPS	PERCENTAGE OF TOTAL DISTANCE	AVERAGE ROUND TRIP DISTANCE PER TRIP	
			Miles	Km
"Purposeful" trips[a]	**50.7**	**28.1**	**6.8**	**11.0**
Work/school	33.1	21.7	8.1	13.0
Others	17.6	6.4	4.5	7.2
"Nonpurposeful" trips[a]	**49.3**	**71.9**	**18.0**	**29.0**
Touring/recreation	29.3	52.4	22.2	35.7
Exercise	16.1	13.3	10.2	16.4
Racing	3.9	6.2	19.9	32.0

[a] "Purposeful" refers to trips for which the bicycle was the transportation means while another purpose motivated the trip. "Nonpurposeful" refers to trips for which cycling was an end in itself.

SOURCE: Reference (4), Table 8.

Patterns of Use

Bicycling is somewhat constrained by weather factors and by the physical limitation of cyclists as well as their willingness to exert themselves. In the survey of regular bicycle riders, referred to above, one-third of the respondents stated that they do not ride during hours of darkness and one-quarter that they do not cycle in the rain. The average round-trip length of trips made by this specialized group within the general population is shown in one column of

Table 3-5. It is safe to assume that the reluctance to use bicycles in the dark or the rain is greater, and distances traveled are less for the average cyclist in the United States.

However, cycling can make a major contribution to the total transportation system in selected circumstances. Young people are more likely to use bicycles than other adults. Thus one finds that a substantial part of commuting to and from colleges and universities located in relatively flat terrain and mild climates is by bicycle.

Bicycle Operating Characteristics

Bicycles generally occupy a space that is 2 ft (0.6 m) wide and 5.75 ft (1.75 m) long. However, to allow for clearances from obstructions and for maneuvering, a minimum path width of 3.3 ft (1.0 m) is recommended. The height of erect adult cyclists above the road surface is about 7.5 ft (2.5 m).

Average cycling speeds on level ground have been found to be about 10 mph (16 kph) both in the United States and Europe. Grades exceeding ±1 percent affect speed, but the amount of reduction on uphill grades varies considerably with the physical condition of the cyclist and the gear ratios and weight of the bicycle.

Groups of bicycles interact in a manner analogous to the stream characteristics of motor vehicle traffic. In the following chapter, Fig. 4-10 indicates the basic relationship between speed, density, and flow rate of bicycle traffic.

TRAFFIC ACCIDENTS

Traffic accidents constitute one of the major public health problems facing advanced societies. Their occurrence indicates a failure by the road facility, vehicle, and vehicle operator separately or jointly. Traffic engineers possessing the necessary analytic skills and practical experience can play important roles in dealing with this problem. They do so by identifying general accident trends and specific hazards, and developing suitable countermeasures.

The costs in human suffering and economic resources exacted by highway traffic accidents are truly immense. Figure 3-11 shows the U.S. fatality record since the automobile became widely used. The long-term trend toward more deaths is apparent. However, short-term reversals of

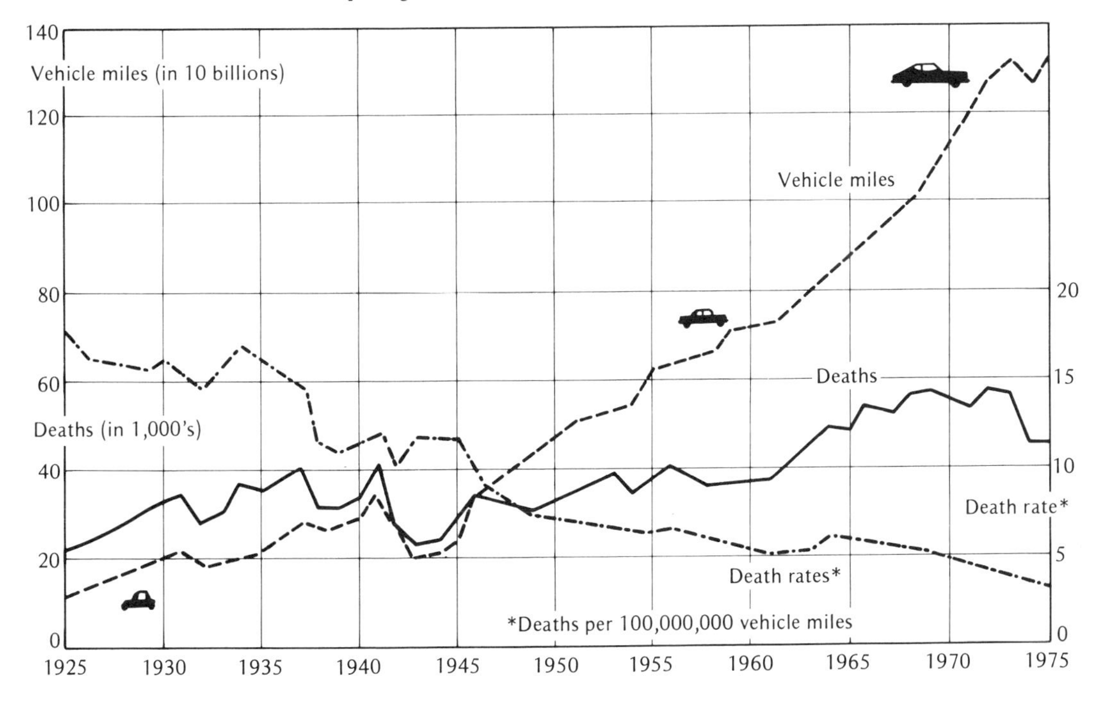

SOURCE: ACCIDENT FACTS, 1976 edition, National Safety Council.

FIGURE 3-11
TRENDS IN TRAFFIC FATALITIES

this trend have occasionally occurred, most recently in 1974 as a result of the fuel shortage and imposition of a nationwide 55 mph (88.5 kph) speed limit. However, the slower upward trend in deaths after 1965 indicates that new safety programs were having a beneficial effect. The accident rate or standardized exposure index (deaths per 100 million vehicle miles) has decreased steadily since 1945.

Traffic Accident Characteristics

In this section some significant statistical characteristics of traffic accidents are summarized. The sources of these data are reports prepared by police, participants, witnesses, and others. Despite efforts by the National Highway Traffic Safety Administration to promote uniform accident records, disparities in reporting policies and practices still exist among the various states. For this reason, some aspects of traffic accidents are illustrated with data from one state whose data processing quality is high.

Figure 3-12 shows that two-thirds of traffic fatalities occur on rural highways. Pedestrian fatalities account for almost 20 percent of the total. Traffic deaths at night comprise more than half the total, although night exposure (traffic volume) is only about half that of day exposure. The variation in accidents by time of day is further illustrated by California statistics in Table 3-6. This shows the increasing accident experience as the day passes, the disproportionately large number of accidents on weekend evenings, and the general variation by day of the week.

Seasonal effects on traffic accident experience is shown in Table 3-7. The summer travel peak is accompanied by a peak in accidents. Other studies have shown that alcohol plays an increasing role in accident causation toward the end of the year.

The age distribution of persons killed or injured in accidents is not representative of the age distribution of the entire population or of that part of the population holding drivers' licenses. Instead, it reflects the relatively unsafe driving strategies of young male drivers, the inexperience of young children and the limited physical ability of old persons as pedestrians, and variations in exposure rates. Table 3-8 shows that accident victim rates vary widely by age and sex in California, which is representative of the entire U.S.

Alcohol is an important factor in traffic accidents. In California one-third of all drivers in fatal accidents and one-eighth of those in injury accidents are reported to have been drinking, with the highest proportion in the age

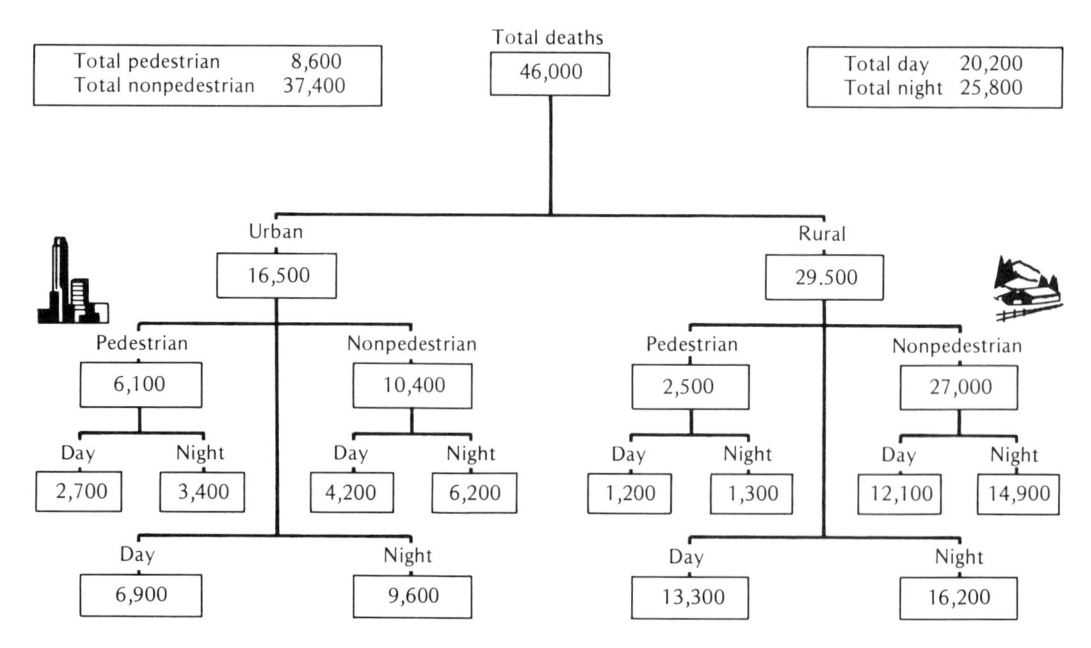

SOURCE: ACCIDENT FACTS, 1976 edition, National Safety Council.

FIGURE 3-12
TYPES OF FATAL TRAFFIC ACCIDENTS—1975

TABLE 3-6
FATAL AND INJURY ACCIDENTS BY TIME OF DAY AND DAY OF WEEK (CALIFORNIA, 1975)

TIME OF DAY	ACCIDENTS AS PERCENTAGE OF AVERAGE 3-HOUR PERIOD			
	Mon-Thurs Average	Friday	Saturday	Sunday
Midnight-3 am	48%	59%	151%	138%
3 am-6 am	16	18	49	44
6 am-9 am	82	79	36	25
9 am-noon	84	90	85	61
Noon-3 pm	131	146	146	122
3 pm-6 pm	193	225	168	157
6 pm-9 pm	113	154	148	133
9 pm-midnight	76	143	127	86
Daily Totals	93[a]	116[a]	115[a]	97[a]

[a]Daily totals are shown as percentage of average weekday.

SOURCE: Derived from California Highway Patrol, *Fatal and Injury Motor Vehicle Traffic Accidents 1975*. Sacramento, Cal.: 1976.

range from 20 to 24 years. Among adult pedestrian victims, one-sixth are listed as having been drinking prior to the accident.

For further analysis, accidents are classified into several types based on the major event involved. The numbers of fatalities occurring in urban and rural areas, classified in this manner, are shown in Table 3-9. Pedestrian deaths are by far most frequent in urban areas, whereas rural accidents cause most deaths when two vehicles collide and in noncollisions involving a vehicle either overturning in the roadway or leaving the roadway before hitting any person or object. Since total vehicle miles traveled in the United States are divided almost equally between rural and urban areas, these death figures also indicate the relative sizes of the fatality rates.

There has been a considerable shift in the relative frequency of different accident types. In the last 20 years the fatality rate overall has almost halved, but collisions with fixed objects have dropped only 12 percent. This can be attributed to the increase in freeway networks and the installation of median barriers, thanks to which many pedestrian accident opportunities have been eliminated and head-on collisions have been converted into sideswiping of median barriers. The increase in the rate of bicycle fatalities shows that a new problem is challenging the abilities of transportation and safety experts.

TABLE 3-7
MOTOR VEHICLE DEATHS BY MONTH (UNITED STATES, 1975)

MONTH	DEATHS AS PERCENTAGE OF AVERAGE MONTH	MONTH	DEATHS AS PERCENTAGE OF AVERAGE MONTH
January	83%	July	117%
February	77	August	119
March	89	September	106
April	88	October	107
May	104	November	103
June	107	December	99

SOURCE: National Safety Council, *Accident Facts 1976*, p. 51.

TABLE 3-8
VICTIMS OF FATAL AND INJURY ACCIDENTS BY AGE AND SEX (CALIFORNIA, 1975)

AGE GROUP	VICTIMS AS PERCENTAGE OF ALL VICTIMS		VICTIMS PER 100,000 POP. IN SAME AGE-SEX CATEGORY	
	Males	Females	Males	Females
0-14 years	6.8%	5.4%	611	508
15-24 years	23.3	15.4	2802	1900
25-34 years	12.2	8.7	1773	1254
35-44 years	5.4	4.6	1108	954
45-54 years	4.0	4.0	848	794
55-64 years	2.7	2.8	725	688
65 and over	2.1	2.5	617	500
All Ages[a]	56.5	43.5	1313	973

[a]Excludes victims whose sex and/or age were not reported.

SOURCE: Derived from California Highway Patrol, *Fatal and Injury Motor Vehicle Traffic Accidents 1975*. Sacramento, Cal.: 1976.

Transit System Accidents

The safety of transit systems is analyzed in terms of two types of occurrences. *Traffic accidents* involve the collision of the bus, streetcar, or rapid transit train with another vehicle, a pedestrian, or an object. *Passenger accidents* include events causing injury on board the vehicle (caused by sudden change of speed or direction), or in

TABLE 3-9
MOTOR VEHICLE TRAFFIC FATALITIES BY TYPE OF ACCIDENT (UNITED STATES)

TYPE OF ACCIDENT	NUMBER OF FATALITIES—1975				CHANGE IN RATE[a] FROM 1955
	Urban	Rural	Total	Rate[a]	
Collision with—					
Pedestrian	6,100	2,500	8,600	0.64	−52%
Other motor vehicle	6,000	14,300	20,300	1.52	−36%
Railroad train	500	500	1,000	0.08	−70%
Bicycle	600	400	1,000	0.08	+10%
Animal	b	100	100	0.01	−50%
Fixed object	1,800	1,300	3,100	0.23	−12%
Noncollision	1,500	10,400	11,900	0.89	−55%
Totals	16,500	29,500	46,000	3.45	−46%

[a]Rate expressed as deaths per 100 million vehicle-miles.
[b]Less than 5.

SOURCE: National Safety Council, *Accident Facts 1976*, pp. 45 and 58.

boarding and alighting; in rail systems an additional category of "in-station" accidents summarizes injuries on stairs, escalators, and platforms and in other station areas.

Table 3-10 indicates transit accident rates for the United States in 1972. These are based on partial data, and neither details nor information on long-range trends are available. However, it is reported that the most common bus traffic accident involves colliding with other motor vehicles, and that three-quarters of bus passenger accidents occur on board, with the remainder involving boarding or alighting. On rail systems, pedestrian collisions are the most frequent cause of traffic accidents; however, since information on suicide jumps is not known, not all these events deserve to be considered failures of the transit system. Among rail passenger accidents, two-thirds to three-quarters occur in stations (the majority of these on fixed stairs), the remainder on the trains or when getting on and off, in about equal proportions.

Accident Prevention

The driver perception analysis described earlier in this chapter, and accident statistics such as outlined here, point to the conclusion that accidents may be caused by human failure, mechanical malfunction of vehicles, design deficiencies of roadways or pedestrian facilities, or a combination of these factors. Only rarely, in fact, does an accident, in the basic sense of the word, occur; most events are incidents occurring in patterns that are predictable and, therefore, susceptible to mitigation.

The minimization of human failure can be achieved by raising the minimum standards for drivers' licenses, better education and communication between traffic engineers and drivers and pedestrians, and good enforcement of the rules of the road. Thanks to the government standards for new motor vehicles and periodic inspection programs in many states, vehicle failure is already a relatively minor cause of accidents. Facilities design, however, remains a

TABLE 3-10
MASS TRANSIT SYSTEM ACCIDENT RATES (UNITED STATES, 1972)

TYPE OF ACCIDENT	MODE	ACCIDENTS PER 10^6 PASSENGERS	ACCIDENTS PER 10^6 VEHICLE-MILES
Traffic accidents	Bus	17.7	60.8
	Rail	0.38	1.75
Passenger accidents	Bus	7.0	24.2
	Rail	11.7	53.4

SOURCE: Battelle Columbus Laboratories, *Safety in Urban Mass Transportation: Research Report.* Washington, D.C.: U.S. Urban Mass Transportation Administration, 1976, Tables B-19 and B-22.

major problem, since resources are not available to bring every stretch of city street and rural highway to optimum safety levels. In succeeding chapters dealing with design and operation of facilities, it will be seen that the concern for enhancing safety is always one of the basic motivations of the transportation engineering task.

REFERENCES

1. Baerwald, John E., editor, *Transportation and Traffic Engineering Handbook.* Englewood Cliffs, N.J.: Prentice-Hall, Inc., 1976. Chapters 2, 3, and 4.
2. Forbes, T.W., editor, *Human Factors in Highway Traffic Safety Research.* New York: John Wiley & Sons, 1972.
3. Fruin, John J., *Pedestrian Planning and Design.* New York: Metropolitan Association of Urban Designers and Environmental Planners, Inc., 1971. 206 pp.
4. Kaplan, Jerrold A., *Characteristics of the Regular Adult Bicycle User.* Washington, D.C.: U.S. Federal Highway Administration, July 1975.
5. Motor Vehicle Manufacturers Association, *Parking Dimensions.* Detroit: annual.
6. National Safety Council, *Accident Facts.* Chicago: annual.
7. U.S. Government, *Code of Federal Regulations,* Title 49, Part 571. Washington, D.C.: U.S. Government Printing Office. Contains Federal Motor Vehicle Safety Standards. Updated regularly.
8. U.S. Federal Highway Administration, *Highway Safety Program Manuals.* Washington, D.C.: U.S. Government Printing Office. Updated as appropriate.

6

4

TRAFFIC STREAM CHARACTERISTICS

Ultimately transportation engineers are concerned with the flow of groups of vehicles on the roadway system that they are planning, designing, or operating. The characteristics of this flow are defined by the constraints imposed by human behavior and vehicle dynamics, described in Chapter 3, the demand for movement, which will be dealt with in Chapter 13, and the physical parameters of the highway system.

Three stream flow characteristics are of prime interest; each of these is related to the other two:

- *Speed,* and its reciprocal, *travel time.*
- *Volume* or *rate of flow,* and its inverses, *headways* and *gaps* between vehicles.
- *Density* of traffic streams, and its reciprocal, *spacing of vehicles.*

In the following sections, these phenomena will first be described and their interrelationships explored, leading to definition of the concepts of capacity and levels of service. The analogous behavior of streams of bicycles and pedestrians will be mentioned. Finally, reference will be made to the application of these concepts in more advanced analytical studies of traffic flow.

SPEED AND TRAVEL TIME

Speed or its reciprocal, travel time, is a simple measure of how well a street, highway, or network is operating. Individual drivers partly measure the quality of their trip by their ability to maintain their desired speed. Many behavioral studies have indicated that minimization of delay is the single most important element in selection of one route from among several alternatives. In the long run, the selection of a home is determined in part by the separation (travel time) from work, and selection of a shopping site is conditioned by travel time to alternate retail areas. The engineer considers speed when planning a freeway network, designing a highway for safe stopping distance, or timing the yellow interval duration at a traffic signal.

Speed varies not only with the rate of flow and traffic stream density, as will be discussed later, but also with the characteristics of drivers and vehicles, the time of day, location, and environmental conditions. The terms *speed* and *travel time* are used in different contexts and must be clearly defined. A driver reporting that he made a trip in "good" time without delay is thinking in terms of *running* time and speed, that is, the relation between the time in motion and the distance traveled. If the same driver considers also the time lost in delays, he is evaluating *overall* travel time and speed. But if he looks at his speedometer, he will read his *spot* speed, the reciprocal of which is his *rate of motion*. In traffic stream analysis, the principal concern is with spot speed characteristics, the instantaneous speed of vehicles at a specified location during a limited time period, when traffic flow rates and environmental conditions remain approximately constant.

In a traffic stream there is a distribution of speeds, which is usually assumed to be represented by a normal curve. The central tendency of this distribution is described by the mean, and the dispersion of the individual speeds about this mean by the standard deviation. They are computed from the following formulae:

$$\bar{v} = \frac{\sum_{i=1}^{m} (f_i \cdot v_i)}{n} \qquad (4.1)$$

and

$$s = \sqrt{\frac{\sum_{i=1}^{m} (f_i \cdot v_i^2) - n \cdot \bar{v}^2}{n - 1}} \qquad (4.2)$$

where

f_i = the number of vehicles in the *i*th speed group.
v_i = the speed of the *i*th group.
m = the number of speed groups.
n = the total number of vehicles in the sample $\left(= \sum_{i=1}^{m} f_i\right)$.
$\bar{v}$ = the mean speed of the traffic stream.
s = the standard deviation of the speed distribution.

There are two distinctly different types of speed distributions, depending on the method of collecting data in the field. As described in Chapter 5, one can sample the speeds of the vehicles that are on a stretch of highway at a given moment (by means of aerial photography, for example) and thereby obtain a distribution "in space" and a *space mean speed*. Or one can measure the speeds of vehicles passing a particular location for a period of time (using radar or flash boxes), which will produce a distribution "in time" and a *time mean speed*. It will later be shown that the space mean speed is the appropriate measure to be used in traffic stream analysis. Since it is easier to obtain time mean speeds in the field, the space mean speed can be calculated as follows:

$$\bar{v}_s = \frac{n}{\sum_{i=1}^{m} \frac{f_i}{v_i}} \qquad (4.3)$$

where

$\bar{v}_s$ = the space mean speed.
v_i = speeds obtained "in time."
f_i, n, and m have the same meanings as before.

The difference between the two mean speeds can be understood if a section of highway is visualized. A spot speed sample taken at the end of this section over a finite period of time will tend to include some fast vehicles that had not yet entered the section at the start of the survey, but will exclude some slower vehicles that were within the highway section when the sample was started. An aerial photograph, however, would include all vehicles within the highway section at the moment of exposure and no others. Therefore, the time mean speed is always somewhat higher than the space mean speed.

Although the space mean speed is used for traffic flow analysis, the time mean speed, being more simply calculated, is usually used in speed zoning studies, analysis of accident locations, and similar work.

A speed distribution may be shown as an approximately normal curve, Fig. 4–1(a), even though not all the data points may lie on it. One assumes that, although the sample shows some irregularities, ever larger samples would lead to the normal curve that so often describes human characteristics. The dispersion is visually depicted by the relative squatness of the curve. More precisely, if one plots one standard deviation to the left and right of the mean, the area under the part of the curve so defined will contain approximately 68 percent of all speeds in the distribution. Moving two standard deviations to each side of the mean encompasses about 95 percent of the speed distribution, and three standard deviations on each side include about 99 percent of the area under the curve.

Commonly the cumulative speed curve is also drawn [Fig. 4–1(b)], and again it is smoothed out. The dispersion is not readily visible here, but is roughly gauged by the steepness of the curve. On the other hand, the cumulative curve shows clearly the median (fiftieth percentile) speed as well as other percentile speeds that traffic engineers often use. For instance, the eighty-fifth percentile is generally accepted as a safe speed under prevailing roadway conditions on the assumption that less than 15 percent of the drivers in any traffic stream are unsafe.

While the central tendency of a speed distribution—the mean speed—can be looked upon as a "quality" measure of traffic flow to the user, the dispersion indicates the efficiency and safety of the traffic stream. A large standard deviation signifies a great variety of speeds with associated frequency of catching up and overtaking, involving speed adjustments and lane changing. Low standard deviations imply a "disciplined" traffic stream, which is, other factors being equal, a desirable objective.

Travel time is frequently used to determine minimum paths through a transportation network. For this purpose, the overall time required to travel between key points on the system, usually called nodes, is determined, rather than the instantaneous rate of motion. As a result, spot speed data cannot be used, but travel times studies, which are relatively more time consuming and expensive, must be conducted. Therefore, much smaller samples are obtained, and results are correspondingly less precise.

Under traffic conditions of low flow and low density, travel time on highways is a function of the geometric design standards and speed limits. As volume and density increase, travel time also rises until it eventually approaches infinity under jam conditions (Fig. 4–9). In the analysis of transit networks, travel time also includes delays when one is waiting for a vehicle, which is a function of the frequency of service.

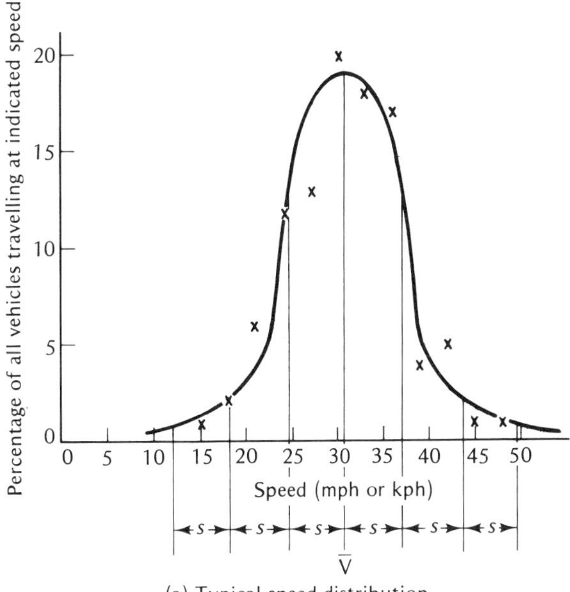

(a) Typical speed distribution

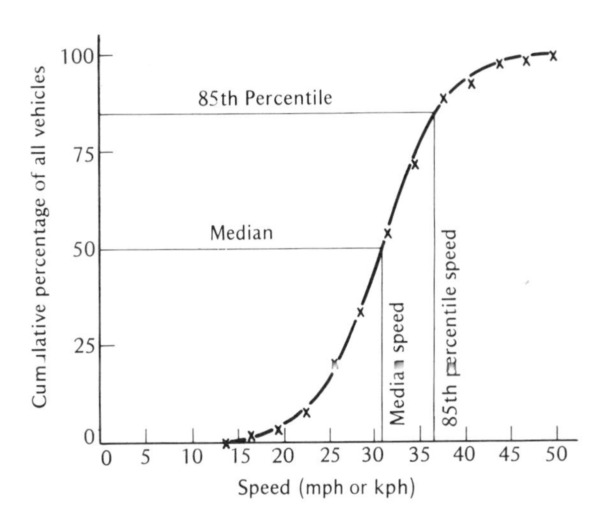

(b) Typical cumulative speed distribution

FIGURE 4–1
TYPICAL SPEED AND CUMULATIVE SPEED DISTRIBUTIONS

VOLUMES, HEADWAYS, AND GAPS

The volume of traffic passing a point, or the flow rate, is the measure of quantitative productivity of the highway or street. It results from the combination of demand for travel and the traffic stream interactions. Especially important in this regard is the capacity of the facility (discussed further below) and the relationship of demand to available capacity. The traffic volumes observed by the traffic engineer show how the system is being used within these capacity constraints.

The time variations in traffic volume reflect the total economic and social demands for transportation. There is less traffic on an urban highway at 3 A.M. than at 8 A.M. Sunday volumes on a street in the retail district may be lower than any other day of the week, whereas a highway serving a major resort or recreational area may have its greatest volumes on that day. Similarly, the volumes observed in December at a regional shopping center exceed those of any other month.

Daily time patterns vary between routes and for different days of the week and months of the year. Figure 4-2 depicts hourly volume patterns for the average weekday for rural Wisconsin trunk highways and urban streets in Milwaukee. The pronounced morning peak flow, typical of urban patterns, is not evident on the rural highways, but the preponderance of travel is during the daylight hours on both types of highway.

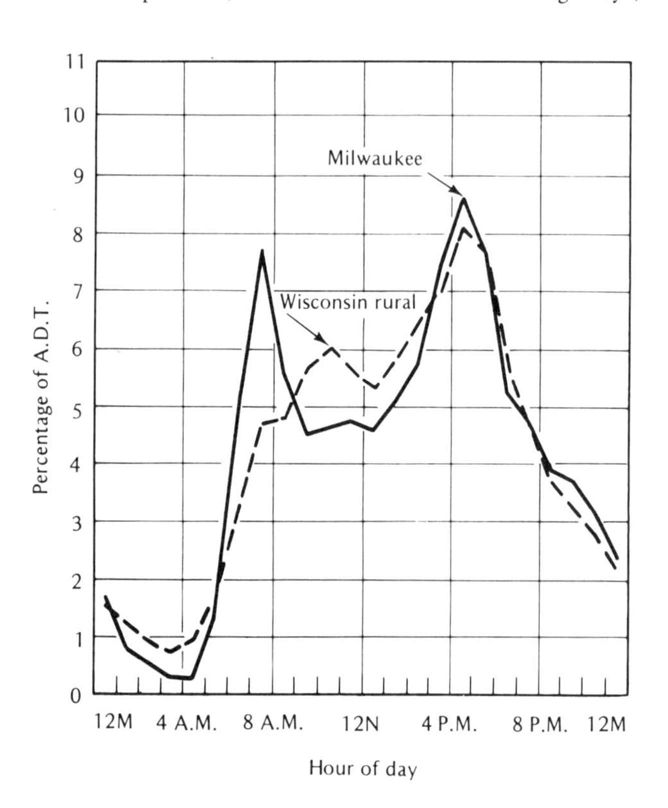

SOURCE: Wisconsin State Highway Department

FIGURE 4-2
HOURLY VARIATIONS OF VOLUME OF AVERAGE WEEKDAY

Weekly volume patterns are shown in Fig. 4-3 for urban streets in Nashville, Tennessee, and for a rural recreational route in Connecticut. For the urban case, Sunday is the day on which the least travel occurs, whereas for the rural-recreational route the weekends are the busiest. These variations in patterns between different types of roadways provide a method of matching similar roadways for volume-sampling purposes. Knowledge of volume-pattern similarities and dissimilarities, particularly for monthly time patterns, enables engineers to match time-pattern profiles in a systematic way so that volume counts of one- or two-day durations may be expanded to Average Daily Traffic estimates.

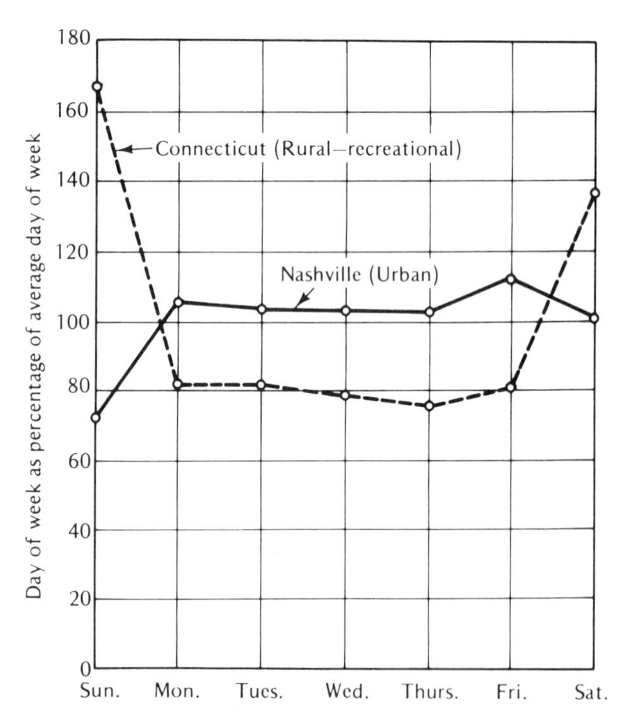

SOURCE: Connecticut State Highway Department and the Nashville Metropolitan Area Transportation Study.

FIGURE 4-3
VOLUME FLUCTUATION BY DAY OF WEEK

Monthly volume patterns are shown in Fig. 4-4 for three Connecticut counting stations. The rural-recreational route serves summer recreational travel, so there are great differences between summer and winter months. The rural interstate station is located on the principal highway between New York and Boston. There is a summer peak

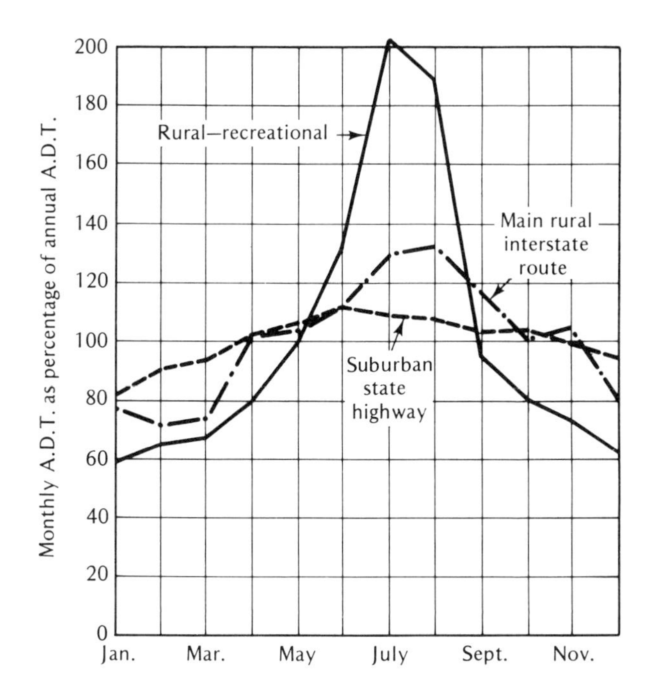

SOURCE: Connecticut State Highway Department

FIGURE 4-4
MONTHLY VARIATION IN TRAFFIC VOLUME

on that route, but it is decidedly less pronounced than for the recreational route. The final location is on a state highway at a suburban location near Hartford. Travel for all months of the year is nearly uniform, reflecting the patterns of work, shopping, and business trips with only a relatively minor increase of automobile travel in the summer months.

Volume patterns are of great significance in determining the capacity of a roadway or intersection. For each time period considered, a capacity value should represent a rate of traffic flow that takes into account fluctuations in traffic demand within that period. For instance, a daily capacity for a particular route or intersection will be based on ensuring that the traffic demand during the peak hours will not exceed a given hourly capacity, while an hourly capacity in turn is based on acceptable shorter peak rates of flow. This emphasizes the necessity of evaluating traffic facilities in terms of their ability to function under specified peak loads rather than average loads.

If all of the hours of the year are ranked from the highest observed volume to the lowest, it will be noted that many hours have volumes substantially exceeding the average hourly volume as shown in Fig. 4-5. Although data for individual facilities are not shown in the figure, there is little dispersion from the average values shown.

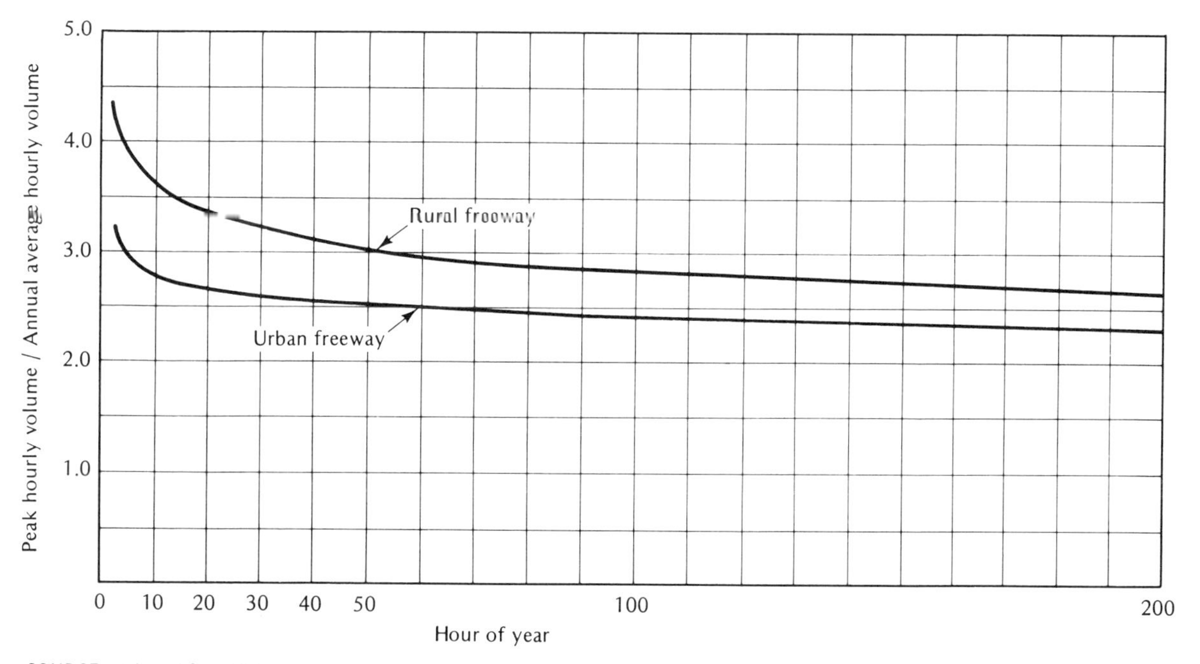

SOURCE: Adapted from Highway Research Board, *Highway Capacity Manual—1965, Special Report 87*, Copyright 1966, p. 40 and Appendix.

FIGURE 4-5
PEAK HOURLY VOLUMES

This is particularly true of the thirtieth-highest hour, which is often used as a design flow. The ratios shown in the figure are also very stable over time. This makes it possible to use these ratios confidently in estimating peak hourly flows when average hourly volumes are known.

Not only are traffic volumes concentrated in certain hours of the day or days of the week, but there is at the same time concentration of traffic on parts of the highway system. Figure 4-6 shows that in both urban and rural areas of the United States traffic is concentrated on the major highway systems, while local streets and roads, which represent over 80 percent of the total mileage, carry less than one-third of all urban and one-fourth of all rural travel.

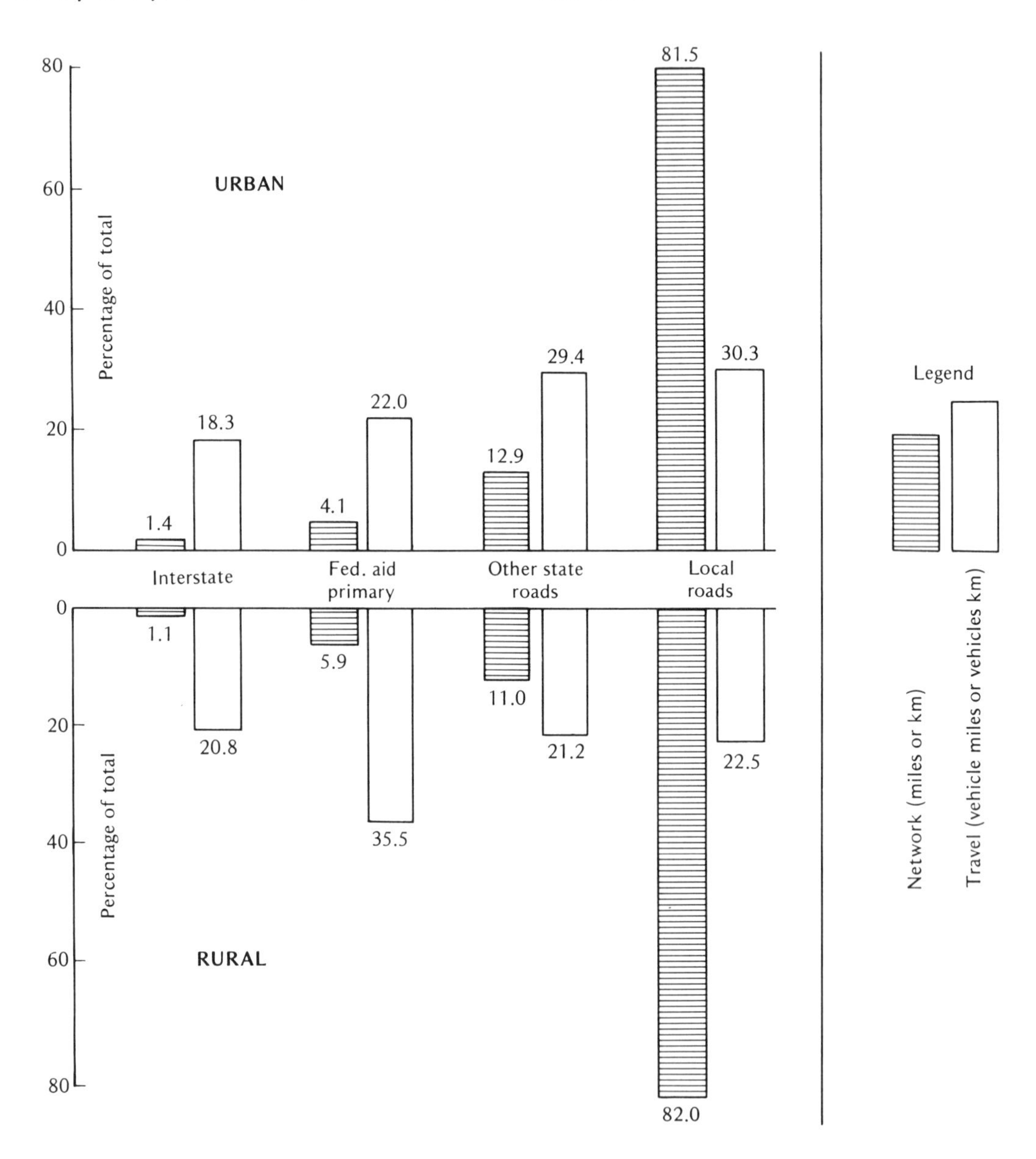

SOURCE: Adapted from data in Federal Highway Administration, *Highway Statistics 1975*, Tables VM2 and M12.

FIGURE 4-6
VEHICLE MILEAGE BY HIGHWAY SYSTEM (UNITED STATES, 1975)

Further, only slightly more than five percent of all rural segments of state highway systems have volumes in excess of 10,000 vehicles per day, whereas 36 percent of urban extensions of state highway systems have volumes exceeding 10,000 vehicles per day.

There is also a directional distribution of traffic volumes on two-way facilities which must be taken into account in design and control. During long intervals of time, a day or a week, the volumes in both directions are approximately equal because of the cyclic patterns of travel. However, for specific hours of the day there is often an imbalance in traffic flow. These directional distributions are most critical during peak periods of traffic flow, as, for example, on major commuter and recreational routes. Table 4-1 shows typical examples of unbalanced directional distributions in peak periods.

Traffic volume lane distribution is an observable characteristic of a multilane highway. The distribution is affected mainly by total volume, and the location and design of the roadways, intersections, and ramps. The geometric design of the highway, together with composition of traffic and driver practices, also affects the lane distribution. At low volumes the left lane carries only a small percentage of the total volume. As volumes increase, the left and middle lanes of a six-lane freeway begin to be utilized more until each will be serving more than one-third of the total flow. On urban arterial facilities with three moving lanes and one parked lane for each direction of movement, the second lane usually carries the highest volume of traffic and the lane next to the parking lane the lowest volume.

Headways and gaps in the traffic stream are of considerable importance in traffic operations. A headway is defined as the time between the arrival of successive vehicles at a given point; the average headway is, therefore, the reciprocal of the average flow rate. A gap is the "open time" between vehicles, or the headway minus the time required for the first vehicle to pass the point of observation. Pedestrians and drivers wishing to cross a stream of traffic are concerned with gaps rather than headways.

Headways and gaps in a traffic stream are distributed about the mean value but, unlike spot speeds, the distributions are skewed toward the lower values. Some types of distributions that approximate headway patterns are given in Chapter 8 of Drew [Ref. (1)] and Chapter 3 of Gerlough and Huber [Ref. (3)]. The fundamental relationships between mean flow rates, headways, and gaps are

$$\bar{h} = \frac{3600}{q} \qquad (4.4)$$

$$\bar{g} = h - \frac{\bar{l}}{\bar{v}} \qquad (4.5)$$

where

$\bar{g}$ = average length of gap in seconds.
$\bar{h}$ = average headway in seconds.
$\bar{l}$ = average length of vehicles, in feet or meters.
$\bar{q}$ = average flow rate in vehicles per hour.
$\bar{v}$ = average speed of vehicles in feet or meters per second.

The information concerning the gap acceptance distribution of drivers or pedestrians (discussed in Chapter 3) can be combined with the distribution of gaps actually available to analyze the quantity of crossings of or merges into an uninterrupted traffic stream which are likely to take place. Such an analysis for merging at freeway on-

TABLE 4-1
DIRECTIONAL IMBALANCE OF TRAFFIC FLOW

FACILITY	TYPE OF TRAFFIC	PEAK HOUR TRAFFIC VOLUME		
		Peak Direction	Both Directions	Percentage in Peak Direction
Caldecott Tunnel	Bedroom towns to center of the urban region.	8,330 vph	10,670 vph	78%
Bay Bridge	Commute to San Francisco; some reverse commuting.	9,000 vph	12,860 vph	70%
Carquinez Bridge	Interstate recreational. Sunday evening peak.	3,180 vph	5,290 vph	60%

SOURCE: Traffic counts by Institute of Transportation Studies, University of California, and by California Department of Transportation, 1975-1976.

ramps is given in Chapter 9 of Drew [Ref. (1)], and for intersection delays in Chapter 8 of Gerlough and Huber [Ref. (3)].

DENSITY AND SPACING

The density of a traffic stream is a measure of congestion. However, since it is more difficult to quantify in the field than are speed or flow rates, it is not used as commonly. It can be obtained by means of aerial photography or by systems of vehicle detectors that report both presence and speed of vehicles. Both methods are relatively sophisticated and expensive, and are, therefore, used most commonly in research studies. For frequent use, as in freeway metering, it is simpler to obtain mean flow rates and speeds and then to calculate the density.

Similarly, spacing of vehicles, which reflects the way a driver evaluates traffic stream congestion at any moment, is seldom measured directly, but usually computed.

Density, or concentration as it is called in some of the literature, is a most important statistic in evaluating traffic streams. Through research it has become known at what densities traffic flow becomes relatively unstable and liable to break down into "forced flow" or bumper-to-bumper conditions. Schemes for metering access to freeways, bridges, and tunnels are based on this knowledge and are designed to maintain traffic stream density on the facility at some point below the region of instability.

Figure 4-7 is an example of a density contour chart for a short section of freeway, representing conditions from 7:00 to 8:15 A.M. An analyst notes locations at which densities exceed 50 vehicles per lane-mile (30 vehicles per lane-km) and then looks for the cause downstream of this location. It is a characteristic of traffic streams that the density at bottlenecks is lower than in the region upstream where vehicles are queueing up to get past the trouble spot. Figure 4-8 shows a density contour diagram for a 30-mile section of a freeway system; this can be used as a general visual tool to show peak period conditions in a system.

RELATIONSHIP BETWEEN FLOW, SPEED, AND DENSITY

The most basic relationship governing traffic streams is that the rate of flow is the product of mean density and

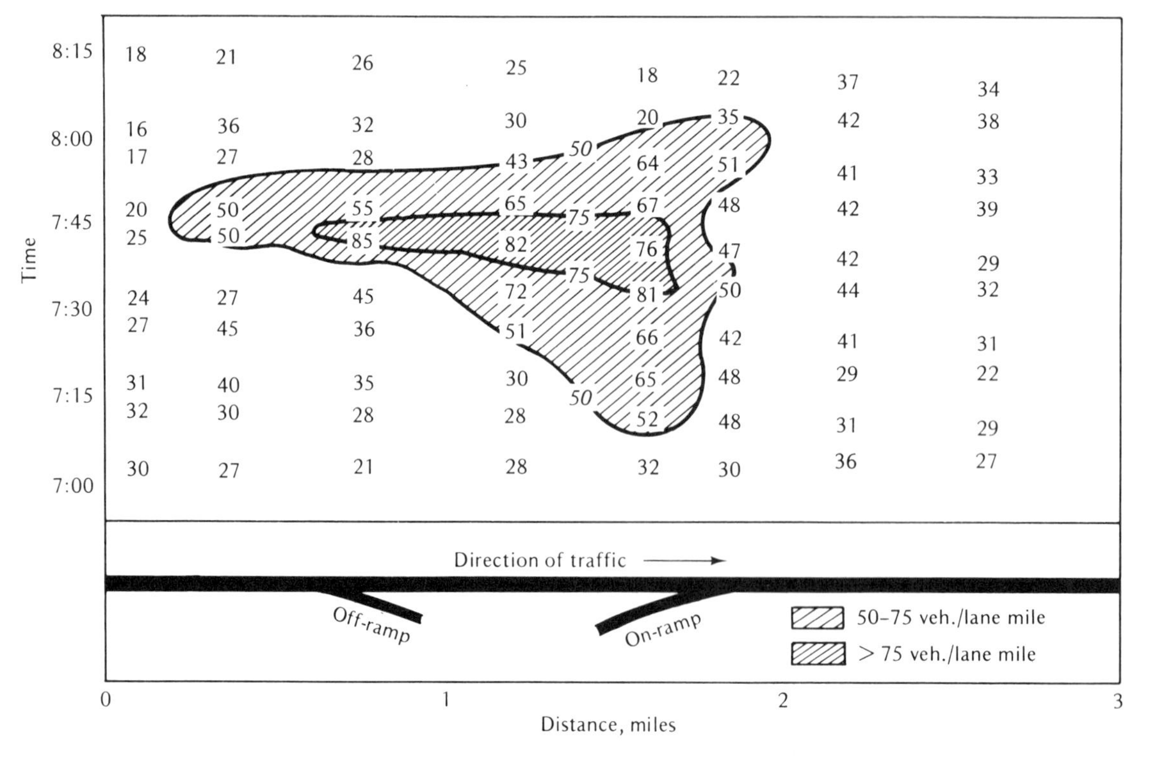

FIGURE 4-7
TYPICAL DENSITY CONTOUR CHART

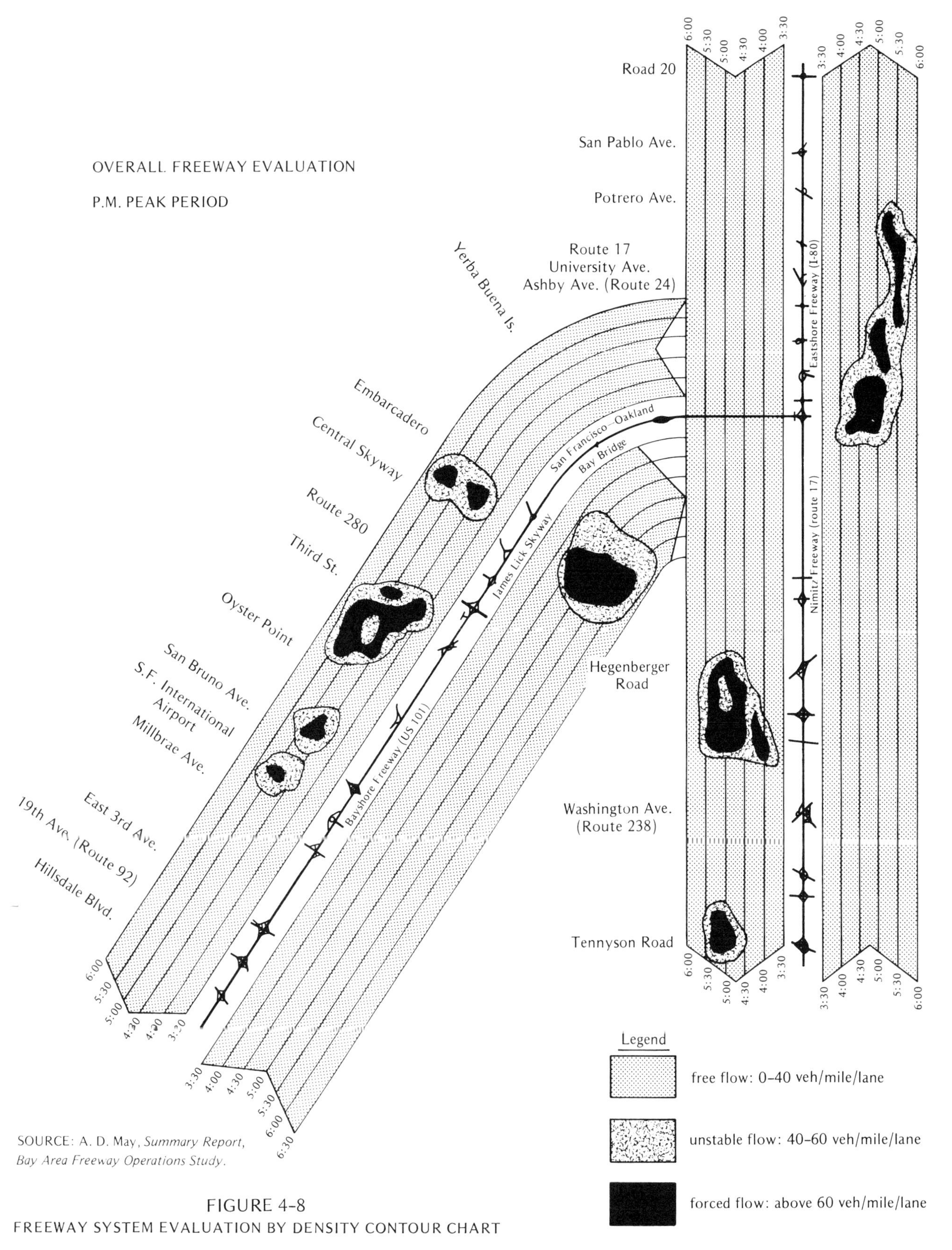

SOURCE: A. D. May, *Summary Report, Bay Area Freeway Operations Study*.

FIGURE 4-8
FREEWAY SYSTEM EVALUATION BY DENSITY CONTOUR CHART

mean velocity, or

$$q = \bar{v} \cdot \bar{k} \qquad (4.6)$$

where

q = the rate of flow in vehicles per unit of time.
$\bar{v}$ = the mean velocity in distance traveled per unit of time.
$\bar{k}$ = the mean density in vehicles per unit length of road.

However, in any stream, vehicles are traveling at a variety of speeds. If one considers only a group of vehicles that travel at the same speed, the above relationship holds for such a group:

$$q_x = v_x \cdot k_x \qquad (4.7)$$

where x refers to a group of vehicles all traveling at v_x.

By definition, the flow rate and the density of the entire traffic stream are the sums of the flow rates and densities, respectively, of each group within the stream which travels at the same speed; or

$$q = \sum_{x=1}^{m} q_x \qquad (4.8)$$

$$k = \sum_{x=1}^{m} k_x \qquad (4.9)$$

where m = the number of speed groups within the stream.

The average speed of the entire traffic stream is then found from

$$\bar{v} = \frac{q}{k} = \frac{\Sigma q_x}{k} = \frac{\Sigma (k_x \cdot v_x)}{k} \qquad (4.10)$$

The value of $\bar{v}$ in Eq. (4.10) is obtained by weighting each group of vehicles traveling at the same speed by their contribution to total density (k_x). The resulting value calculated is, therefore, the space mean speed, as mentioned earlier.

The basic traffic stream relationships are shown in Fig. 4-9. In each graph the solid line represents normal conditions ranging from minimal traffic at low densities to capacity flow. The broken line indicates forced flow, which occurs when traffic densities exceed k_m, the value corresponding to capacity flow v_m. The region near capacity is unstable; some researchers show discontinuity between the solid and broken lines of these graphs.

For a freeway lane, typical values of k_m are about 60 vehicles per mile (40 vehicles per km). The mean speed varies on different lanes at all conditions (Table 4-2), including capacity; thus v_m is perhaps 35 mph (55 kph) in the lane nearest the median and 25 mph (40 kph) in the shoulder lane because of heavy truck traffic. Grade also influences the value of v_m. The resulting capacity flow, q_m, can be calculated from Eq. (4.6) to be approximately 2000 and 1500 vehicles per hour in the median and shoulder lanes, respectively.

CAPACITY AND LEVEL OF SERVICE ANALYSIS

The traffic stream characteristics sketched in Fig. 4-9 include a set of conditions under which maximum flow (q_m) occurs. This flow, the *capacity*, is defined as "the maximum number of vehicles which has a reasonable expectation of passing over a given section of a lane or a roadway in one direction . . . during a given time period under prevailing roadway and traffic conditions" [Ref. (5), p. 5]. The same term can be applied to streams of bicycles and pedestrians (and, for that matter, to trains, aircraft, and ships). Capacity is expressed in the same units as volume or traffic flow. Volume represents an actual rate of flow and responds to variations in traffic demand, whereas capacity indicates a capability or maximum rate of flow with a certain level of service characteristics that can be carried by the roadway.

The need for estimating the capacity of both existing and proposed highways is twofold. First, in the design of street and highway improvements, a capacity analysis indicates the ability of the improvement to carry the expected traffic under satisfactory operating conditions. The second is in the evaluation of existing street and highway networks to determine inadequacies and priority of needs. However, capacity is not the sole criterion used by engineers in their efforts to provide the public with an adequate highway system. Safety, economics, directness of routing, public policy, level of service, land use considerations, and environmental elements must also be considered.

The term "level of service" is used to describe operations where the actual volumes are below the capacity of a given highway. Level of service is a general term that de-

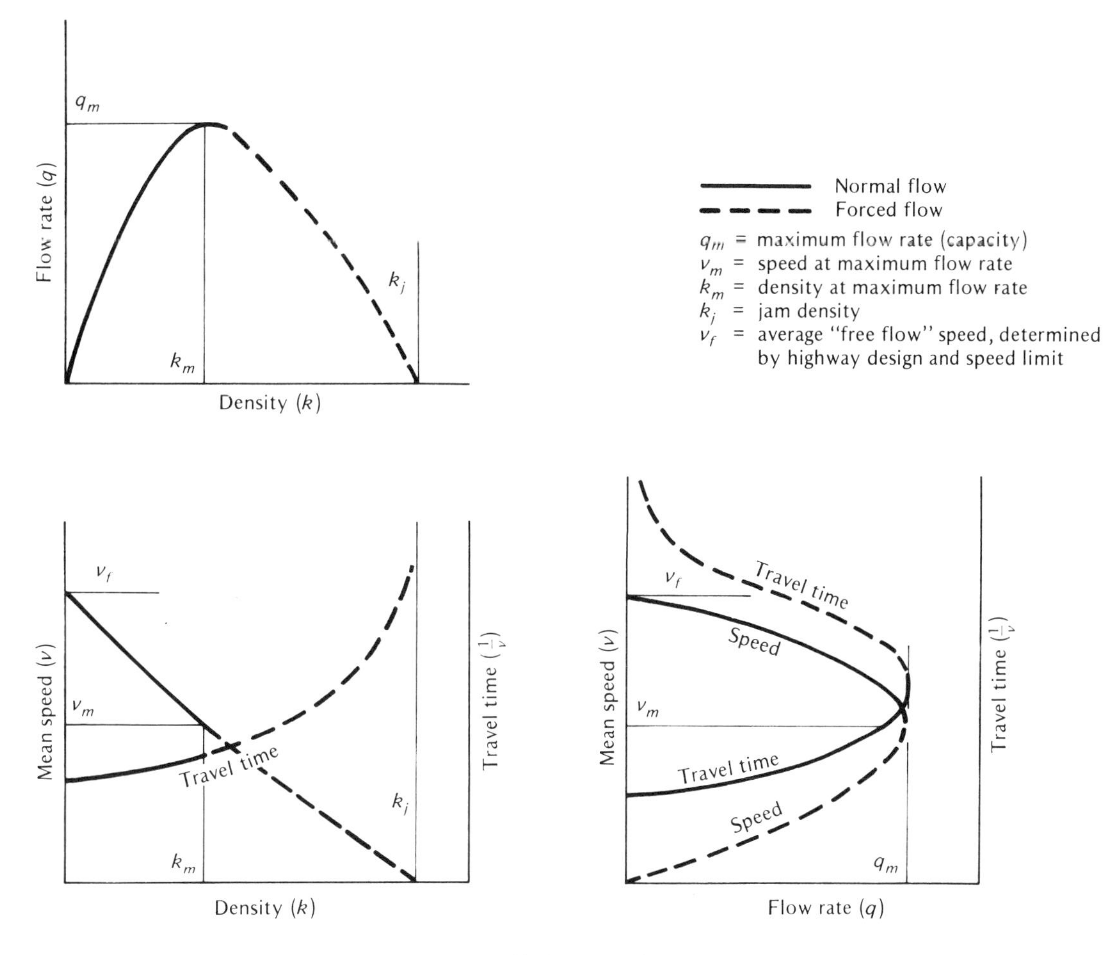

FIGURE 4-9
BASIC RELATIONSHIP BETWEEN FLOW RATE, DENSITY, AND SPEED

TABLE 4-2
MEAN SPEEDS BY LANE (MPH)

LOCATION		LANE 1*	LANE 2	LANE 3
New Jersey Turnpike	(a)	46	55	60
Eastshore Freeway	(b)	46	51	53
Pasadena Freeway	(b)	42	44	46
Ford Expressway	(c)	42	46	47
Davidson Expressway	(c)	45	48	47
Santa Ana Freeway	(b)	44	45	—
North Sacramento Freeway	(b)	46	50	—
Merritt Parkway	(d)	47	51	—

*Lane 1 = Right-hand lane in direction of travel

SOURCE: (a) Ricker, Edmund R., "Monitoring Traffic Speed and Volume," *Traffic Quarterly*, Vol. 13, No. 1, January, 1959, p. 139.
(b) Webb, George M., and Karl Moskowitz, "California Freeway Capacity Study—1956," *Proceedings*, Highway Research Board, Vol. 36, 1957, p. 618.
(c) Wagner, Frederick A., and Adolf D. May, "Volume and Speed Characteristics at Seven Study Locations," *Bulletin 281*, Highway Research Board, 1961, p. 65.
(d) Huber, Matthew J., "Effect of Temporary Bridge on Parkway Performance," *Bulletin 167*, Highway Research Board, 1957, pp. 66-70.

scribes the operating conditions that a driver will experience while traveling on a particular street or highway. In addition to the physical characteristics of the highway, the traffic volume or presence of other vehicles affects level of service. Since roadway conditions are fixed, level of service on any particular highway varies primarily with volume. Measurable items of level of service include frequency of stops, operating speed, travel time, traffic density, and vehicle operating costs.

By far the most commonly used measures of level of service are operating speed for uninterrupted flow and amount of delay for interrupted flow. The validity of these measures is shown by the following generalized description of traffic characteristics for a roadway section.

Capacity has little meaning when traffic demand is low, since there is no direct constraint on the amount of traffic using the highway. The major controls on drivers are inherent roadway characteristics, since other traffic causes little interference with their movements. Horizontal and vertical curvature, location and magnitude of fixed traffic interruptions (usually traffic signals), speed limits, and personal preference control the speed at which the highway is traveled. There is little or no restriction in maneuverability due to the presence of other vehicles, and drivers can maintain their desired speeds with little or no delay. This condition is frequently called "free flow," or level of service A. Traffic density is low.

Level of service B is in the zone of stable flow, with operating speeds beginning to be restricted somewhat by traffic conditions. Drivers still have reasonable freedom to select their speed and lane of operation. Reductions in speed are not unreasonable, with a low probability that traffic flow will be restricted. This level of service has been associated with service volumes used in the design of rural highways.

Level of service C is still in the zone of stable flow, but speeds and maneuverability are more closely controlled by the higher volumes. Most of the drivers are restricted in their freedom to select their own speed, change lanes, or pass. A relatively satisfactory operating speed is still obtained, with service volumes perhaps suitable for urban design practice.

Level of service D approaches unstable flow, with tolerable operating speeds being maintained, though considerably affected by changes in operating conditions. Fluctuations in volume and temporary restrictions to flow may cause substantial drops in operating speeds. Drivers have little freedom to maneuver, and comfort and convenience are low, but conditions can be tolerated for short periods of time.

Level of service E cannot be described by speed alone, but represents operations at even lower operating speeds than in level D, with volumes at or near the capacity of the highway. At capacity, speeds are typically, but not always, in the neighborhood of 30 mph (50 kph). Flow is unstable, and there may be stoppages of momentary duration.

Level of service F describes forced-flow operation at low speeds, where volumes are below capacity. These conditions usually result from queues of vehicles backing up from a restriction downstream. The section under study will be serving as a storage area during parts or all of the peak hour. Speeds are reduced substantially, and stoppages may occur for short or long periods of time because of the downstream congestion. In the extreme, both speed and volume can drop to zero. Densities exceed 60 vehicles per lane per mile (40 vehicles per lane per km).

These levels of service are depicted conceptually in the top graph of Fig. 4-10, and photographically in Fig. 4-11. At what point traffic conditions have lowered the level of service to a minimum that is still acceptable is a matter to be determined by the engineer. The maximum volume or volume/capacity ratio considered acceptable will vary with each type of route, its prevailing roadway characteristics, and the average length of trips made on the facility, longer trips warranting higher levels of service.

Operating conditions within a roadway section are seldom uniform, but tend to vary with the ratio of traffic volume to capacity, which itself is subject to fluctuations in time and place. For example, traffic may enter a freeway in platoons from a ramp, so that the level of service will drop while the platoon is entering and then rise until the next group of entering vehicles appears. Or the presence of a slow-moving vehicle in the traffic stream may be akin to a capacity bottleneck that is shifting its location. Obviously, special incidents, such as stalled vehicles or maintenance work in progress, will cause temporary and fluctuating traffic stream conditions.

Although the discussion here has focused on motor vehicle traffic, flows of bicycles and pedestrians can be analyzed in an analogous fashion. The center and lower graphs of Fig. 4-10 show the speed-flow relationships for these modes of travel. Both the quantities shown and the suggested designations of levels of service are based on recent research into the behavior of bicycle riders and pedestrians [Refs. (2) and (7)]. Again, level of service A represents very low density and complete freedom for the bicyclist or pedestrian to maneuver, level of service E is close to capacity, and level of service F is a stop-and-go condition occurring upstream of bottlenecks.

The information about pedestrian stream characteristics is particularly useful in evaluating the adequacy of sidewalks and crosswalks, and in designing improvements or new facilities. Engineers are also involved in determining the requirements for walkways, stairs, escalators, and moving ramps or belts in transportation terminals, such as airports, transit stations, and parking garages. Design is based on handling large surges in demand at an acceptable level of service. Such demand surges may be caused by the arrival of a train or an aircraft, by the end of an athletic or cultural event which has attracted a large number of spectators, or perhaps just by the normal rush hour travel patterns. In some cases, queueing analysis, as described in the next section for vehicles, can also be useful in understanding the effect of bottlenecks on a system of pedestrian facilities.

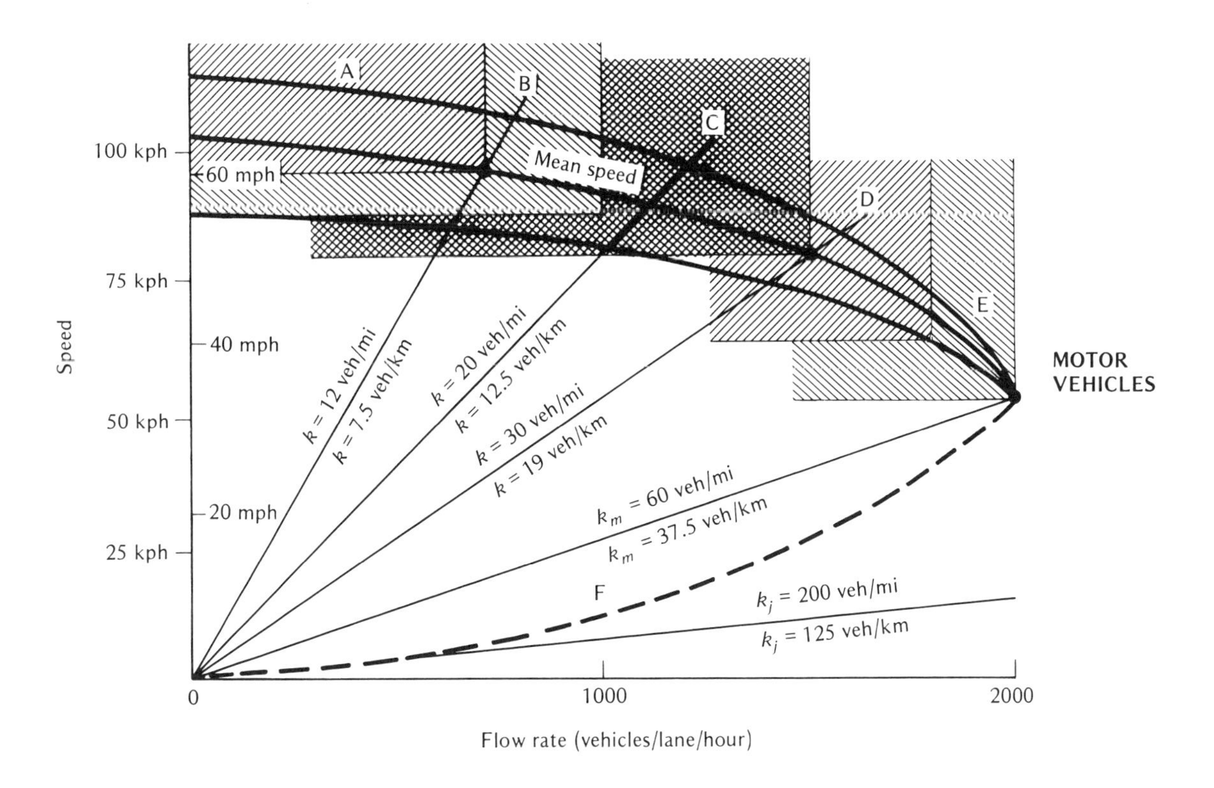

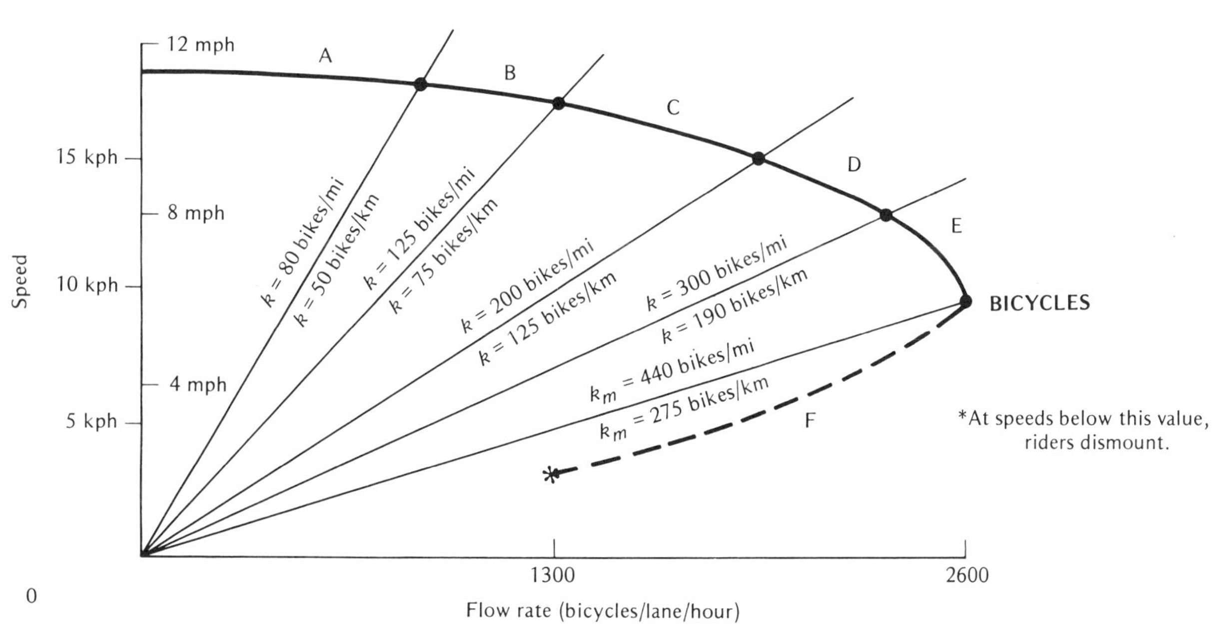

FIGURE 4-10
SPEED FLOW RELATIONSHIPS AND LEVELS OF SERVICE FOR UNINTERRUPTED STREAMS OF MOTOR VEHICLES, BICYCLES, AND PEDESTRIANS

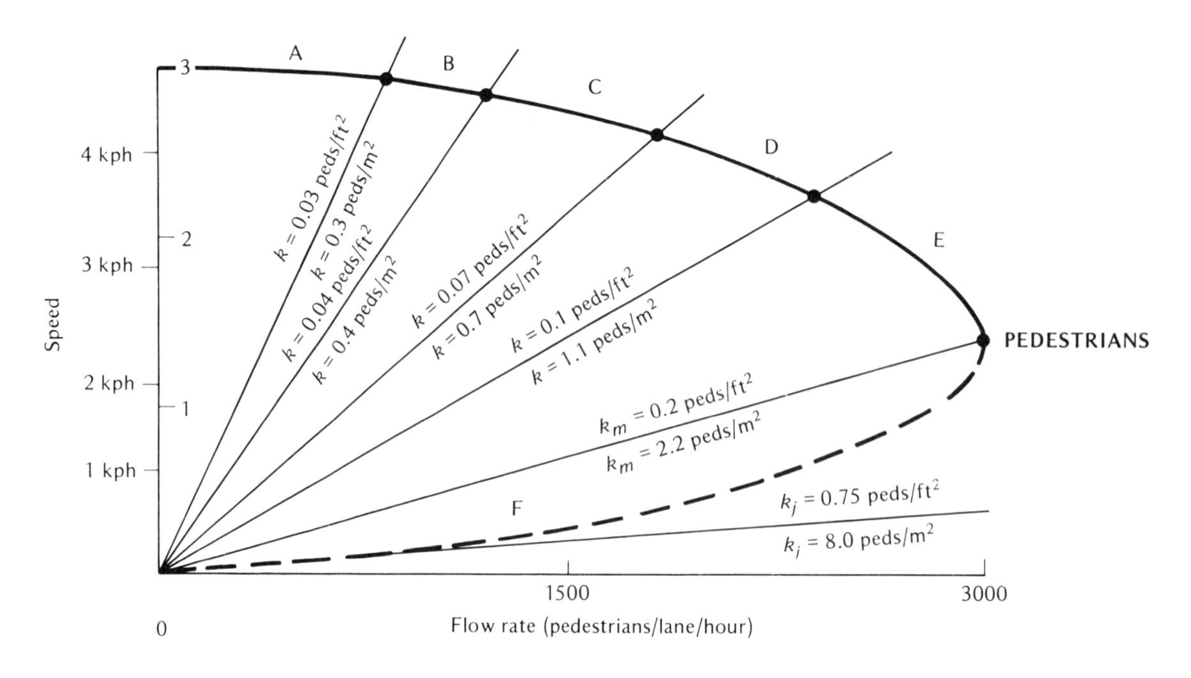

SOURCE: Based on data from References (2), (5), and (7).

FIGURE 4-10
(CONTINUED)

Level A
Free flow, no restrictions on maneuvering or operating speed

Level B
Stable flow, few restrictions

FIGURE 4-11
LEVELS OF SERVICE CONCEPT: ONE DIRECTION OF A MULTILANE FREEWAY

Level C
Stable flow,
more restrictions

Level D
Approaching
unstable flow

Level E
Unstable flow,
some stoppages

Level F
Forced flow,
many stoppages

FIGURE 4-11
(CONTINUED)

QUEUEING ANALYSIS

Queues of vehicles develop upstream of any location where demand approaches or exceeds capacity. The cause may be either a temporary increase in demand or a transitory decrease in capacity. If the condition is caused by demand increases, queues are built up until the demand drops sufficiently of its own accord to cause the backups to dissipate. However, demand can be held to a level that will prevent the formation of queues by such operational techniques as access metering. Temporary reductions in capacity can range from the red phase interruption of traffic streams at signals to the momentary delay caused by toll booth transactions, and include unforseeable incidents and accidents that block or otherwise affect part of the roadway.

A detailed discussion of queueing analysis is beyond the scope of this book. It involves formulation of the headway distribution of traffic upstream of the bottleneck—the arrival pattern—and of traffic passing the bottleneck—the service or departure pattern. These two distributions are combined in equations from which the number of vehicles queueing, the length of queues, and the delays can be calculated.

A graphical analysis of some bottleneck situations is fairly simple and gives a good approximation of queue parameters. The arrival and departure patterns are assumed to be uniform, which may be realistic for some situations, but which will produce results that are too low in others. Figure 4-12, for example, illustrates a case in which demand exceeds capacity temporarily. The arrival and departure curves are drawn on a graph with time on the horizontal and cumulative vehicles on the vertical axis. The arrival curve can be obtained by counting traffic at a point far enough upstream so as not be affected by the bottleneck, or it can be assumed from demand analysis. The departure curve can also be obtained from field studies, or, if the capacity at the bottleneck is known, it can be drawn as a straight line with the slope corresponding to this capacity. From the diagram it can be seen that:

- A queue of vehicles begins to form at time t_1, and does not disappear until t_3.
- The length of the queue at any time t between t_1 and t_3 is $Q(t) = A(t) - D(t)$.
- A vehicle that arrives at t does not depart until $W(t)$ later.
- The queue is longest at t_2 when the arrival rate is just equal to q_m.
- The total queueing delay is equal to the area between the arrival and departure curves.

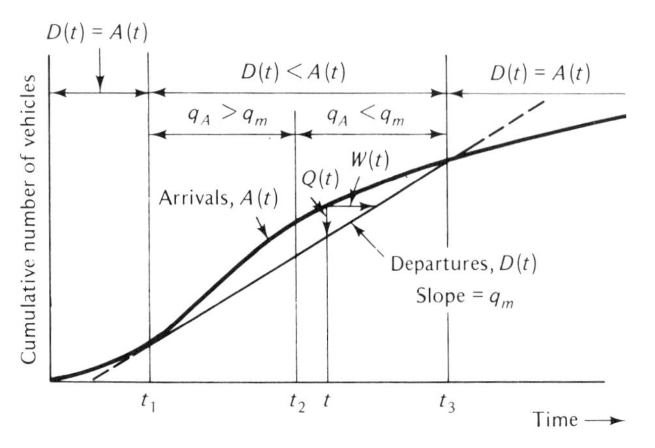

FIGURE 4-12
QUEUEING DIAGRAM WITH VARYING DEMAND

Figure 4-13 illustrates a case where the demand (arrival rate) is constant but the service rate varies. This diagram may be thought of as representing flow on one approach to a signalized intersection. The red phase occurs between time t_1 and t_2, and again from t_3 to t_4. During this phase no traffic leaves the front of the queue, so the departure curve $D(t)$ is drawn horizontally. During the green phases cars depart at the rate representing the capacity of the intersection approach (slope q_g) provided there is a queue waiting to enter, or at the arrival rate $A(t)$ if the queue has dissipated. Again, queue length, amount of delay per vehicle, and total delay can be read off the figure, as was the case in Fig. 4-12.

This diagram, though greatly simplified, is a useful introduction to the study of capacity at signalized intersections, which is described in Chapter 11.

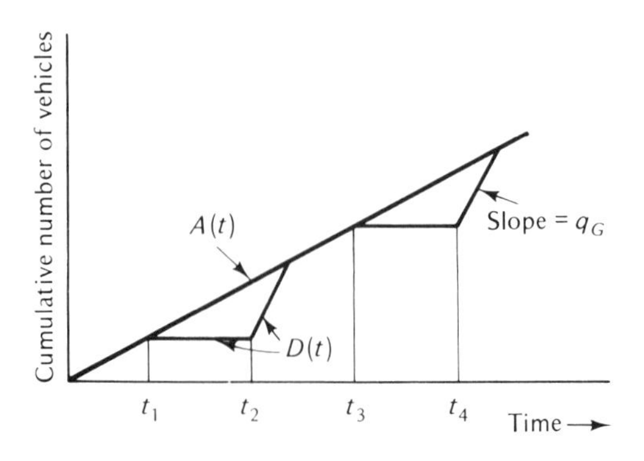

FIGURE 4-13
QUEUEING DIAGRAM WITH VARIABLE OUTPUT CAPACITY

REFERENCES

1. Drew, Donald R., *Traffic Flow Theory and Control.* New York: McGraw-Hill Book Co., 1968, 467 pp.
2. Fruin, John J., *Pedestrian Planning and Design.* New York: Metropolitan Association of Urban Designers and Environmental Planners, Inc. (MAUDEP), 1971. 206 pp.
3. Gerlough, D.L. and M.J. Huber, *Traffic Flow Theory.* Washington, D.C.: Transportation Research Board, Special Report 165, 1975. 222 pp.
4. Haight, Frank A., *Mathematical Theories of Traffic Flow.* New York and London: Academic Press, 1963.
5. Highway Research Board, *Highway Capacity Manual, 1965.* Washington, D.C.: Special Report 87, 1966. 397 pp.
6. Newell, Gordon F., *Applications of Queueing Theory.* London: Chapman and Hall Ltd., 1971. 148 pp.
7. Smith, Dan T., Jr., *Safety and Location Criteria for Bicycle Facilities.* Washington, D.C.: U.S. Federal Highway Administration. Report FHWA-RD-75-112, Feb. 1976.

5

TRAFFIC MEASUREMENTS

The engineer cannot embark on planning systems, designing facilities, or developing operating strategies without first obtaining an understanding of the transportation environment. One must learn the scope and nature of problems that exist at present and must develop a knack for anticipating trends and future situations. For this one needs a solid base of information about current transportation system conditions and a file of data from the past. The data must be accurate or, if the nature of the information does not guarantee precision, then the confidence with which it can be used must be known.

However, traffic measurements are often time-consuming and expensive, and unnecessary data can clutter up files and conceal the valuable information. The engineer, therefore, develops a systematic data acquisition and storage program. Such a program will include some studies carried out at regular intervals to monitor system performance and trends. Other measurements will be taken only to analyze specific problems and to evaluate the impact of solutions applied to these problems.

Traffic measurements are useful in almost every phase of transportation engineering. They provide the information required for determination of supply, demand, and quality of transportation service. Measurements of existing conditions and time trends provide the basis for forecasting the essential input required for effective planning. Measurements provide information on characteristics of flow and geometry that make it possible to design a traffic facility that will operate in a predicted way under the loads that will use it. In the operation of an extensive transportation system, measurements are used to learn the location of trouble spots that need attention. Measurements also make it possible to predict and verify the effectiveness of traffic control improvements introduced at specific locations. Results of studies give information on the level of service users of the system are receiving. Measurements also provide bases for developing warrants for installation of traffic control devices and comparing operations with these warrants. Measurements are also used to assist in the programming and administration of operating departments. The flow of data is invaluable in developing priorities and schedules and assists the administrator in evaluating the performance of the organization, personnel, and equipment.

Most engineers dealing with traffic spend part of their time in planning, making, and using measurements. Although the actual measurements in many cases are made by subprofessionals and unskilled personnel, the engineer must understand the basic principles applicable and guide the execution of the work to obtain results within the limits of accuracy required. In this chapter an introduction to the principles of traffic measurement is presented, and many typical engineering studies are briefly described. The references at the end of the chapter give more detailed descriptions of the studies.

PRINCIPLES OF TRAFFIC MEASUREMENTS

The principles of good traffic measurements are the same as those for studying human behavior and the performance of engineering systems in any situation. The two principal tasks—data collection and interpretation—must be carried out accurately and scientifically, and sound statistical procedures must be followed.

A discussion of the necessary statistical tools is beyond the scope of this book. The reader is referred to Appendices A and B of Reference (2) for a concise summary of the necessary procedures, to References (4) and (5) for texts written with traffic engineers in mind, and to References (3) and (6) for general books on statistical methods. Of the latter, Reference (6) is written for the layman in a very readable form, and Reference (3) is an example of a standard text for engineering students.

Data Collection and Sampling

The type of data collected and the methods used in obtaining them vary with the type of study being conducted, as will be described later in this chapter. However, standards of accuracy and freedom from bias must be adhered to in all field measurement. These are determined by the type and size of samples used. The type of sample reflects which "population" (i.e., which set of total activities or phenomena) is represented by the information. The sample size establishes the level of accuracy and the reliability of the results.

Three sampling situations may arise. In inventory studies, the entire population (e.g., traffic control devices, bus routes, or parking facilities) is counted, and sampling is not involved. In many counting studies and accident analyses, a sampling time period is chosen, such as an entire day, peak periods, hours of darkness, or weekends. Samples are also selected geographically. Within the selected time period and for the chosen location all data are then obtained and tabulated. A third condition occurs when it is impossible to measure all traffic or interview all persons involved in a sample period at a sample location; examples include spot speed studies, origin-destination surveys, and vehicle occupancy counts. In such cases a sample of the sampled fraction of the population must be used, and this must be obtained in a randomly unbiased way.

If it is assumed that the techniques of study used in the field are appropriate, the accuracy of a study sample depends on two factors. First, as the number of measurements in the sample is increased, the accuracy of the sample increases if the population being measured does not change. Second, the method of sampling affects the accuracy. If the sample is drawn from a different population than that intended, the findings may be meaningless. An example of this type of error occurs in a speed study of vehicles passing a point at which it is very difficult for an observer to measure a representative sample of the speeds of all vehicles passing the location. There is a demonstrated tendency to measure more of the high-speed vehicles in the stream, thus giving a biased and meaningless result. In questionnaire studies in which the response is voluntary, the failure of less well-educated groups to respond at all is a consistently observed phenomenon. This must be kept constantly in mind when attempting to use the results of the study.

It is necessary that users of the measurements be able to assess the quality of the result. The engineer can assist the user by describing completely the techniques of measurement and sampling and by indicating the probable effects of uncontrolled and unstudied variations.

The engineer should use every opportunity to check the accuracy of the measurements. This can be done by comparing the results with those of other studies. An example of this can be found in the Origin-Destination survey, described later in the chapter, in which the results of the home interviews are compared with screenline vehicle counts.

Data Analysis and Interpretation

The value of traffic data depends in part on the way they are used. By careful data reduction, summary, and study, the engineer can discover trends and characteristics not necessarily apparent from the original data or from observation of the phenomenon. This data analysis permits the confirmation of hypotheses or rejection of certain prior suspicions. To accomplish these ends the analyst requires an appreciation of and a familarity with tests of reliability and confidence used by statisticians. Most observations are made on systems not suitable for controlled experiments. Without statistical checks one cannot know whether measured events are related to the transportation or traffic situation or to extraneous events, or whether they occurred purely by chance. A good example of this is the impact study, often called the "before-and-after" study. Some characteristics are measured before and again after the introduction of a change in the transportation network, method of operation, level of pricing, or other feature. In a complex system in which the before and after periods are of substantial duration, it is impossible to attribute with confidence any changes observed only to the effect of the change being studied; long-term or unrelated changes may have occurred and may have masked the true impact.

Finally, engineers must not lose sight of the fact that their observations sometimes represent *usage* of the transportation system rather than *demand*. They must be aware of the constraints within the system that may affect the behavior being measured and incorporate this information in their data files. Constraints include both long-range conditions, such as deficiencies in capacity or levels of service, and short-range events, such as poor weather, transportation strikes, or even temporary closures of an artery due to an accident.

TYPES OF TRAFFIC MEASUREMENTS

Engineering measurements of traffic and related transportation studies can be classified by the type of activity involved into six categories: inventories, observation studies, interview studies, accident record analyses, statistical studies, and experiments. Some studies do not involve traffic measurements at all. To understand the human and social factors that define the scope of the transportation mission, the engineer may be involved in studies that cover such areas as economics, geography, psychology, and sociology.

Inventory studies provide the background information about the current state of the transportation network, the highway and transit links, traffic control devices, parking facilities, the transit vehicle fleet, and similar elements. Once a thorough inventory has been completed, a good program of updating the data will make a second complete survey unnecessary.

Numerous observation studies are made to evaluate the performance of the transportation system, detect problems such as congestion and potential hazards, and measure the impact of new or altered traffic facilities or controls. Since traffic patterns and behavior are subject to many influences, observation studies must be repeated at intervals ranging from perhaps every month to every four years.

The only way to obtain information on the motivation for travel, including the way in which choices among alternate modes are made, is by means of interviews or self-administered surveys of samples of the population. Such studies are usually quite expensive and therefore are scheduled only as needed to update older information.

Accident records must be studied on an almost continuous basis. This work is generally performed by office staff, tabulating and analyzing records as they become available. As locations with high frequencies of accidents are identified, field investigations are undertaken.

Statistical studies are also undertaken in the office. These produce information on long-range trends in transportation uses, economic indicators, population and land use changes, and in other factors that are required for planning purposes.

Controlled experiments are sometimes needed to evaluate certain physical characteristics of the transportation system or its components. Examples include the testing of traffic paints for physical characteristics important to the selection of the most visible and durable type, skidding characteristics of pavements, and effectiveness of different types of vehicle detectors.

TRAFFIC DATA SYSTEMS

The transportation engineer accumulates an immense amount of information from measurement studies. Advanced data processing techniques are needed to accomplish a smooth flow of information into and out of data files and the efficient correlation of the various types of data in the files. A method for purging outdated information from the files is also necessary. Computers play a significant role in the operation of such a system.

INVENTORIES

The engineer must have readily available information on the amount and condition of the physical elements of the transportation system. One needs a listing of all the links in the network—freeways, streets, alleys, transit routes—together with data from which one can estimate the capacity and level of service offered. Similarly, an inventory of major terminal points, such as curb and off-street parking facilities, is needed. For system management, a complete file of traffic control devices and local ordinances is needed.

The information gathered is used to program improvements, to establish effective maintenance and replacement programs, and to formulate detailed plans. It is important that as much of this inventory information as possible be kept as machine records to facilitate the speedy retrieval and the correlation of data for purposes of analysis. These inventories are described in the following sections.

Highway Network Inventory

An inventory of all highways in the network should be made and maintained. The roadway inventory is kept current through "change reports" prepared by the highway organization responsible for making the changes. Field checks are made periodically to insure that the inventory is complete and accurate. In jurisdictions with responsibility for major networks, "photologging" (a series of photographs taken through the windshield as a car moves along a highway) is used both for the initial inventory and for updating.

The collection of data is designed to permit drafting of maps and compiling of statistics on the mileage of the several types of streets and highways, the kinds of structures, and other data.

Items usually included in such an inventory are

1. Location and boundary lines of all governmental units.
2. Description of roadway, right-of-way and surface width, surface type, section length, and information on pavement design.
3. Roadway condition: riding quality, condition of surface, and drainage facilities.
4. Roadway structures.
5. Railroad crossings.
6. Critical features of roadway geometry.
7. Services performed on the highway, such as transit, mail, school bus, and other special service routes.
8. Abutting land use information, including driveways and other "marginal friction."

This information is obtained from field surveys, road plans, pertinent maps, aerial photographs, and construction plans. On roads where plans are not available, information on critical features, such as sight distance, curvature, superelevation, gradients, and clearances must be obtained in the field.

A series of maps, charts, and lists is prepared from these data.

Parking Space Inventory

One part of this inventory, the tabulation of curb parking spaces, can be combined with the inventory of other street features. The entire length of curb is classified into parking and other uses, and the parking spaces are identified as unrestricted, subject to time limitations, and metered. Off-street parking facilities are surveyed to obtain a count of the spaces available, types of fees charged, if any, and hours of operation. Curb and off-street information is combined in maps that show the total number of parking spaces available in each block of the study area. Further details on the conduct of parking inventories is given in Chapters 2 and 10 of the *Manual of Traffic Engineering Studies* [Ref. (2)].

Traffic Control Device Inventory

An inventory of all in-place traffic control devices should be made and maintained. This inventory record is essential for planning future needs, programming maintenance, and insuring that all devices conform to current standards. The inventory should include such items as:

1. Type of control device.
2. Description of device—size, model number, special physical features, etc. as appropriate.
3. Location information that can be referenced to the lo-

cation identification system used for other highway records.

4. Authority for erection.
5. In-place history—initial installation date, dates and record of maintenance and repair work, dates of routine maintenance checks and trouble calls, and dates and record of any changes or modifications.

This inventory information may be kept on a manual system when an agency such as a small city or rural county is responsible for only a limited number of traffic control devices. However, for larger street and highway systems, easily retrievable computer tabulations are desirable to provide the flexibility needed for full use of the inventory data.

Special maps are prepared to show the location of street lights, including data on the type and output of luminaires used and the location of overhead or underground wiring. The average level of illumination can be indicated by color codes. These maps can often be prepared by the local utility company or the municipal electrical department.

Transit Network Inventory

An inventory of transit service provides the essential background information for other transit studies and for evaluation of the service being provided. Data gathered include maps showing the routes of all transit operators in the study area, schedules indicating frequency and hours of service on each route as well as trip times between key points on the system, a summary of the rolling stock used showing its capacity, age and condition, and the schedule of fares charged.

TRAFFIC OBSERVATION STUDIES

Most of the studies made on a day-to-day basis are observations of the characteristics of traffic that are made without interfering with the flow. The simplest kind of observational study would obtain information on the behavior of individual vehicles at a given location. An example of such a study would be the investigation of the gap in a through traffic stream that drivers require to cross safely at a given location. Studies of the time interval required for a vehicle stopped on an upgrade at a signal to start and accelerate into the intersection would also be of this type. The following sections describe the most important traffic observation studies.

Measuring Traffic Volume

The engineer's need for traffic flow data is widespread. Almost every facet of operational highway transportation engineering requires knowledge of the traffic volume in the area being studied. When other information is available on conditions at a site, knowledge of the traffic volume makes it possible to develop a reasonable estimate of the general quality of service to users. Different locations can be compared by using the traffic volume to develop rates. This is usually done in accident and delay studies.

The usage of the data determines the base period for which volume must be estimated. The average annual flow on a highway is used as a means of estimating gross annual user revenues, determining total travel, and, in conjunction with information on the length of the section of highway, in computing annual vehicle miles for use in accident and other rate studies. However, since few locations in a network can be counted continuously for a year, the annual total is often calculated from weekly or other short-period traffic counts adjusted for seasonal variations.

The Annual Average Daily Traffic (AADT or ADT), 1/365th of the total annual flow, is a universally accepted measurement. Most standards for improvements and evaluation of facility adequacy are expressed in terms of the average daily traffic flow.

Since average daily traffic does not adequately represent the short-term peaking conditions found on many facilities, a third level of measurement, the hourly flow, is used. Generally, capacity analysis and detailed geometric design require that volume be expressed as an hourly rate. Warrants for devices, regulations, and controls are often expressed in terms of the average hourly volumes that must exist for a certain number of hours. Traffic flows based on counting periods of less than one hour are usually developed to describe unusual surges at localized problem spots.

The most common form of volume measurement is the count of all vehicles passing a point. This process is mechanized. Passage of vehicles is detected by a pneumatic hose or by a magnetic or induction loop. The simplest version of the counter in which the number of vehicles is accumulated has only a display window through which a dial is read at desired intervals. More elaborate machines in common use print the counts on paper tape at preset intervals, usually every five or fifteen minutes, and can also record these data in machine-readable form. The most sophisticated installations, used for permanent counting stations, are connected to a central computer by telephone; volume totals are transmitted automatically and are stored and processed electronically.

For certain kinds of detailed data, manual counting methods must be used. For example, intersection design and operation strategies are based on data that include volumes of turning vehicles, pedestrians, and perhaps bicycles. Capacity analyses require information on the classification of vehicles by type. For policy and planning, the total number of persons traveling on a facility is often desired, necessitating manual counts of vehicle occupants.

Traffic volume measurements are usually taken at single locations, either as a part of an areawide traffic counting program or as a part of a special study. Sometimes the engineer wishes to know the total traffic moving between two parts of the region or into and out of a particular area, such as the central business district. In such cases one locates a screen line or a cordon line on the map and establishes volume measurement stations on all highways crossing such a line.

Continuing estimates of traffic volume and changes in the average flow on extensive highway systems must be kept for a wide variety of planning needs. It is not feasible to equip each counting station with a permanent recording counting machine. Therefore, the use of sophisticated sampling techniques based on statistical principles is required if adequate information is to be obtained at minimum cost.

The key to the efficient measurement of traffic volume is to take advantage of the demonstrated regularities in human activity, time of trip making, and trip route selection. These regularities result in cyclical patterns of remarkable consistency. One studies time rates of flow only long enough to identify the pattern with adequate accuracy, relates the patterns at different types of locations with each other, and then takes samples at other locations for short periods of time, making use of these patterns to expand the counts.

For these continuing measurement programs, therefore, highways are grouped by the cyclical pattern of their traffic flow. In rural areas roads are classified as general-purpose highways, recreational routes, farm service roads, etc. In cities, the principal groups are freeways, other major arterials, collector streets, residential streets, and business/industrial district streets, with the first two groups subdivided into those which are radial or tangential with respect to the city center.

A few *permanent counting stations* are established on each type of rural highway and urban facility. To get greater precision on seasonal, daily, and hourly volume variations, additional *key count stations* are operated on each group of highways for a seven-day period monthly or quarterly. The rest of the network is measured by *coverage* stations, which are operated for a 24-hour period annually or, in the case of very minor roads, every two to four years. The results are usually plotted as a traffic flow map of the area (Fig. 5-1), and are used to obtain both general trends in traffic flow and specific information that might identify a new or imminent capacity problem. An example of a pictorial representation of a cordon count result is shown in Fig. 5-2.

Travel Time and Delay Studies

The importance of travel time as a measure of system performance has been mentioned in Chapter 4. Travel time represents the trip maker's view of the quality of transportation service received.

Travel time varies with time of day; substantial changes can occur at the beginning and end of peak flow periods. In planning travel time measurements, one must, therefore, decide whether both peak and off-peak conditions are to be studied, and plan the sampling procedures accordingly.

If travel time between only a few key points is required, the most efficient procedure is the license plate method. Observers are placed at each observation point with synchronized stop watches. They record three digits of each license plate onto forms or onto a tape, together with the time expressed to within two to five seconds. If the traffic stream is at all substantial, it will be impossible to obtain all numbers. Therefore, all observers sample by reporting only plates ending in a set of agreed-to digits; for example, if it is decided that a sample of 20 percent should be taken, plates ending in any two of the ten digits chosen at random would be included. Usually a sample of sufficient size for statistical accuracy can be obtained in 30 to 60 minutes. One team can, therefore, collect data from several highways on the same day.

The license plate data are later matched manually or by a computer program. Not all vehicles seen at one point will be found at another, since some will have turned onto or off the road between observation points. However, enough matching numbers will be found to permit the calculation of mean travel time, standard deviation, and other descriptors of the data distribution.

However, there is often a need also to obtain data on the delays along a road that are reflected in the total travel time. This information can be obtained only by traveling along the route. Test cars are used for this purpose. Drivers

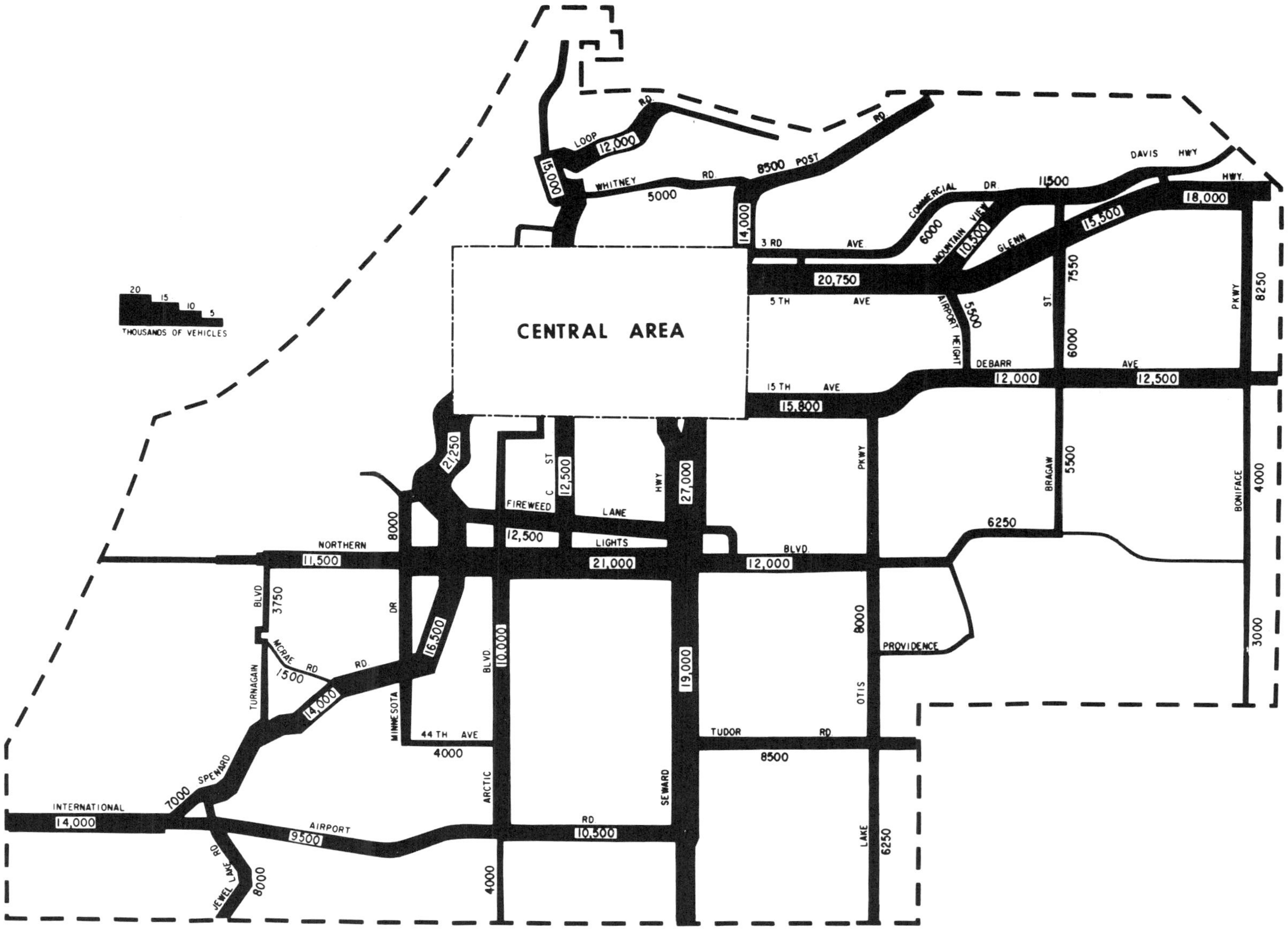

SOURCE: Wilbur Smith and Associates, Anchorage Metropolitan Area Transportation Study. Columbia, S.C.: 1969.

FIGURE 5-1
TRAFFIC VOLUME MAP

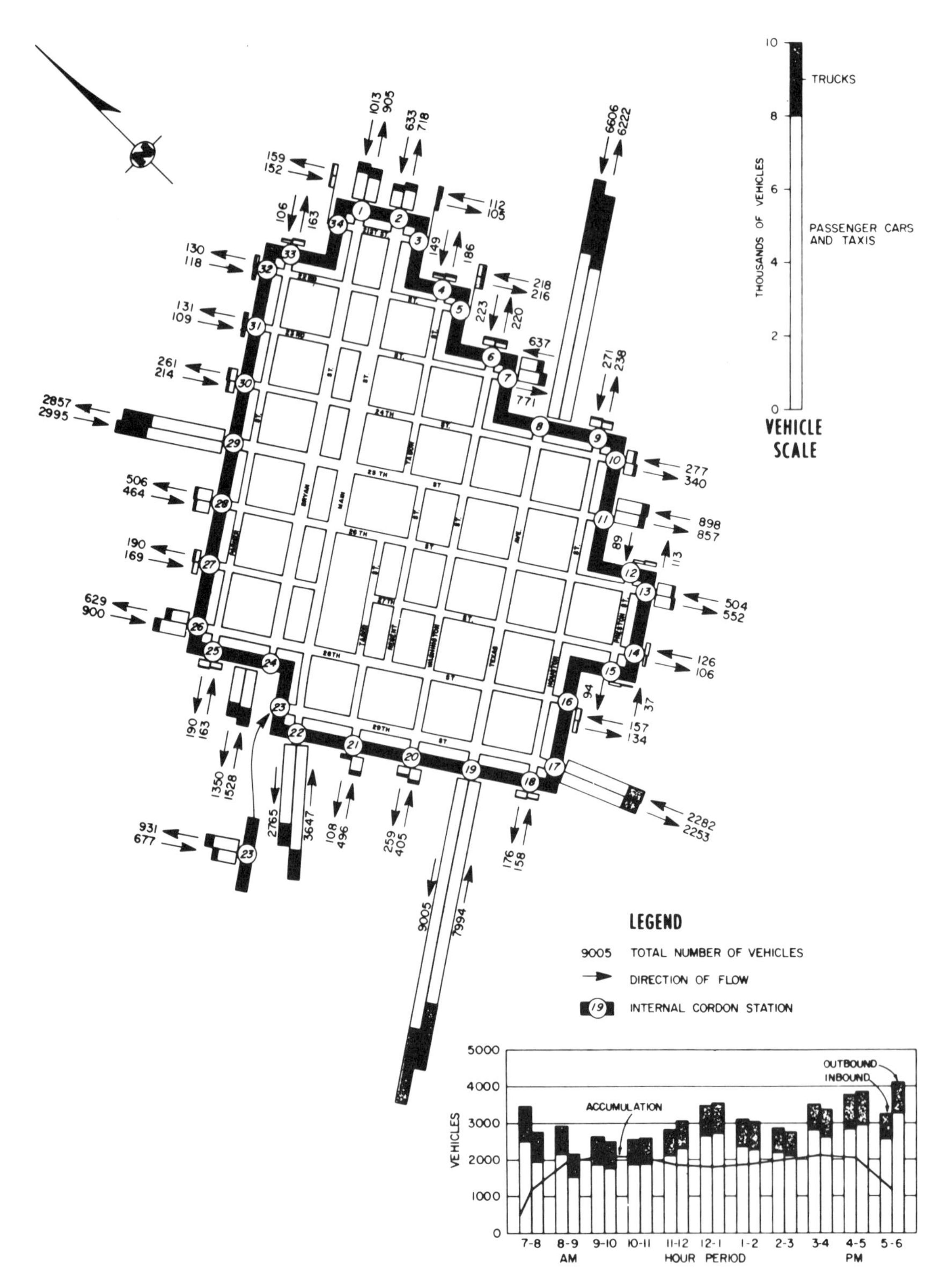

SOURCE: Texas Highway Department, Bryan—College Station—Brazos County Origin—Destination Survey. Austin, Texas: 1970.

FIGURE 5-2

CORDON COUNT MAP OF A CENTRAL BUSINESS DISTRICT FOR A WEEKDAY, 7 AM TO 6 PM are instructed to drive either at what they consider the average speed of traffic or at such speed that they overtake just as many vehicles as overtake them. An observer records the time when the vehicle passes key points along the route, the length of delays caused by traffic signals, conflicts with other vehicles or pedestrians, and other causes, and the location of these conflicts. Thus a complete picture of travel conditions along the route emerges.

The disadvantage of the test car technique is, of course, the time and personnel required to obtain samples of adequate size. For example, if a section of road some 10 minutes long is studied, a round trip will require about 20 minutes. Should the peak period last only an hour, a sample of only three observations can be obtained per car per day. The engineer usually must settle for sample sizes of from 6 to 10, and a permitted error in the mean speed obtained from the study of ±3-5 mph (5-8 kph).

Travel time studies are usually summarized for control sections of the major arterial and freeway network and for representative sections of the minor system. Information is generally determined for both peak and off-peak conditions to give guidance in the study of immediate and long-range improvements. It is often needed for the network analysis models described in Chapter 13.

Spot Speed Studies

The engineer, in addition to learning about the speed performance along a route for the average driver, is also interested in studying the variation in speeds at a given location throughout the day and night as environmental conditions and the characteristics of the drivers using the road change. This type of information is obtained from the spot speed study.

Spot speed studies can be used to detect changes in average speed. Information on the dispersion of speeds is valuable in establishing maximum and minimum speed limits, in determining the need for posting advisory speed signs, in establishing no-passing zones, in analyzing the need for school zone protection, and as a means of evaluating the performance of a geometric improvement or traffic control device.

Since it is often impractical to observe all vehicles passing the study point, it is necessary to resort to sampling techniques based on statistical principles. Average conditions are most often of interest, and if a study is conducted on an average day the result can safely be assumed to be typical of other "average" days. Since absolute precision is not critical in speed measurement, often a relatively small sample (50 or so) of vehicles may be fully adequate. A formula for calculating the specific sample size for each study is given in Reference (2) on pp. 80 and 81.

The measurement of the speed of vehicles can be made with the aid of several different devices. Frequently traffic is timed with stop watches as it passes through a "trap" measured precisely along the road. Pavement marks may be used to mark the ends of the trap, but precision is gained if mirrors are used. Each mirror is placed at an angle of 45 deg to the roadway centerline. An observer standing along the shoulder or sidewalk anywhere between the mirrors can see exactly when a vehicle enters and leaves the trap. Alternatively, a radar meter can measure speeds instantaneously, with the observer recording individual readings from an output dial or using a separate graphic recorder. By any of these methods a *distribution in time* (see Chapter 4) is obtained.

Occasionally speeds are measured by aerial photography as a part of a larger project, perhaps including density measurements. Two photographs are taken at a precisely controlled short interval in the range of 1 to 3 seconds. The distance traveled in this interval by all vehicles appearing in both pictures is measured and converted to the equivalent speeds. A *distribution in space* results from this method.

The data are processed, as are any group of data points of this type, by calculating the mean, standard deviation, and standard error of the mean. The distribution and the cumulative distribution are plotted (see Chapter 4, Fig. 4-1). From the latter one can read the value of the eighty-fifth percentile speed, which is often used in analyzing the adequacy of speed limits.

Traffic Density Studies

The density of traffic or instantaneous concentration of vehicles on a section of road may be measured for only a short distance but is generally expressed in vehicles per mile or km. Since average density, average speed, and average traffic volume are related mathematically, the measurement of two of the characteristics may be easier to accomplish than the direct measurement of the third characteristic. For example, an effective method of estimating the average time that vehicles spend passing through a signalized intersection is to count the volume of traffic and sample the density of traffic at points in time rather than to clock a small sample of vehicles through the intersection one at a time with a stop watch.

Traffic density studies require little more than a vantage point from which to view the section under study. If the section is large, a still photograph may be used to record the instantaneous density, and the counting may be done at a later time. Aerial photographs are often made to determine density. Study results can be represented graphically by the use of contour charts (see Chapter 4, Fig. 4-7).

Traffic Conflict Studies

Traffic conflict measurements are used to obtain an estimate of the accident likelihood of such potentially dangerous locations as intersections and merging and weaving areas. They can provide a faster evaluation of the efficacy of an improvement at such locations than the study of accident records, since for the latter an "after" period of a year is usually required.

Two types of conflicts are counted. The first comprises all infractions of traffic regulations whether or not another vehicle was in the vicinity. The second is defined as evasive actions by drivers—applications of the brakes or lane changing—to avoid a potential collision. At the same time total traffic through the area is counted.

The study is summarized in terms of the ratio of traffic conflicts observed to total vehicles using the facility. Standard statistical tests are made to determine the reliability of the results. The data can be shown in tabular form, with the principal types of conflicts shown separately. "After" data are listed with corresponding "before" figures, and the statistical significance of the differences is shown.

Parking Usage Studies

Parking usage studies are made to measure parking characteristics and demand. Parking facilities exhibiting high occupancy, heavy usage, or frequent turnover indicate the location and characteristics of significant parking demand. These studies provide information on the use being made of alleys and illegal spaces by both cars and trucks as well as the use of spaces reserved for special types of vehicles, such as truck loading zones, taxi stands, and bus stops.

The primary information obtained in parking usage studies is parking duration and facility occupancy. Some studies are designed to obtain information only on occupancy. Other studies provide data on occupancy and duration, permitting a determination of accumulation during the course of a day. In their simplest form these studies consist of continuously watching the parking space, noting the time of entry and exit of a vehicle from the parking space, and determining the parking duration by noting the time difference. Since the ability of an observer to watch more than a few (15-20) spaces of typical curb operations is limited, a periodic check study is often used. In this technique it is assumed that a vehicle is unlikely to enter and leave a parking space between two observations made at a specified time interval apart. The observer follows a designated route, noting the license plate number of each parked vehicle on his route each trip.

In the case of a large off-street parking facility it is unnecessary to observe each space unless detailed information of that type is required. A notation as to the time and identification (license plate number) can be made for each vehicle observed entering and leaving the facility through each entrance and exit. The duration can be determined once the information on entering and leaving for each vehicle has been merged. In commercial off-street parking lots and garages it is frequently possible to obtain from the operator the parking tickets, which have the necessary information for a duration and occupancy study of transient parkers.

In some cases information only on the extent of occupancy of parking facilities is required. Changes in occupancy throughout the day are usually needed. Such information can be obtained by establishing a route along which the observer drives or walks. As each facility is passed, the number of vehicles parked in it is counted. The same route is followed several times during the day to obtain time variations in occupancy. Figure 5-3 shows how usage and supply can be reported geographically.

Transit Usage Studies

Observational studies of the use of transit routes and vehicles assist the transportation engineer in planning and operational decisions. The number of passengers using transit can be measured in two ways. *Load check studies* are conducted at a transit stop at or near the point where the maximum passenger volumes along the route are known or suspected to occur. At bus stops, a single observer records the time of arrival of each vehicle, its route and schedule number, and the number of passengers on board. An attempt is made to count all vehicles (100 percent sample) to avoid data biases caused by arrivals of buses in platoons. Either at the time of the study or after returning to the office, the observer enters the time at which the vehicle was scheduled to pass the observation point, and the number of minutes by which it deviated

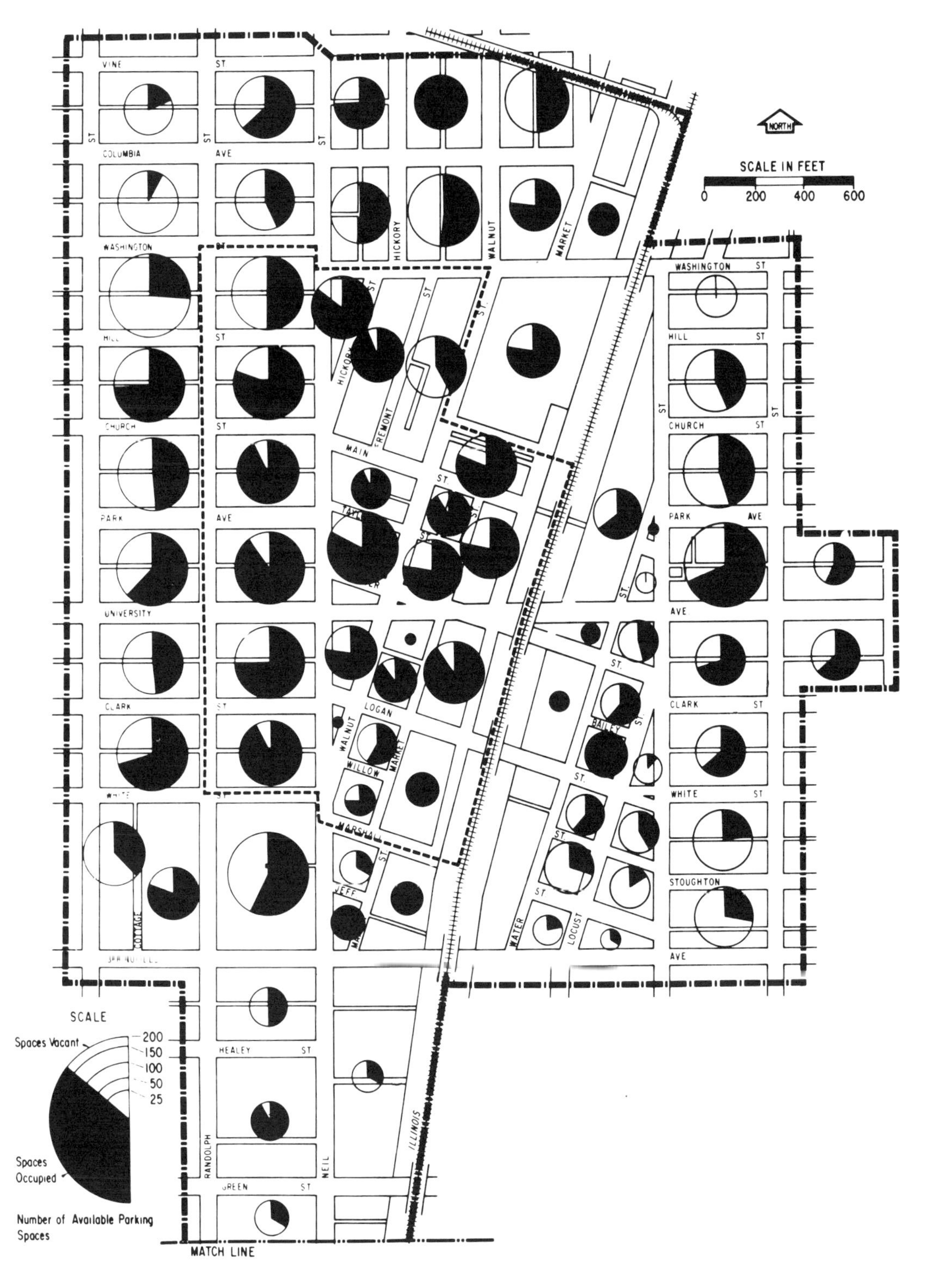

SOURCE: De Leuw, Cather & Co. Report on the Motor Vehicle Parking System. Champaign, Ill.: 1964.

FIGURE 5-3
CURB PARKING SPACE USAGE MAP—CENTRAL BUSINESS DISTRICT—PEAK HOUR

from the schedule. At rapid transit stations, the procedure is similar, but a team of observers is required, with each member responsible for counting two cars. The counts are plotted against time; a vertical line or bar can show the seating capacity of each bus or train, the number of passengers, and the number of empty seats or standees. The data are also summarized by intervals of 15 or 30 minutes.

Load check studies produce a complete count for transit riding on a route with minimal effort, but do not indicate anything about riding patterns along the route. To obtain such data, a *boarding and alighting check* is carried out. Observers ride the transit vehicles from one terminal to the other, recording the number of passengers entering and departing at each stop. A running total of the passengers on board is also kept. If information on the types of fares (adult, children, monthly passes, etc.) used is desired, the entering passengers can be classified accordingly.

If sufficient personnel are available, all vehicles on a route will be studied on the same day. Should this not be possible, a sample of all vehicles is observed, but care must be taken on routes showing a tendency to platoon that the sample is not biased in favor of vehicles that always lead or trail in such a group.

The data from such a study are shown in a route profile, with distance plotted along the horizontal axis and key locations shown, and passengers boarding, alighting, and on board plotted vertically. Chapter 11 of the *Manual of Traffic Engineering Studies* (2) illustrates such a profile.

Because of train lengths and crowded conditions, boarding and alighting studies cannot be conducted on rail systems. Instead, separate counts must be made at each station of the number of passengers entering and leaving each train.

Transit Travel Time and Delay Studies

This type of measurement is analogous to travel time and delay studies for the general vehicle stream. However, the information is collected by observers riding on buses or trains. (On buses, this study can be combined with the boarding and alighting check.) In addition to recording the progress of the vehicle or train along its route and the causes and duration of traffic and signal delays, the time required to allow passengers to board and alight is measured. See Chapter 12 of the *Manual of Traffic Engineering Studies* [Ref. (2)] for further details.

Other Observation Studies

Another frequently used observance study involves the determination of driver route selection. This is accomplished by stationing observers at key locations and having them record the last few digits of the licenses of the vehicles seen passing their vantage point. The information can be correlated in the office and the route of individual vehicles determined. If time information is kept, this method can also be used to measure overall travel time.

Studies to determine the observance of regulations by drivers are frequently conducted at high accident-frequency locations and are made at the time of day and week at which the effectiveness of a particular traffic control device is in question. An example would be a study of the reaction to stop signs by drivers.

The criteria for school crossing protection have evolved from observational survey studies of potential hazards based on street widths, pedestrian volumes, vehicular speeds and volumes, and gaps in the traffic stream. In a school crossing study, the number of pedestrians, the pavement width (curb to curb), and the actual pedestrian delay time (as a percentage of the total survey time) created by the traffic flow are determined. Measurements of gaps in the traffic stream complete the needed information. Data are obtained on a typical school day during the peak crossing hours in both the morning and afternoon.

INTERVIEW STUDIES

The information about transportation that can be obtained through observation is necessarily limited. Therefore, interview studies are used to obtain needed information from the transportation users themselves. The primary data concern the recent trip-making patterns of the population surveyed and certain indicators that describe the individual's social, economic, and mobility status. If the study is intended to assist in the formulation of new transportation policies, attitudinal questions may also be included. However, extensive surveys can be conducted only by interviews in the home or by self-administered questionnaires. When travelers are surveyed at roadside locations, in parking facilities, or within the transit system, questions must be limited to the characteristics of the trip being undertaken at that time and, possibly, one or two background items.

The Home Interview Study

Travel by residents of a metropolitan area is quite regular in character. Travel patterns on typical weekdays are enough alike to permit studies made on several weekdays to be averaged to represent the daily travel. A sampling technique recommended by the Federal Highway Administration [Ref. (7)] makes it possible to estimate total travel made on an average day by residents of small areas of a region called *zones*. Within these areas a small sample (ranging from 2 to 20 percent) of the dwelling units is selected for interviewing. Interviewers visit each selected residence and through direct questioning learn the details of trips made by all residents of the home on the previous day.

The information obtained includes data on household characteristics, such as number of residents, occupation, number of vehicles owned, and other socioeconomic data significant for transportation planning. The travel information obtained for each trip made by every resident older than five years of age on the previous day includes:

1. Mode of travel (auto driver or passenger, transit user, taxi passenger, etc.).
2. Origin of trip.
3. Destination of trip.
4. Time of trip.
5. Purpose of trip.
6. Number of occupants in car.
7. Kind of parking.

This study is expensive but is one of the major ingredients of the comprehensive origin-destination study, a vital input in the urban transportation planning process, as described in Chapter 13.

Roadside Interview Studies

The roadside interview study is designed to obtain information on a particular trip. The study is used most frequently to obtain information on those travelers who pass through an area, but live outside it and would, therefore, not be sampled in the home interview survey.

The study is carried out by stopping drivers on the road at a check point, usually at a cordon or screen line, and asking questions concerning the trip currently being made. Origin and destination of trip and trip purpose are obtained from the driver. The time the trip is made, the type of vehicle, and occupant characteristics can be easily determined.

The survey party often cannot interview every driver, and it is necessary to sample from the group of drivers passing the point. The best way to exercise this control is to sample carefully, to count the number of vehicles, classified by type, passing the location, and to expand the interviews by hourly periods on the basis of the total vehicles of the same class observed. In studies of this type it is necessary to be particularly alert to the danger resulting from attempting to stop vehicles moving on a rural highway.

A variation of the driver interview technique used under extremely heavy traffic conditions is to hand out a postal questionnaire card to drivers as they are stopped briefly at the interview location. The drivers are requested to complete the cards and mail them.

Truck and Taxi Surveys

As a part of the comprehensive origin-destination study, information on typical daily movements of trucks and taxis is obtained. A relatively large sample of the registered trucks and taxicabs is selected, and the travel for the preceding day is determined from documents kept by the vehicle operators as a matter of course or at the request of the interviewing agency.

Other Interview Studies

In a parking usage study the driver of the parked vehicle is interviewed or is requested to return a postcard left with the vehicle. In addition to the observer's obtaining information on usage from direct observation, the parker is generally requested to supply information on the purpose of the trip, the origin of the trip for which the parking location is the destination, and the parker's destinations to be visited after parking the vehicle.

In smaller cities, where motorist trip information is obtained by roadside interview, the usage of the local transit system may not be large enough to be of concern. In larger cities special studies of transit travel may be made by interviewing passengers on transit vehicles. In some cases a postcard questionnaire is used instead of actually interviewing the riders. The returns are correlated with total bus usage on that route for that trip.

ACCIDENT RECORD STUDIES

Accident records are a valuable source of information on the effectiveness of a highway system in providing safe transportation. The engineer, the police department, the courts, educational groups, and safety organizations all use accident records. The objective of all interested groups with regard to accident analysis is to reduce accidents and to evaluate the effectiveness of safety improvement programs. Accident summaries are an important input in design of improvements and in determining priorities for the allocation of funds for construction or for enforcement efforts.

In the study of accidents, data are gathered on the location, date, day of week, time of day, severity and type of accident, the types of vehicles involved, the condition of the road surface, and a general chronological description of the occurrence of the accident and the events leading up to it.

In the simplest record systems, reports are filed manually by location of accident occurrence. Spot maps are then prepared to show visually the locations at which engineering studies are warranted. Different-colored pins are used to illustrate descriptive information about each accident, such as property damage, personal injury, or fatality.

However, a manual records system for accident information has deficiencies and limitations, particularly when large numbers of accident reports must be processed. The manual operations for spot map preparation or report retrieval for analysis purposes require many man-hours. Because of this, large-volume production and detailed statistical analysis are difficult to achieve.

Electronic data processing provides a means for maintaining a large-volume accident records system that is capable of providing the accident data needed for effective traffic safety improvement programs. In the simplest automated systems the accident report information is coded and put on punch cards. In this form the accident information can easily be put into a variety of meaningful statistical summaries, and specific information items, such as high accident location listings, can be quickly prepared. In more advanced systems the coded accident data on the cards are put onto magnetic tape for use in computers. This permits a more refined and detailed analysis of the data and their correlation with other data files such as traffic volumes, traffic control device inventories, and driver and vehicle records.

The safety conditions at specific locations are analyzed by drawing condition and collision diagrams (Fig. 5-4) from the accident reports in the file. The condition diagram, which is drawn to scale, shows all features that could play a role in safety, including sight-obstructing buildings and trees, light poles, hydrants, pavement markings, and the like. In the collision diagram, not to scale, accidents are plotted in clusters, each group depicting one type of collision. Such a diagram facilitates detection of safety problems, and reference to the condition diagram will aid in finding the causes. However, traffic volumes (including turning movements if appropriate), speed distributions, and observance of control devices must also be measured and incorporated in the anlaysis.

STATISTICAL STUDIES

Basic statistical information on the scope and condition of all transportation systems is essential for sound planning and effective administration. The data are fundamental for proper analysis of highway and transit use in various studies such as system classification, cost allocations, fiscal needs, and physical and financial program development.

Mileage statistics are developed to provide information on the physical layout of the transportation network, and trends in system development, system improvement, and rate of overall progress in plant improvement.

Highway statistics should be collected and analyzed on an areawide basis and then identified so that they can be studied separately for rural and urban areas. Data should cover the several road types, surface widths, traffic volume groups, and administrative systems.

Statistics on the physical extent of the highway systems are assembled from data collected as part of the inventory and from travel estimates made in the traffic counting and classification program.

Tables of general and specific interest on motor fuel, motor vehicles, population, highway user and nonuser taxation, as well as finance, mileage, and inventory information are generally available in publications such as *High-*

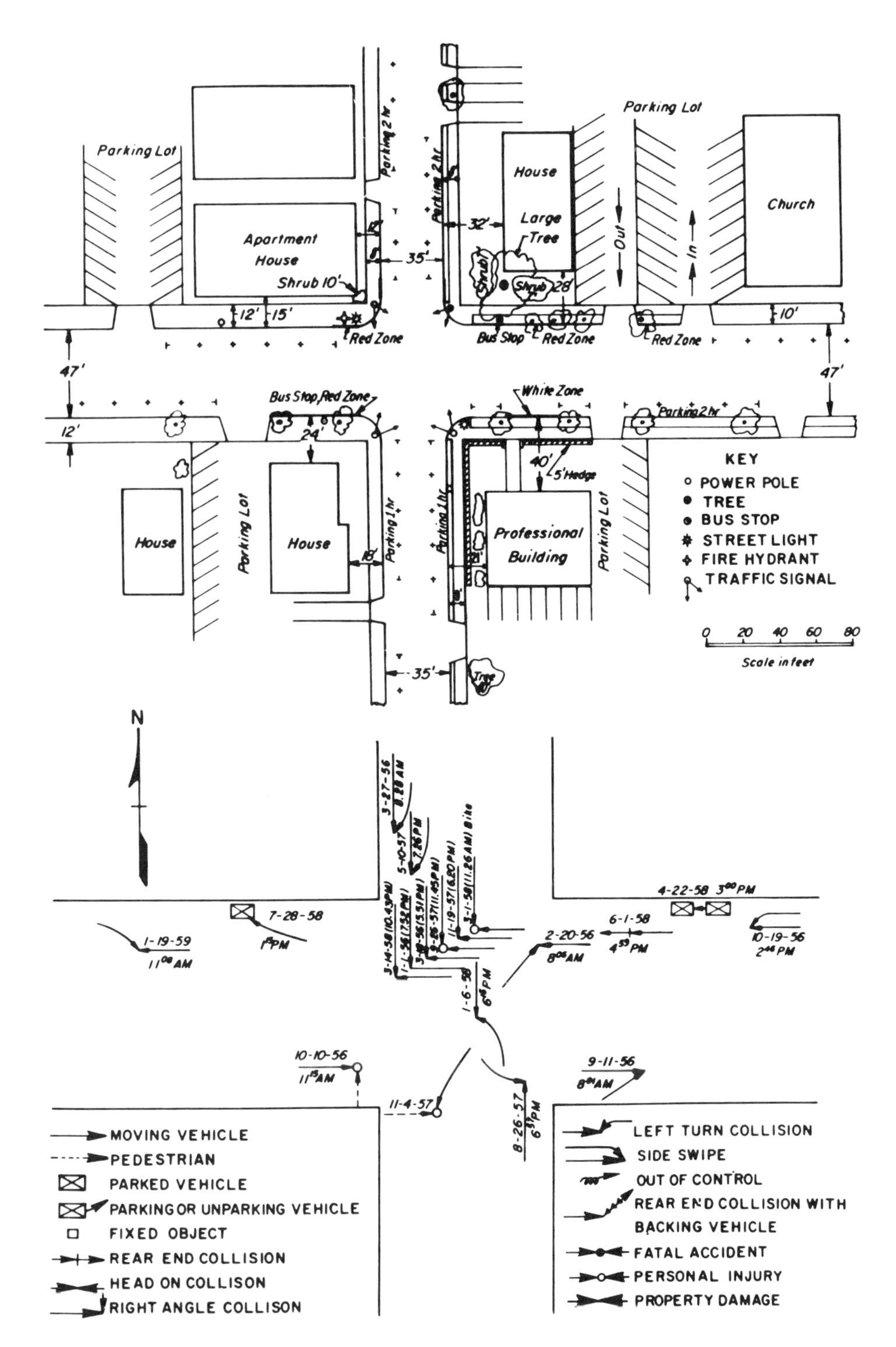

SOURCE: W. S. Homburger and J. H. Kell, Fundamentals of Traffic Engineering, 9th ed. Berkeley, Calif.: University of California, Institute of Transportation Studies, 1977.

FIGURE 5-4

TYPICAL CONDITION DIAGRAM (TOP) AND COLLISION DIAGRAM (BOTTOM) *way Statistics,* published annually by the Federal Highway Administration.

EXPERIMENTAL STUDIES

The engineer frequently tests new traffic control devices to determine their effectiveness in serving traffic needs. This type of measurement study is similar to the laboratory experiments used by other disciplines. For example, advanced technology may be used to test the performance of traffic markings for durability, or to test reflective devices for their brightness and other visual characteristics.

Sometimes the characteristics of the vehicle-road system where the driver does not play an important role must be studied. The most frequently encountered situation of this type is the study of skidding. The analysis of skidding at accident scenes can assist the traffic engineer in estimating speeds and other characteristics of the operational situation that prevailed at the time of the accident. In order to estimate the tire-road coefficient of friction at the time of the accident, it may be necessary to conduct a test at the scene.

REFERENCES

1. Baerwald, John E., editor, *Transportation & Traffic Engineering Handbook,* Englewood Cliffs, N.J. Prentice-Hall, Inc., 1976. Chapter 10.
2. Box, Paul C. and Joseph Oppenlander, *Manual of Traffic Engineering Studies,* 4th ed. Arlington, Va.: Institute of Transportation Engineers, 1976.
3. Dixon, W.J. and F.J. Massey, *Introduction to Statistical Analysis,* 3rd ed. New York: McGraw-Hill Book Co., 1969.
4. Greenshields, Bruce D. and Frank Weida, *Statistics with Application to Highway Traffic Analysis.* Saugatuck, Conn.: Eno Foundation, 1952.
5. Leeming, J.J., *Statistical Methods for Engineers.* London: Blackie, 1963.
6. Moroney, M.J., *Facts from Figures,* 3rd ed. Harmondsworth, England, and Baltimore, Md.: Penguin Books, 1956 (paperback).
7. U.S. Federal Highway Administration, *Urban Origin-Destination Surveys.* Washington, D.C.: U.S. Government Printing Office, 1973.

TRUCKERS

6

GEOMETRIC DESIGN

Geometric highway design pertains to the design of the visible features of the highway and may be thought of as the tailoring of the highway to the terrain, to the controls of urban land space usage, and to the requirements of the highway user [Ref. (1)]. It deals with such roadway elements as cross section, curvature, sight distances, and clearances and thus depends directly on traffic flow characteristics.

In earlier years, the geometrics of an individual road were developed primarily in relation to right-of-way, physical controls, and economic feasibility. Factors such as the road's importance, terrain, and availability of funds were used in establishing design standards. The designer then applied these or higher standards to the terrain to produce a plan and profile of the highway.

However, because the construction of new facilities vitally affects traffic operations, design engineers are giving more and more consideration to traffic requirements. As a result, geometric design is sometimes referred to as traffic design. In traffic design the principles of location and design are developed with reference to the needs of through and local traffic and their effect on the community and other traffic facilities in the system.

There are many factors that influence geometric design. It is imperative that the abilities and limitations of the driver and the vehicle described in earlier chapters be considered. Who will use the road, and how often, is also of extreme importance. Thus, traffic composition, volume, and speed are significant. The geometry of the highway facility must be related to traffic performance and traffic demand to achieve safe, efficient, and economic traffic operations.

Today's concepts of geometric design have emerged both from research and experience. Two publications of the American Association of State Highway and Transportation Officials (AASHTO), *A Policy on Geometric Design for Rural Highways* and *A Policy on Arterial Highways in Urban Areas,* outline the design standards of accepted practice in the United States [Refs. (2) and (1)]. These policies have standardized design practices in the various transportation agencies and by so doing have established a consistency in highway design. The 1978 ITE publication, *Urban Street Design Standards,* presents more recently accepted standards and practices. [Ref. (3)].

DESIGN CONTROLS AND CRITERIA

Highways should be designed to provide mobility, convenience, economy of operation, safety, and to promote desirable land-use relationships. Highways of high standards tend to fulfill these goals, whereas highways of low standards may compromise them. High design standards are warranted where there are sufficient benefits to justify the additional cost. For design purposes, highways and streets are grouped into freeway, primary, and secondary road classes in rural areas; and freeway, major arterial, collector, and local street classes in urban areas. Considerations in highway classification are the travel desires of the public, access requirements, and the continuity of the system. Thus, the service function of a controlled access facility (freeway) is movement, whereas the service function of a local street is to provide access to abutting property. Chapter 14 further describes highway classification.

Each class of highways has a "design designation," which basically expresses the major design controls, reflecting the relative importance of that class. The four major controls are (1) traffic volume and its composition, (2) design vehicles, (3) design speed, and (4) level of service. Two additional controls are driver and pedestrian characteristics. The characteristics associated with these criteria are discussed in earlier chapters. The application of each to geometric design is discussed in the following sections.

Traffic Volume

The high cost of right-of-way and construction dictates the need for thorough engineering analyses in designing a facility to accommodate traffic. Future volumes and type of traffic indicate the service to which a highway improvement is being made and largely determine the type of highway and the geometric features of design. In rural areas, traffic data must include traffic volumes for an average day of the year, volumes by hour of the day at strategic points on the system, and the distribution of vehicles by type, weight, and performance. In urban areas, the need for traffic data is similar, but must be more comprehensive because of the high concentration of traffic.

Volume data needs are determined by the purpose of their intended use. Traffic estimates for a particular route are usually expressed as average daily traffic (ADT) volume. Although, the ADT volume is important in classification studies, in programming capital improvements, and in the design of the structural elements, the sole use of ADT in geometric design is not practical, because it does not indicate the cyclical traffic patterns. A time interval shorter than a day is, therefore, used as a basis for design. In most cases, the interval used is one hour. However, periods of less than one hour are now being considered in urban design.

The peak hour volume, representing conditions that will not be exceeded too often or too much, is the generally accepted criterion for use in geometric design. It is the traffic volume expected to use a facility at some designated future date and is called the design hourly volume (DHV). In rural areas, the *thirtieth highest hourly volume* (30 HV) of the year is usually taken as the design hour volume (DHV). The determination of the 30 HV and justification for its use is explained in *A Policy on Geometric Design of Rural Highways* [Ref. (2), pp. 54-56]. A brief explanation is also presented in an earlier chapter.

In urban areas an appropriate hourly volume may be determined from the study of traffic during the normal daily peak periods. One approach for determining peak-hour volume is to use the highest afternoon peak traffic flow for each week and then to average these values for the year to represent an hourly volume for use in design. If all the morning peak-hour volumes are less than the afternoon peaks, the average of the 52 weekly afternoon peak-hour volumes would be about the same value as the twenty-sixth highest hourly volume of the year. If the morning peaks are equal to the afternoon peaks, the average of the afternoon peaks would be about equal to the fiftieth highest hourly volume. The volumes represented by the twenty-sixth and fiftieth hours are not sufficiently different from the 30-HV value to affect design. Therefore. for use in urban design the thirtieth highest hourly volume can be accepted, since it is a reasonable representation of daily peak hours during the year.

The DHV is expressed as a percentage (K) of the ADT. Traffic data show that the K value generally ranges from 7 to 18 percent in urban areas and 11 to 20 percent in rural areas.

In urban areas, the peak rates of flow within the peak hour often exceed the average hourly rate of flow. These peak flow conditions vary, and this variation is related to the size of the city. For cities of one million population, the peak five-minute flow is approximately 20 percent higher than an average five-minute flow during the peak hour. This percentage is even larger for smaller cities. Therefore, in urban design, the five-minute volume is expanded to a peak hourly volume to accommodate the higher rates of flow that exist over shorter intervals within the peak hour. The *Highway Capacity Manual* supports the use of the highest 15-minute rate of flow as the basis for determining a "peak hour factor" for design purposes at signalized intersections.

A knowledge of the traffic load in each direction is essential in geometric design. Although the traffic volume in each direction on two-way facilities tends to be equal for periods such as a day, an unbalance of flow frequently exists during the peak periods. Two-way design hourly volumes are adjusted by the percentage (D) of traffic that is moving in the heavier direction. Typical values of D for rural and suburban highways range from 60 to 80 percent, with an average of about 67 percent. In and near central business districts, D approaches 50 percent. As with peak hour percentage, the directional splits can be developed and applied to the entire transportation system or applied by area and facility type.

For design purposes, the composition of traffic must also be known; it is normally expressed as the percentage of trucks during the design hours, referred to as T. Vehicles have different operating characteristics. Trucks are heavier, slower, occupy more roadway space, and have an overall effect on traffic operation equivalent to that of several passenger vehicles. Thus, a larger number of trucks (all buses, single-unit trucks, and truck combinations that have operational characteristics different from those of passenger vehicles) in the traffic stream creates a greater traffic load and, therefore, they require more highway capacity.

The estimated values for the average daily traffic, design hourly volume, directional distribution, and composition of traffic for the design year are determined from the travel surveys that are conducted during the planning phase. This information is essential for the design of the highway.

Design Vehicles

As indicated in the AASHTO design manuals [Refs. (1) and (2)].

> The physical characteristics of vehicles and the proportions of various size vehicles using the highways are positive controls in geometric design. Therefore, it is necessary to examine all vehicle types, select general class groupings, and establish representative size vehicles within each class for design use. A "design vehicle" is a selected motor vehicle the weight, dimensions, and operation characteristics of which are used to establish highway design controls to accommodate vehicles of a designated type. For purposes of geometric design, the design vehicle has physical dimensions and a minimum turning radius larger than those of almost all vehicles in its class. Freeways nearly always are designed to accommodate the largest of the several design vehicles. However, on other urban highways, which may include intersections with small turning radii and narrower pavements with curbs, a careful determination should be made as to which design vehicle or vehicles control the design.

Consideration of two general classes of vehicles, passenger cars and trucks, is usually sufficient to determine the required number of traffic lanes. More detailed information on vehicular characteristics is needed for individual design elements. As noted in a previous chapter, the width of the vehicle affects the width of traffic lanes; the length of vehicle affects the turning radii; and the height of the vehicle affects the clearance of the grade separation structures.

The American Association of State Highway and Transportation Officials has determined that six design vehicles are required for design. The six vehicles and their dimensions are shown in Table 6-1, and range from a passenger vehicle to a semitrailer-full trailer combination.

The minimum turning paths of two representative vehicles, the passenger car and a wheel base 40 (WB-40) semitrailer are shown in Figs. 6-1 and 6-2, respectively. A more detailed discussion can be found in Reference (1).

TABLE 6-1
DESIGN VEHICLE DIMENSIONS

DESIGN VEHICLE TYPE	SYMBOL	WHEELBASE	DIMENSION IN FEET (METERS)				
			Front Overhang	Rear Overhang	Overall Length	Overall Width	Height
Passenger car	P	11 (3.4)	3 (0.9)	5 (1.5)	19 (5.8)	7 (2.1)	...
Single unit truck	SU	20 (6.1)	4 (1.2)	6 (1.8)	30 (9.1)	8.5 (2.6)	13.5 (4.1)
Single unit bus	BUS	25 (7.6)	7 (2.1)	8 (2.4)	40 (12.2)	8.5 (2.6)	13.5 (4.1)
Semitrailer combination, intermediate	WB-40	13 + 27 = 40 (12.2)	4 (1.2)	6 (1.8)	50 (15.2)	8.5 (2.1)	13.5 (4.1)
Semitrailer combination, large	WB-50	20 + 30 = 50 (15.2)	3 (0.9)	2 (0.6)	55 (16.8)	8.5 (2.6)	13.5 (4.1)
Semitrailer fulltrailer combination	WB-60	9.7 + 20.0 +9.4[a] + 20.9 = 60 (18.3)	2 (0.6)	3 (0.9)	65 (19.8)	8.5 (2.6)	13.5 (4.1)

[a]Distance between rear wheels of front trailer and front wheels of rear trailer.

SOURCE: *A Policy on Design of Urban Highways and Arterial Streets,* American Association of State Highway Officials, 1973, p. 269.

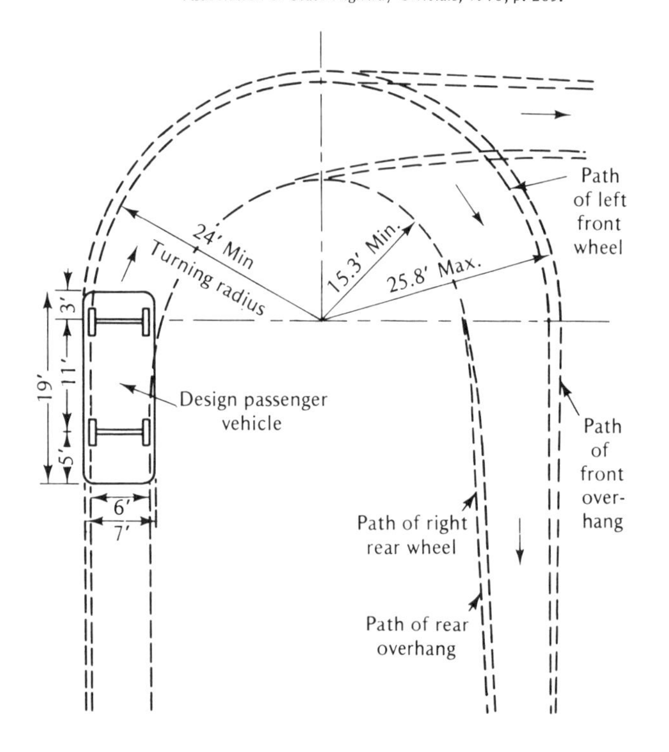

SOURCE: American Association of State Highway Officials, *A Policy on Geometric Design of Rural Highways,* 1965, p. 73.

FIGURE 6-1
TURNING PATH OF PASSENGER DESIGN VEHICLE

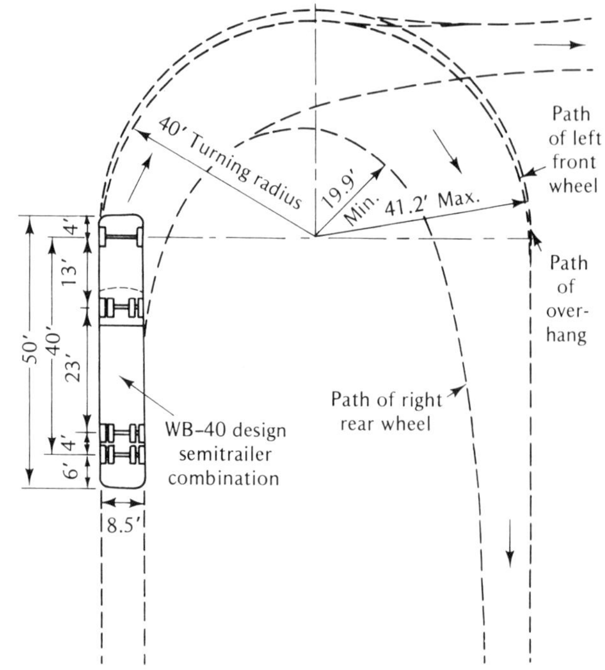

SOURCE: American Association of State Highway Officials, *A Policy on Geometric Design of Rural Highways,* 1965, p. 83.

FIGURE 6-2
TURNING PATH OF WB-40 DESIGN VEHICLE

Design Speed

Speed is one of the most important factors in the user's choice of alternate routes or transportation modes. It is also very important in highway design. The maximum safe speed that can be sustained over a specified section of highway provides another standard design determinant. This speed is called the *design speed.* Such design elements as horizontal and vertical alignment, sight distance, superelevation, shoulder and lane widths, and clearances depend on design speed and vary with a change in design speed.

The choice of design speed is based principally on the function of the roadway. For example, a highway in a rural area carrying a large volume of interstate traffic justifies a higher design speed than a city arterial that serves only local traffic. Average trip length should be considered in selecting a design speed, with a larger value for longer trips. Facilities should be designed with all elements in balance and consistent with a proper design speed. Drivers tend to adjust their speed to the characteristics of the roadway and traffic, not to its system designation.

Design speeds are usually expressed in 10-mile-per-hour (16 kmh) increments between 30 and 60 miles per hour (48 to 97 kmh), and five-mile-per-hour (8 kmh) increments between 60 and 80 miles per hour (97 and 129 kmh). It should be emphasized that design speed is used to establish minimum geometric standards. Higher standards should be used whenever they can be attained at little or no additional cost.

A consistent design speed is most important in the design of a substantial length of highway. Changes in design speed on short sections are sometimes required because of changes in terrain and physical controls. In such cases, the design speed should be gradually reduced over a sufficient distance to permit a safe change in operating speed before the restricted section of highway is reached. A more detailed discussion of design speed can be found in the AASHTO red book, pp. 281-289 [Ref. (1)] .

Design Designation

From the discussion of design criteria, it is evident that in a broad sense there are four major controls—traffic volume and its composition, the design vehicle, design speed, and level of service. These controls should be formally incorporated in a set of plans for a proposed facility—preferably on the title sheet. A typical design designation should include the following:

- *ADT* (current year)
- *ADT* (future design year)
- *DHV* (future design year)
- *D* (directional distribution)
- *T* (percentage of trucks)
- *V* (design speed)
- Level of service (usually B or C)

In addition, modern practice includes designating the degree of control of access (i.e., full or partial) and the design vehicle (i.e., P, SU, etc.).

ELEMENTS OF DESIGN

Stopping Sight Distance

Safe vehicle operation demands that a clear line of sight of suitable length be provided in the design of highway alignment. A primary feature in highway design is the arrangement of the geometric elements so that there is adequate sight distance for safe and efficient operations. The sight distances to be provided are either "stopping sight distance" or "passing sight distance." A safe stopping sight distance is the minimum distance required for a driver to stop his vehicle while traveling at or near the design speed before reaching an object seen in his path.

Stopping sight distance is made up of two elements: the distance traveled from the time the object can be seen to the instant that the brakes are applied, and the distance traveled while the driver brakes the vehicle to a stop. For design purposes a value of 2.5 seconds is generally used. This is made up of a perception time of 1.5 seconds and a brake reaction time of one second. The formula for computing stopping sight distance is discussed in the section on vehicle acceleration and deceleration in Chapter 3.

Table 6-2 shows the relationship of sight distance to various design speeds, calculated according to the formula in Chapter 3. The stopping sight distances shown are for level roads. When a highway is on a downgrade, the sight distances shown are slightly modified. The minimum stopping sight distances in Table 6-2 should be provided throughout the length of all streets, highways, and highway connections.

Passing Sight Distance

On two-lane highways with two-way operation, the opportunity to pass slow-moving vehicles must be provided at

TABLE 6-2
MINIMUM STOPPING SIGHT DISTANCES

DESIGN SPEED		ASSUMED SPEED FOR CONDITION		PERCEPTION AND BRAKE REACTION			COEFFICIENT OF FRICTION	BRAKING DISTANCE ON LEVEL		STOPPING SIGHT DISTANCE	
				Time	Distance						
mph	(km/hr)	mph	(km/hr)	sec.	feet	(m)	f	feet	(m)	feet	(m)
Design Criteria—*Wet Pavements*											
30	(48)	28	(45)	2.5	103	(31)	0.36	73	(22)	176	(54)
40	(64)	36	(56)	2.5	132	(40)	0.33	131	(40)	263	(80)
50	(80)	44	(71)	2.5	161	(49)	0.31	208	(63)	369	(112)
60	(97)	52	(84)	2.5	191	(58)	0.30	300	(92)	491	(150)
65	(105)	55	(89)	2.5	202	(62)	0.30	336	(102)	538	(164)
70	(113)	58	(93)	2.5	213	(65)	0.29	387	(118)	600	(183)
75	(121)	61	(98)	2.5	224	(68)	0.28	443	(135)	667	(203)
80	(129)	64	(103)	2.5	235	(72)	0.27	506	(154)	741	(226)
Comparative Values—*Dry Pavements*											
30	(48)	30	(48)	2.5	110	(34)	0.62	48	(15)	158	(48)
40	(64)	40	(64)	2.5	147	(45)	0.60	89	(27)	236	(72)
50	(80)	50	(80)	2.5	183	(56)	0.58	144	(44)	327	(100)
60	(97)	60	(97)	2.5	220	(67)	0.56	214	(65)	434	(132)
65	(105)	65	(105)	2.5	238	(73)	0.56	251	(76)	489	(149)
70	(113)	70	(113)	2.5	257	(78)	0.55	297	(91)	554	(169)
75	(121)	75	(121)	2.5	275	(84)	0.54	347	(106)	622	(190)
80	(129)	80	(129)	2.5	293	(89)	0.53	403	(123)	696	(212)

SOURCE: *A Policy of Geometric Design of Rural Highways*, American Association of State Highway Officials, 1965, p. 136.

intervals in the interest of safety and service to the faster-moving drivers. Passing sight distance is measured from driver's eye, 3.75 feet (1.1 m) above the pavement to the top of an object 4.5 feet (1.4 m) high on the pavement. A passenger car is assumed to be the object sighted. Passing sight distance, to be effective on two-lane highways, should be provided over a high proportion of the highway length. The minimum clear distance ahead to permit safe passing is called the safe passing sight distance. This distance should be long enough to permit a vehicle to safely pass an overtaken vehicle and return to the right lane with reasonable clearance before meeting an oncoming vehicle.

In computing safe passing sight distance, certain assumptions relative to traffic behavior are generally made. These are

1. The overtaken vehicle travels at a uniform speed.
2. The passing vehicle is required to follow at the same speed until there is an opportunity to pass.
3. The driver of the passing vehicle requires a certain period of time to start his maneuver.
4. The passing vehicle accelerates during the passing maneuver, and its average speed during occupancy of the left lane is 10 miles per hour (16 kmh) higher than that of the overtaken vehicle.

The minimum passing sight distance for two-lane highways is determined from the sum of the four distances illustrated in Fig. 6-3 and described as follows:

d_1 = distance traveled during the preliminary delay period (the distance traveled during perception and reaction time and during the initial acceleration to to the point of encroachment on the left lane).

d_2 = distance traveled while the passing vehicle occupies any part of the left lane.

d_3 = distance between the passing vehicle at the end of its maneuver and the opposing vehicle.

d_4 = distance covered by an opposing vehicle for two-thirds of the time the passing vehicle occupies the left lane, or $\frac{2}{3}d_2$. (The implication here is that during the first part of the passing maneuver the driver can still return to the right lane if he sees an opposing vehicle; this uncommitted distance is about $\frac{1}{3}d_2$. Since the opposing and passing vehi-

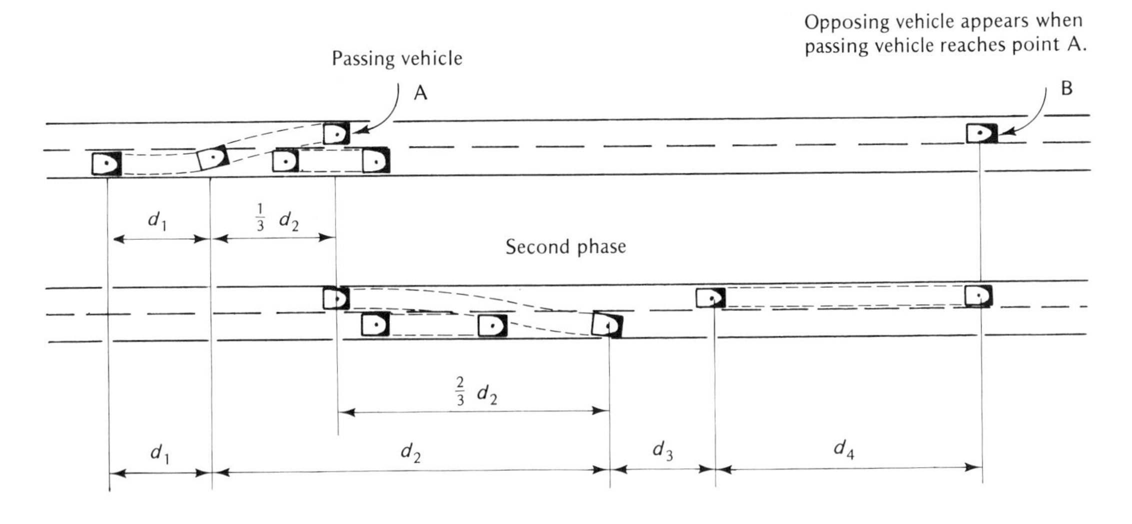

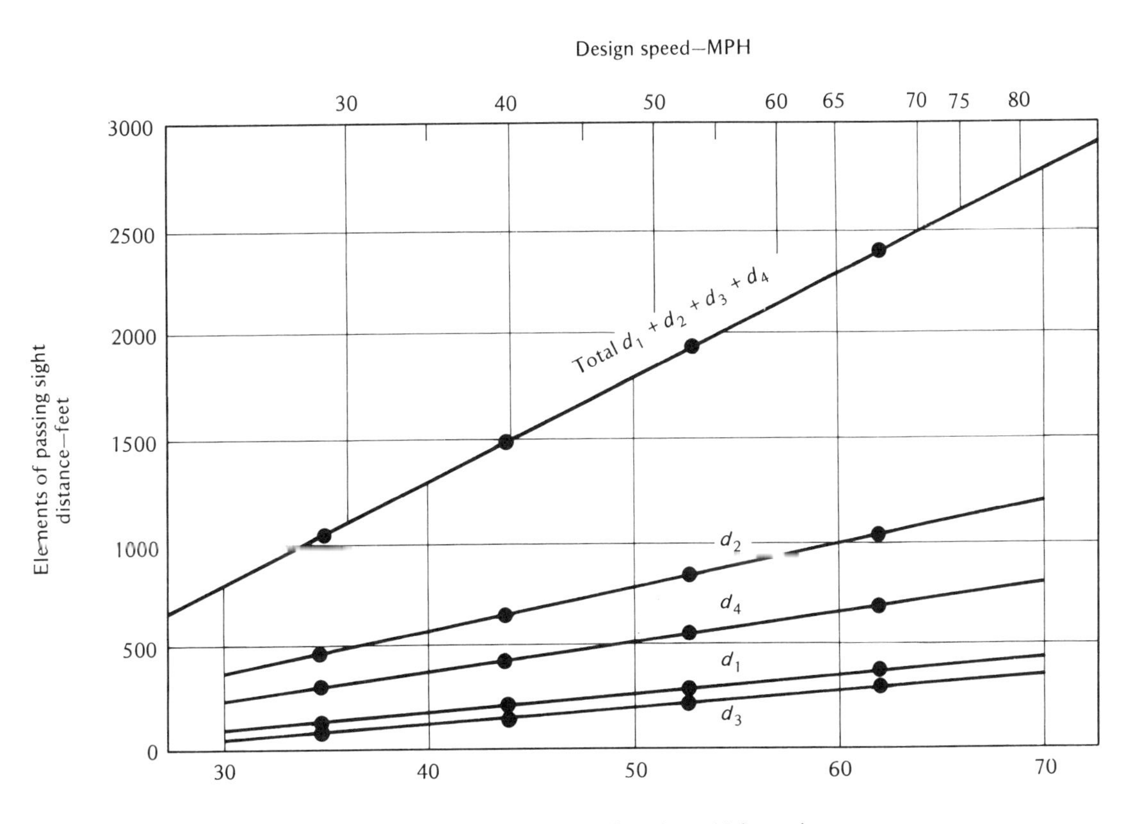

SOURCE: American Association of State Highway Officials, *A Policy on Geometric Design of Rural Highways*, 1965, p. 143.

FIGURE 6-3

PASSING SIGHT DISTANCE FOR TWO-LANE HIGHWAYS

cles are assumed to be traveling at the same speed, $d_4 = \frac{2}{3} d_2$).

Safe passing sight distance for various speed ranges, determined from distance and time values observed in the field during passing maneuvers, are summarized in Table 6-3.

Vertical Alignment

Vertical alignment consists of a combination of grades and vertical curves. As shown in profile, it is a series of straight lines connected by vertical parabolic curves, which represent the changes in elevation along the centerline of the roadway. Where possible, the grade line should follow the general terrain. The most desirable design is one with long vertical curves between grade tangents that provides smooth riding qualities and good visibility with a minimum amount of earthwork.

Grades

Grade affects speed, capacity, and cost of operation. Driving practices with respect to grades vary greatly, but modern passenger vehicles are equipped with sufficient power to ascend long grades up to 10 percent without an appreciable reduction in speed. Since grades in excess of 10 percent are rarely employed, the effect of steep grades on speed is essentially a study of the effect of these grades on truck speeds. Table 6-4 summarizes maximum grade controls employed in terms of design speed for main rural highways and major streets and urban expressways of different topography as suggested by AASHTO.

Maximum grades for secondary highways may be in the

TABLE 6-3
SAFE PASSING SIGHT DISTANCE

Speed group, mph Average passing speed, mph (km/hr)	30–40 34.9 (56.2)	40–50 43.8 (70.5)	50–60 52.6 (84.6)	60–70 62.0 (99.8)
Initial maneuver:				
a = average acceleration, mphps[a]	1.40	1.43	1.47	1.50
t_1 = time, seconds[a]	3.6	4.0	4.3	4.5
d_1 = distance traveled, feet (m)	145 (44)	215 (66)	290 (88)	370 (113)
Occupation of left lane:				
t_2 = time, seconds[a]	9.3	10.0	10.7	11.3
d_2 = distance traveled, feet (m)	475 (145)	640 (195)	825 (251)	1030 (314)
Clearance length:				
d_3 = distance traveled, feet[a] (m)	100 (30)	180 (55)	250 (76)	300 (91)
Opposing vehicle:				
d_4 = distance traveled, feet (m)	315 (96)	425 (130)	550 (168)	680 (207)
Total distance, $d_1 + d_2 + d_3 + d_4$, feet (m)	1035 (315)	1460 (445)	1915 (584)	2380 (735)

[a]For consistent speed relation, observed values adjusted slightly.

SOURCE: *A Policy on Geometric Design of Rural Highways*, American Association of State Highway Officials, 1965, p. 144.

TABLE 6-4
MAXIMUM GRADES AND DESIGN SPEED (MAIN HIGHWAYS)

TYPE OF TOPOGRAPHY	DESIGN SPEED, mph (km/hr)							
	30 (48)	40 (64)	50 (80)	60 (97)	65 (105)	70 (113)	75 (121)	80 (129)
Flat	6	5	4	3	3	3	3	3
Rolling	7	6	5	4	4	4	4	4
Mountainous	9	8	7	6	6	5	–	–

SOURCE: *A Policy on Geometric Design of Rural Highways*, American Association of State Highways Officials, 1965, p. 195.

range of 1.2 to 1.5 times those shown in the table. However, grades in excess of five percent should be avoided, if possible, particularly in snow areas. In addition to the percent grade, the "critical length of grade" indicates the maximum length of an upgrade that a loaded truck can operate without an unreasonable reduction in speed.

It is desirable to provide a separate climbing lane on the upgrade of a two-lane highway with truck traffic where the length of grade causes a reduction in speed of 15 miles per hour (24 kmh) or more or a reduction to a speed of less than 30 miles per hour (48 kmh), provided that the volume of traffic and percentage of trucks warrant the additional expense. As a general rule, a large percentage of trucks will reduce the capacity and level of service of a facility. This is sufficient justification for climbing lanes.

The motor vehicle operating cost is a third factor to be considered in the analysis of grade control. The ideal approach involves balancing the added cost of grade reduction against the cost of vehicle operation without grade reduction.

Vertical Curves

The gradual transition from one grade to another is accomplished by means of a vertical parabolic curve connecting two intersecting tangents. If the point of intersection of the two tangents is above the road surface, the vertical curve is called a *crest*. If it is below, the vertical curve is called a *sag*. Vertical curves should result in a design that provides for safe, comfortable operation and adequate drainage. Factors to be considered in selecting the length of vertical curves are the sight distance along the curve, riding comfort, and the economy of earthwork.

Vertical curves are characterized by their length (L) and the algebraic difference (A) of the intersecting grades (G_1 and G_2). Figure 6-4 illustrates the parabolic curve and gives the important mathematical relations needed to describe its properties.

Crest Vertical Curves

Minimum length of crest vertical curves is usually dictated by the requirement that the driver be able to see an obstacle over the crest of the curve, within the safe stopping distance. In deriving the basic equations for length of parabolic vertical curve L, in terms of algebraic difference of grade A, and sight distance S, the sight distance and the length are considered to be the horizontal projection of the line of sight (see Fig. 6-5.). Illustrated is the situation in which the sight distance is greater than the length of the vertical curve, or $S > L$, and in which the sight distance is less than L, or $S < L$. The height of the driver's eye H_1 and the height of object H_2 are vertical offsets to the tangent sight line.

In case $S > L$ [Fig. 6-5(a)], use is made of the property of the parabola that the horizontal projection of the intercept formed by a tangent is equal to one-half the projection of the long chord of the parabola. Thus, the sight distance S may be expressed as the sum of the horizontal projections $ab + bc + cd$, as shown in Fig. 6-5(a). The solution for L that gives the minimum length of vertical curve necessary to provide the required sight distance S is

$$L = 2S - \frac{200(\sqrt{H_1} + \sqrt{H_2})^2}{A} \tag{6.1}$$

In case $S < L$, use is made of the basic offset property of the parabolic curve, as shown in Fig. 6-5(b). Solving for S_1 and S_2, summing to get S, and then solving for L, we obtain the following relationship:

$$L = \frac{AS^2}{100(\sqrt{2H_1} + \sqrt{2H_2})^2} \tag{6.2}$$

The criteria for height of eye and object used in the geometric design of crest curves and equations based on these values of H_1 and H_2 are summarized in Table 6-5. Stopping sight distance is based on a driver with an eye height of 3.75 feet (1.1 m) seeing a six-inch-(0.15 m) high object.

The minimum length of vertical curve for passing is based on the distance at which a driver whose eye is 3.75 feet (1.1 m) above the pavement can see the top of an oncoming vehicle assumed to be 4.5 feet (1.4 m) above the pavement. Since passing sight distances are as much as four times as long as stopping sight distances, vertical curves based on the passing criteria must obviously be much longer than those based on the stopping criteria. For this reason, if a continuous passing opportunity is desired in rough terrain, a four-lane design with crest curves based on stopping sight distance may be cheaper than a two-lane design based on passing sight distance.

Sag Vertical Curves

There is no single design criterion for establishing lengths of sag vertical curves. Four criteria have been used: (1)

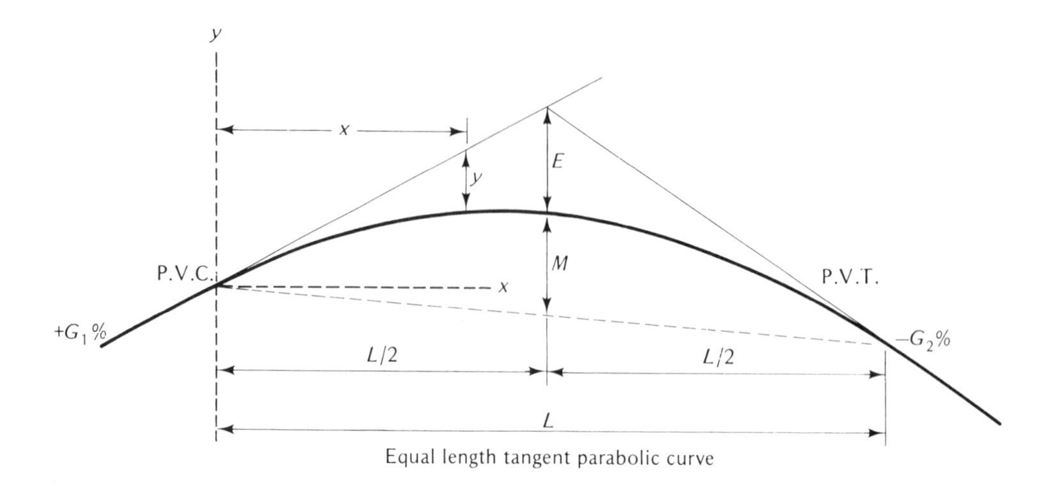

General equation of the parabola	$y = ax^2$
Rate of change in slope of the tangent to parabola	$\frac{dy}{dx} = 2ax$
Rate of change in grade per foot or meter	$\frac{d^2y}{dx^2} = 2a = \frac{A}{E}\,100\,L$

Where A is the algebraic difference in gradients, $G_1 - G_2$, and L, the length of curve in feet or meters is horizontal projection of curve on x-axis

Mid-curve offset $E = \frac{AL}{8} = M$

Other offsets vary as the squares of the distances from P.V.C. (or P.V.T.). For example,

$$\frac{y}{x^2} = \frac{E}{(L/2)^2} \quad \text{or} \quad y = \frac{Ax^2}{200L}$$

FIGURE 6-4
THE VERTICAL CURVE

headlight sight distance, (2) rider comfort, (3) drainage control, and (4) general appearance. Headlight sight distance is the most widely used criterion. When a vehicle traverses a sag vertical curve at night, the portion of highway lighted ahead is dependent on the position of the headlight and the direction of the light beam. Figure 6-6 illustrates the conditions for deriving the minimum length of sag vertical curve, L, in terms of the distance between the vehicle and point where the light ray hits the pavement surface, S; the height of the headlights above the pavement, H; and the angle of the light ray above the horizontal, B.

For case 1 (Fig. 6-6 when $S > L$), the intersection of the light beam with the pavement requires a curve of length:

$$L = 2S - \frac{200(H + S \tan B)}{A} \qquad (6.3)$$

In case 2, when $S < L$, the length of vertical curve is

$$L = \frac{AS^2}{200(H + S \tan B)} \qquad (6.4)$$

For design purposes, headlights are assumed to be mounted 2.0 feet above the highway surface with a 1-deg upward divergence of the beam. Substituting these values for H and B in Eqs. (6.3) and (6.4), we find that the length of vertical curve becomes

A: Case 1 Sight distance greater than length of vertical curve ($S > L$)

$$L = 2S - \frac{200\,(\sqrt{H_1} + \sqrt{H_2})^2}{A}$$

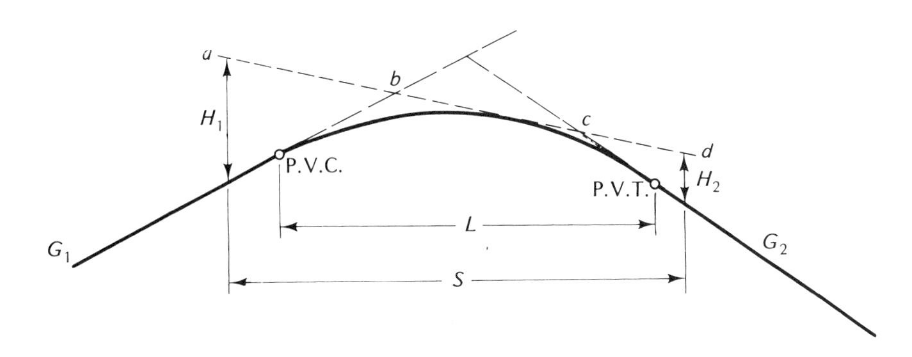

B: Case 2 Sight distance less than length of vertical curve ($S < L$)

$$L = \frac{AS^2}{100\,(\sqrt{2H_1} + \sqrt{2H_2})^2}$$

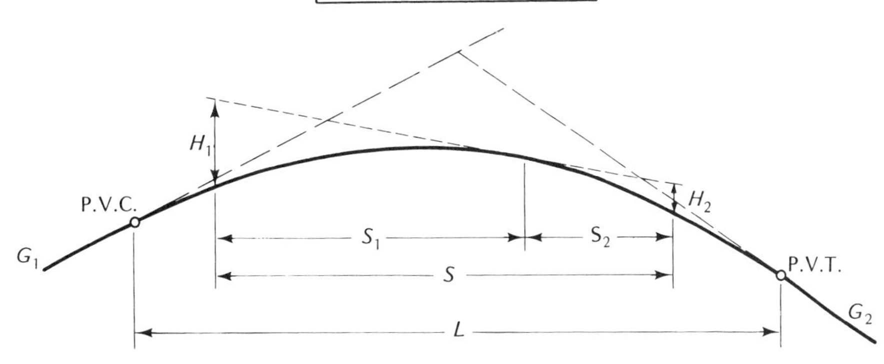

where: L = Length of vertical curve, feet or meters
S = Sight distance, feet or meters
A = Algebraic difference in grads, percent
H_1 = Height of eye above roadway surface, feet or meters
H_2 = Height of object above roadway surface, feet or meters

FIGURE 6-5
SIGHT DISTANCE ON CREST VERTICAL CURVES

$$L = 2S - \frac{400(122\text{ m}) + 3.5S}{A} \quad \text{when } S \geqslant L \qquad (6.5)$$

and

$$L = \frac{AS^2}{400(122\text{ m}) + 3.5S} \quad \text{when } S \leqslant L \qquad (6.6)$$

The discomfort effect of change in vertical direction is greater in sag than on crest vertical curves, because gravitational and centrifugal forces are combined rather than cancelled. The effect is not easily evaluated, but attempts at its measurement have led to the general conclusion that riding on sag vertical curves is comfortable when the centripetal acceleration does not exceed one foot per second [Ref. (2)]. The general expression for lengths of sag vertical curves satisfying this comfort criterion is

TABLE 6-5

CRITERIA IN DESIGN OF CREST VERTICAL CURVES

PARAMETER	STOPPING SIGHT DISTANCE	PASSING SIGHT DISTANCE
Height of eye, H_1	3.75 feet (1.1 m)	3.75 feet (1.1 m)
Height of object, H_2	0.5 feet (0.15 m)	4.5 feet (1.4 m)
$S > L$	$L = 2S_s - \frac{1400}{A}$	$L = 2S_p - \frac{3290}{A}$
$S < L$	$L = \frac{AS_s^2}{1400}$	$L = \frac{AS_p^2}{3290}$

Case 1 Sight distance greater than length of vertical curve ($S > L$)

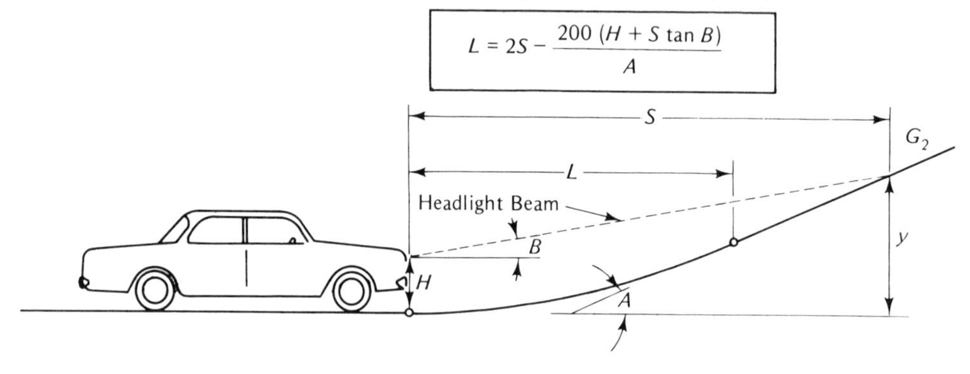

Case 2 Sight distance less than length of vertical curve ($S < L$)

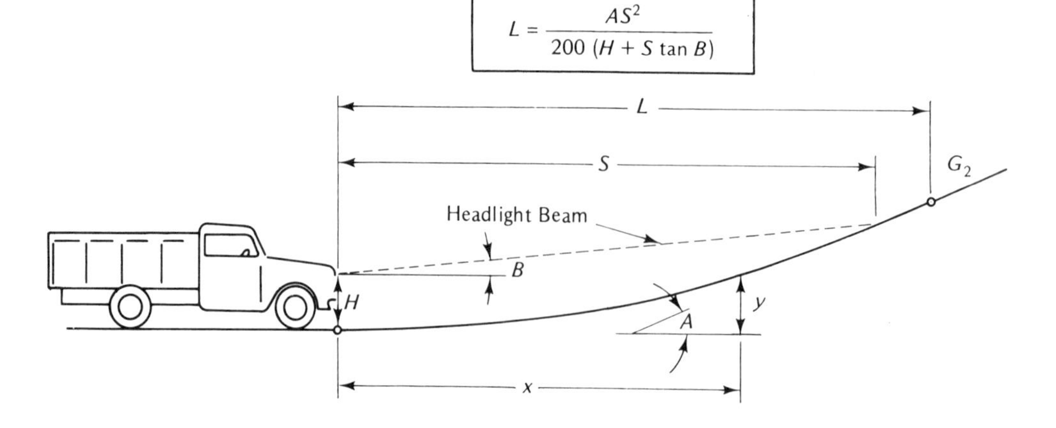

L = Length of vertical curve, feet or meters
A = Algebraic difference in grades, percent $(G_2 - G_1)$
S = Sight distance, feet or meters
H = Headlight height, feet or meters
B = Upward divergence of light beam, degrees

FIGURE 6-6

HEADLIGHT SIGHT DISTANCE ON SAG VERTICAL CURVES

$$L = \frac{AV^2}{46.5} \tag{6.7}$$

When a sag vertical curve occurs at an underpass, the overhead structure may shorten the sight distance somewhat. Under such conditions, the length of vertical curve required for the proper sight distance and adequate clearance may be determined as shown in Fig. 6-7. For further

Case 1 Sight distance greater than length of vertical curve ($S > L$)

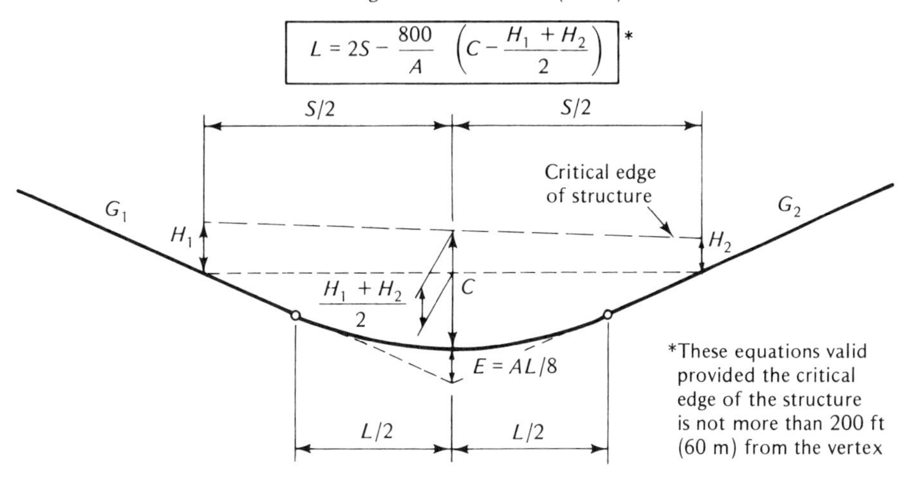

Case 2 Sight distance less than length of vertical curve ($S < L$)

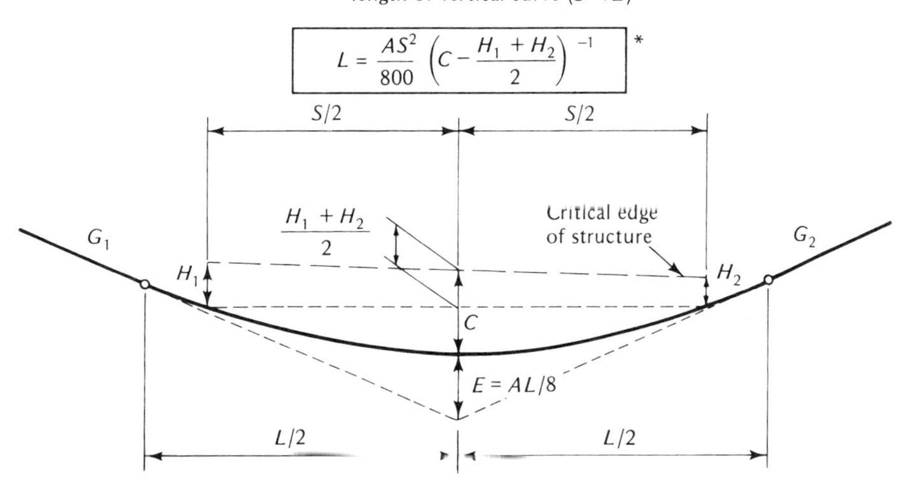

where L = Length of vertical curve, feet or meters
A = Algebraic difference in grades, percent
S = Sight distance, feet or meters
C = Vertical clearance of underpass, feet or meters
H_1 = Vertical height or eye above roadway surface, feet or meters
H_2 = Vertical height of object above roadway surface, feet or meters

FIGURE 6-7
SIGHT DISTANCE AT UNDERPASSES

discussion of vertical alignments as a design control, see the AASHTO red book, pp. 337–346 [Ref. (1)].

Horizontal Alignment

The horizontal alignment of a highway is a series of straight "tangents" joined by curves. The centrifugal force associated with a vehicle moving in a curved path may require that the roadways be "banked" or superelevated to overcome this centrifugal force. To facilitate a smooth change from a straight path to a curved path and the introduction of the superelevation, a transition or spiral curve is used on all modern highways.

The horizontal alignment must be balanced to provide as nearly as possible continuous operation at the speed most likely to prevail under the general conditions existing on that section of road. For example, sharp curves should not be used after long straight tangents on which high-speed operation is likely. Drivers can generally adjust to changes in conditions if these changes are obvious and reasonable. The element of surprise should, by all means, be avoided.

Curvature

In proper geometric design, it is necessary to establish the proper relationship between design speed, curvature, and superelevation. Horizontal curves are circular and are described by either their radius or degree of curve. In highway design the "arc definition" is usually used: i.e., the degree of curve, D, is the central angle subtended by an arc of 100 feet measured along the centerline of the road.

Superelevation

On highway curves, the centrifugal force of the vehicle tends to cause overturning or skidding outward from the center of curvature. If the surface is flat, the only resisting force is side friction between the tires and pavement. Curved sections of modern highways are usually superelevated, utilizing the force of gravity to offset the tendency of vehicles to slide outward. The solution of the equilibrium diagram (Fig. 6-8) yields the following formula for vehicle operation on a superelevated curve:

$$e \pm f = V^2/15R \quad \text{or} \quad v^2/127R \text{ (metric)} \qquad (6.8)$$

where

e is the rate of superelevation.

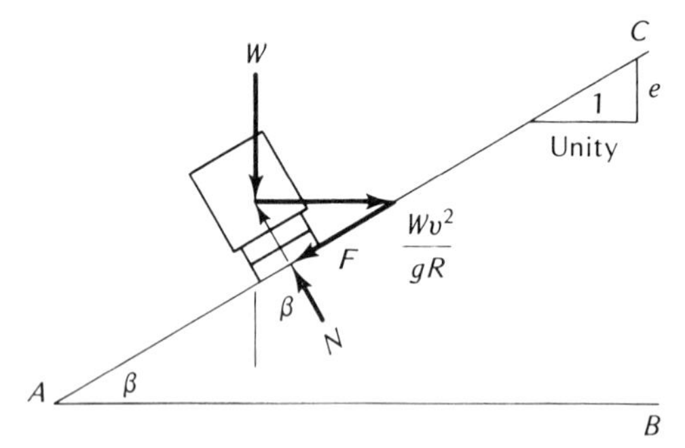

W = Weight of vehicle
β = Angle of pavement slope
e = Superelevation = $\tan\beta$
F or f = Side friction
$F = fN = fW\cos\beta$ where f is the coefficient of friction
g = 32.2 ft/sec² or 9.81 m/sec²
v = Longitudinal velocity, ft/s or m/sec
When all the forces on the vehicle are in equilibrium:

$$W\sin\beta + F = \frac{Wv^2}{gR}\cos\beta$$

Dividing by $W\cos\beta$

$$\tan\beta + F \text{ or } f = \frac{v^2}{gR}$$

or

$$e + F \text{ or } f = \frac{v^2}{gR}$$

Changing v ft/sec to V mph,

and considering case when F acts to left

$$e \pm f = \frac{V^2}{15R} \text{ or } \frac{v^2}{127R}$$

FIGURE 6-8
SUPERELEVATION THEORY

f is the coefficient of friction.
R is the radius of curve in feet.
V is the vehicle speed in miles per hour.

The minimum radius R for a given design speed is determined by the maximum rate of superelevation and the maximum coefficient of friction:

$$R = \frac{V^2}{15(e \pm f)} \quad \text{or} \quad \left(\frac{v^2}{127(e \pm f)}\right) \qquad (6.9)$$

The maximum coefficient of friction depends on the condition of the car tires and the roadway surface. The values employed in curve design are governed by driver comfort, since the side friction force is sensed by the driver. High centrifugal forces tend to cause the driver to slide across the seat of the automobile. Recommended comfortable side friction factors are linearly related to speed, varying from 0.16 at 30 miles per hour (48 kmh) to 0.12 at 70 miles per hour (113 kmh). The designer should select a value that will encourage speeds on curves consistent with those found on the nearby tangent sections on the same highway.

Superelevation values usually range from one to ten percent. Factors affecting the selection of a superelevation value are ice problems, which limit the maximum superelevation, and drainage, which establishes the lower limit of superelevation. Table 6-6 gives values for the minimum radius of curvature for selected design speeds and two values of superelevation.

Transitions

Transition curves (spiral curves) are based on the principle of introducing horizontal curvature and superelevation gradually. This gradual change follows the path of the vehicle as it enters and leaves a circular curve. The ideal design will introduce the unbalanced side friction force on the driver in a smooth progression to the maximum value experienced on the curve.

Some highway departments do not use spiral curves. Where a spiral is not used, the superelevation is transitioned from a normal crown section to a fully superelevated section over some empirically designated proportional lengths of tangent and circular curve. These transitions generally approximate the length of an appropriate spiral curve. A driver, operating on a tangent section on which superelevation has been added, may be forced to steer opposite to the direction of the curvature ahead in order to stay on a straight path. When he reaches the beginning of the circular curve he must immediately steer to the curved path.

At low speeds and small curvatures his path toward the outside of the curve is not critical. However, as speeds become faster and as the degree of curvature increases, a smooth compensation by the driver requires more and more distance. In general, the transition should be such that the relative longitudinal slope of the edge of pavement to the centerline should be no greater than one in 200 for a two-lane highway.

Superelevation may be attained by rotating the pavement about:

1. The centerline

TABLE 6-6
DEGREE OF CURVE AND MINIMUM RADIUS FOR LIMITING VALUES OF *E* AND *F*

DESIGN SPEED mph	(kmh)	MAX. e	MAX. f	TOTAL $(e+f)^a$	MIN. RADIUS ft.	(m)	MAX. DEGREE OF CURVE
30	(48)	0.06	0.16	0.22	273	(83)	21.0
40	(64)	0.06	0.15	0.21	508	(155)	11.3
50	(80)	0.06	0.14	0.20	833	(254)	6.9
60	(97)	0.06	0.13	0.19	1263	(385)	4.5
65	(105)	0.06	0.13	0.19	1483	(452)	3.9
70	(113)	0.06	0.12	0.18	1815	(553)	3.2
75	(121)	0.06	0.11	0.17	2206	(672)	2.6
80	(129)	0.06	0.11	0.17	2510	(765)	2.3
30	(48)	0.12	0.16	0.28	214	(65)	26.7
40	(64)	0.12	0.15	0.27	395	(120)	14.5
50	(80)	0.12	0.14	0.26	641	(195)	8.9
60	(97)	0.12	0.13	0.25	960	(292)	6.0
65	(105)	0.12	0.13	0.25	1127	(344)	5.1
70	(113)	0.12	0.12	0.24	1361	(415)	4.2
75	(121)	0.12	0.11	0.23	1630	(497)	3.5
80	(129)	0.12	0.11	0.23	1855	(565)	3.1

[a] e = rate of superelevation
f = coefficient of side friction

SOURCE: *A Policy on Geometric Design of Rural Highways*, American Association of State Highway Officials, 1965, p. 158.

2. The inside pavement edge
3. The outside pavement edge

Figure 6-9 illustrates the two cases where the pavement is revolved about the centerline and about the inside edge.

Multilane roadways with a median may be rotated individually or as a unit about the master centerline if the median is narrow or if there is a possibility of widening on the inside. Further discussion of horizontal alignment is found in the AASHTO manuals [Refs. (1) and (2)] and in the ITE publication [Ref. (3)].

Sight Distance on Horizontal Curves

Another element of horizontal alignment is the sight distance across the inside of curves. Where there are sight ob-

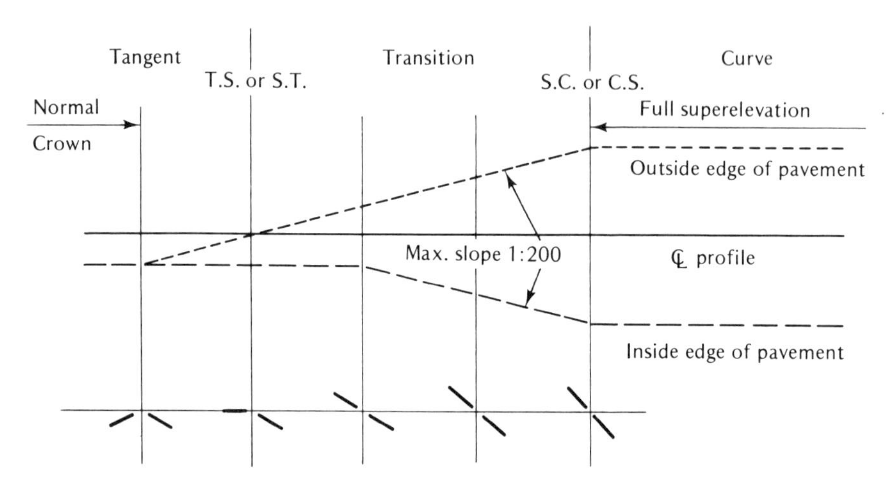

(a) Pavement revolved about center line

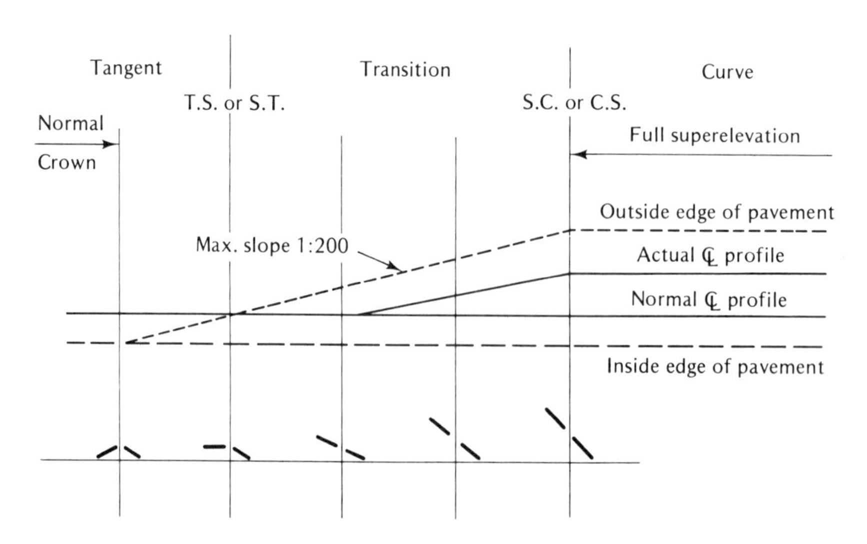

(b) Pavement revolved about inside edge

SOURCE: Joseph Barnett, *Transition Curves for Highways,* U.S. Government Printing Office, 1940, p. 10.

FIGURE 6-9
METHODS OF ATTAINING SUPERELEVATION

structions—walls, cut slopes, buildings, guardrail under certain conditions, etc.—on the inside of curves, a design to provide adequate sight distance may require adjustment in the normal highway cross section or change in alignment if the obstruction cannot be removed [Ref. (2)]. If design speed and minimum (or desirable) stopping sight distance are used, adjustments based on actual conditions are required to provide adequate sight distance. For general use in design of a horizontal curve, the sight line is a chord of the curve, and the applicable stopping sight distance is measured along the centerline of the inside lane around the curve. Figure 6-10 shows the required middle ordinates for clear sight for safe stopping distances for curves of varying degrees.

Coordination of Horizontal and Vertical Alignment

The visibility of the curved roadway is vital to safe operation. If the vertical and horizontal alignment are designed independently or are not properly coordinated, the driver may get a completely inadequate perspective of the road ahead, which will adversely affect his operation.

Horizontal and vertical alignment must be in balance and coordinated not only on the main lanes of the roadway but also at ramps and intersections, where changes in direction are introduced. In those cases where it is necessary to introduce horizontal curves that reduce the safe operating speed, it is generally better to introduce them

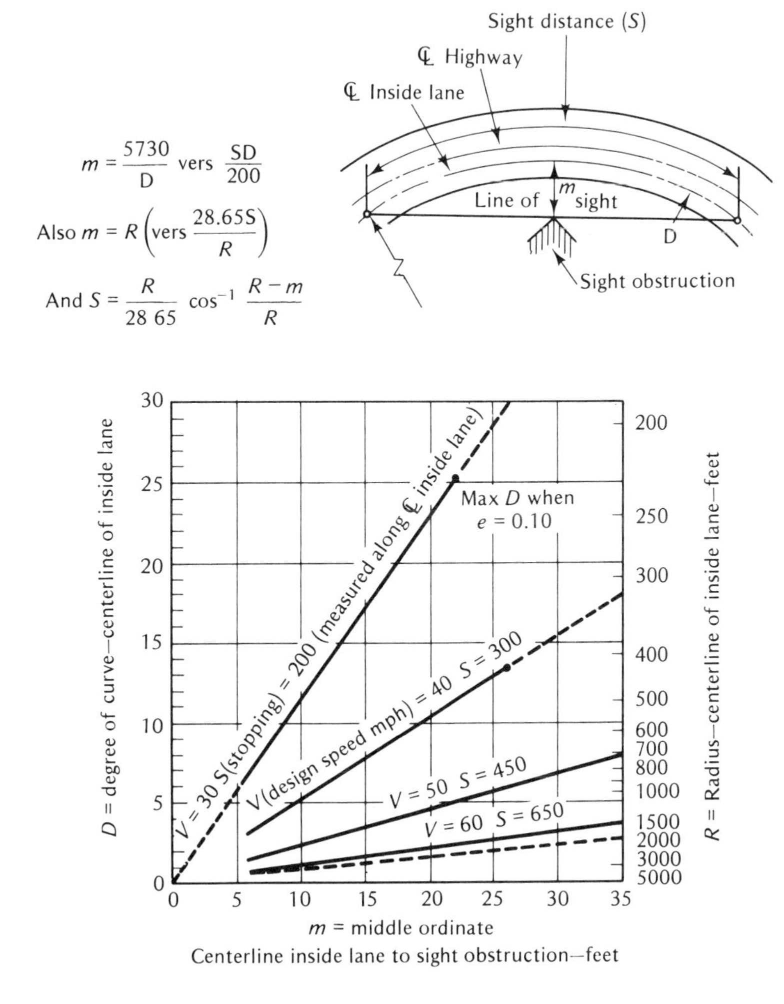

FIGURE 6-10
DESIRABLE STOPPING SIGHT DISTANCE ON HORIZONTAL CURVES

on an upgrade to take advantage of the improved deceleration conditions. In all cases, sufficient sight distance and adequate panoramic view of the beginning of the curved roadway section are essential to safe operation. When curves are introduced at the end of ramps, sufficient view of this condition and adequate deceleration distance are essential. The introduction of a sharp horizontal curve near the top of a crest vertical curve or near the low point of a sag curve can be very hazardous.

It is important that all flow lines of drainage paths be examined to prevent drainage pockets or ditch grades inadequate for proper drainage. In more complex situations, such as interchanges, contour maps of the entire roadway section are helpful in locating those designs having poor drainage. In introducing superelevation on curbed roadways, drainage across the roadway should be carefully considered. This problem becomes even more complex on roadways with continuous curbed medians.

CROSS-SECTION DESIGN

The elements of the highway cross section vary directly with the usage of the facility. Roadways with higher design volumes naturally require more lanes and have greater need for design features such as separate left-turn lanes, wider shoulders, medians, and control of access.

The highway cross section is composed of many design elements. These elements can be classified into three broad groups:

1. The traveled portion of the road: the traffic-bearing lanes.
2. Road margins: the shoulders, ditches, and roadside.
3. Traffic separation: the medians.

Dimensions for each element are based on the evaluation of traffic characteristics and the level of service established for the proposed facility. Figure 6-11 shows two typical urban cross sections.

Lane Widths

Although most design standards still permit lane widths of less than 12 feet (3.7 m), there is general agreement that 12 feet (3.7 m) is the desirable width for roadway lanes. Lane widths below 12 feet (3.7 m) can adversely affect capacity and safety, and their use should be limited to other than high-speed, high-volume facilities. However, stringent controls of right-of-way and existing development in urban areas may permit lanes only 10 feet (3 m) or 11 feet (3.4 m) wide [Ref. (3)].

On high-speed, rural, two-lane roads, lane widths of 13 and 14 feet (4.0 and 4.3 m) have been used. Widths in excess of 14 feet (4.3 m) are not recommended, because of the tendency of some drivers to use such a wide facility as a multilane road. Parking lanes and auxiliary lanes at intersections should be as wide as other lanes but never less than 10 feet (3 m).

Pavement Slope

Pavements are sloped from the center to each edge to prevent water from ponding on the pavement. The effect on steering of slopes up to four percent is barely perceptible. Normally, high-type pavements (e.g. portland cement and asphaltic concretes) are sloped from 1 to 2 percent. Lower types of pavements (e.g. surface treated or gravel) require greater slopes and generally range from 1.5 to 4 percent.

Pavements with three or more lanes inclined in the same direction should have greater slope across the outside lane than across adjacent lanes. On roadway sections where lanes are to be added in the future, the proper relationships should be established between the various cross-sectional elements during the original design process.

Shoulders

Right shoulders should be provided continuously along all highways for emergency stopping. Experience has shown that paved shoulders are desirable for safety and to lessen structural failure along the outside edge of the pavement. Shoulders of 10 feet (3 m) in width should be provided on all high-standard roadways. Shoulders of 4 to 8 feet may be satisfactory on low-standard roadways or may be used on long-span, high-cost structures. Left shoulders are highly desirable on divided urban arterials. Urban freeways should have a left shoulder of at least 4 feet (1.2 m) [preferably 6 feet (1.8 m)].

Shoulders are generally sloped more than the traffic lanes to speed runoff of surface water. Paved shoulders are sloped from 3 to 5 percent, gravel and stone from 4 to 6 percent, and turf shoulders 8 percent.

In most cases, paved shoulders generally result in overall economy. Unpaved shoulders have a low initial cost but are difficult and expensive to maintain. The color and texture of shoulders should provide adequate contrast between the shoulder and the adjacent pavement. Because of

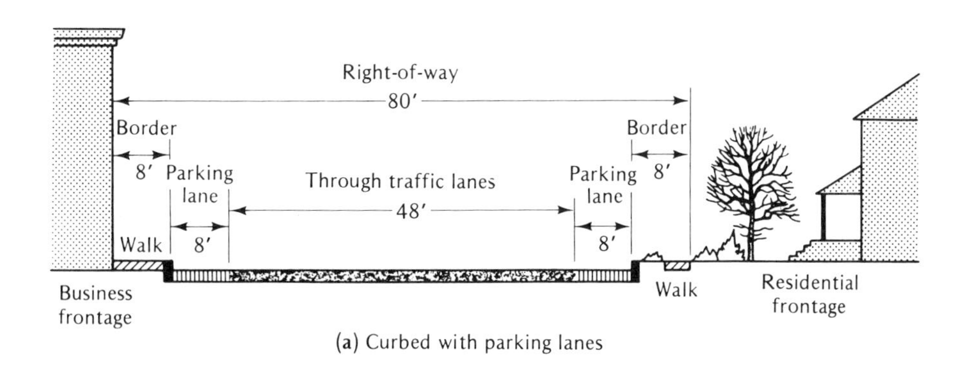

(a) Curbed with parking lanes

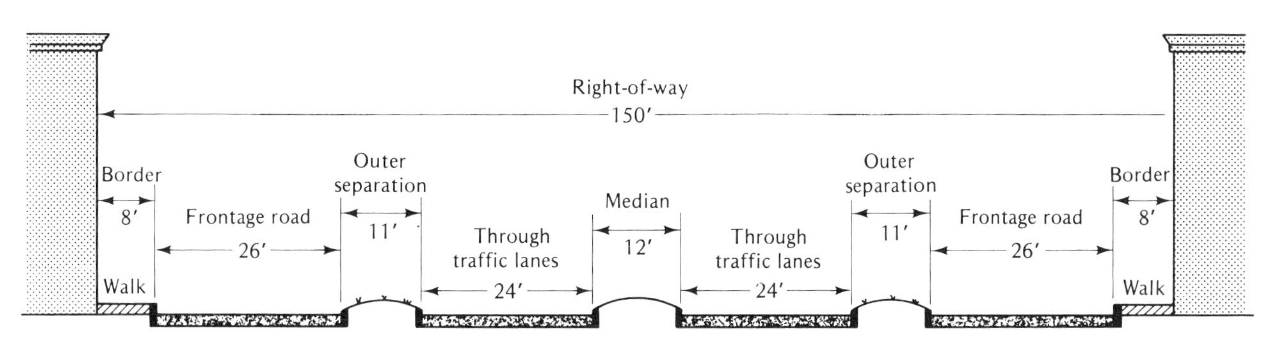

(b) Divided with separated parking and service lanes

SOURCE: American Association of State Highway Officials, *A Policy on Arterial Highways in Urban Areas,* 1957, pp. 201-202.

FIGURE 6-11
TYPICAL URBAN ARTERIAL CROSS SECTIONS

the frequent use of shoulders for disabled vehicles, structural quality must be considered to minimize maintenance problems. A number of agencies use the same structural criteria for shoulders as for pavement, in order to use the shoulders as traffic lanes when needed because of accidents, repairs, multioccupancy vehicles, over-capacity peak flow, etc.

Side Slopes

Safety and economy in maintenance necessitate relatively flat side slopes and rounded ditches. Slopes steeper than 4:1 (four horizontal to one vertical) should be avoided. The driver's chances of regaining control of his vehicle if he leaves the roadway are greatly reduced, and maintenance, such as mowing, becomes difficult.

The intersection of the slope planes should be well rounded, and abrupt changes should be avoided. Back slopes for ditches in cut sections should be a maximum of 4:1 where practical; however, the slopes may be increased in rock cuts or where other special conditions prevail.

Ditches

Roadside drainage ditches should have a minimum depth of 2 feet (0.6 m) as measured from the centerline profile grade and should be a minimum of 6 inches below (0.15 m) the subgrade line to insure adequate drainage of the base course. Wide ditch sections are often required to provide ample waterway. In all cases, dimensions should be determined by the drainage conditions for the particular area.

Curbs

Curbs are used to control drainage, delineate proper vehicle paths, and to deter vehicles from leaving the roadway or encroaching in certain areas. Modern designs use curbs mainly in urban areas.

The two general types of curbs are barrier and mountable. Barrier curbs are relatively high and steep-faced and are intended to prevent or at least deter encroachments, whereas mountable curbs are designed so that they may be crossed easily without discomfort or undue hazard, even at relatively high speeds. Mountable curbs are generally low [less than 6 inches (0.15 m)] and have a flat slope (see Fig. 6–12).

Barrier curbs range from 6 to 20 inches (0.15 to 0.5 m) in height, depending on the nature of encroachment they

Barrier curbs

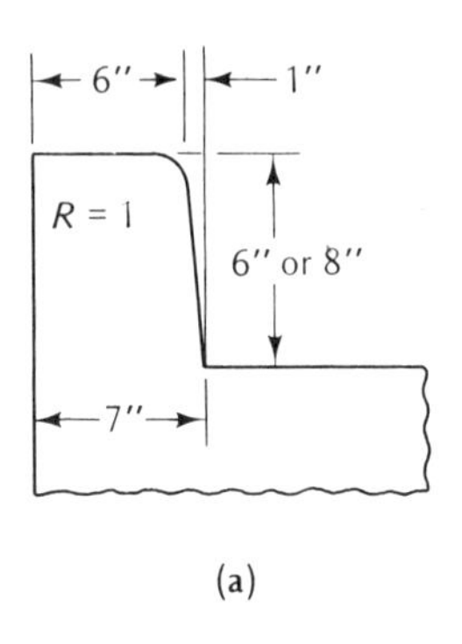

(a)

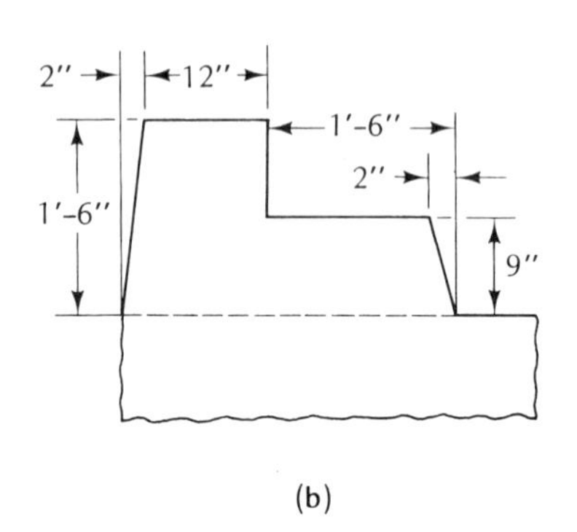

(b)

Mountable curbs

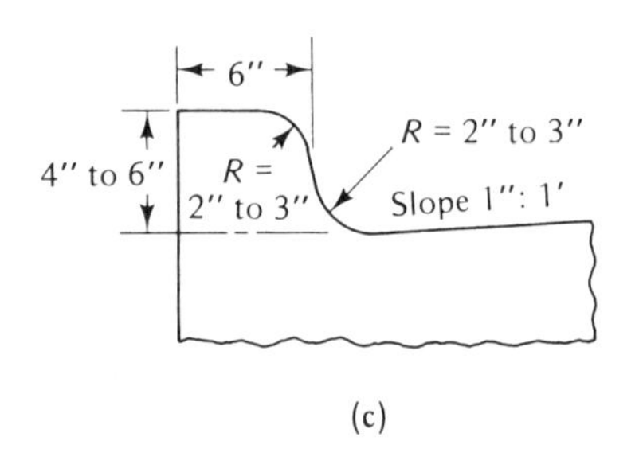

(c)

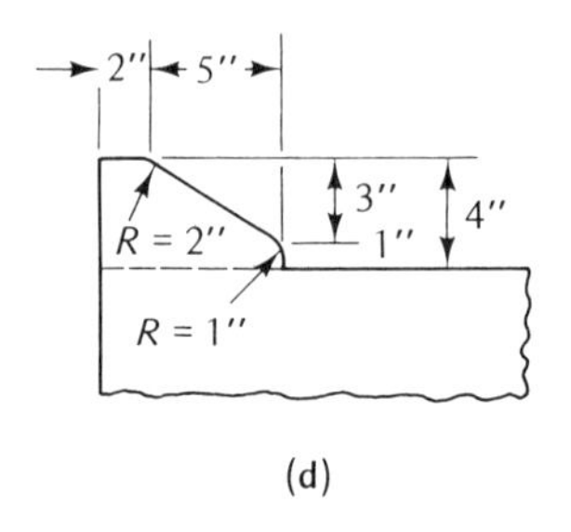

(d)

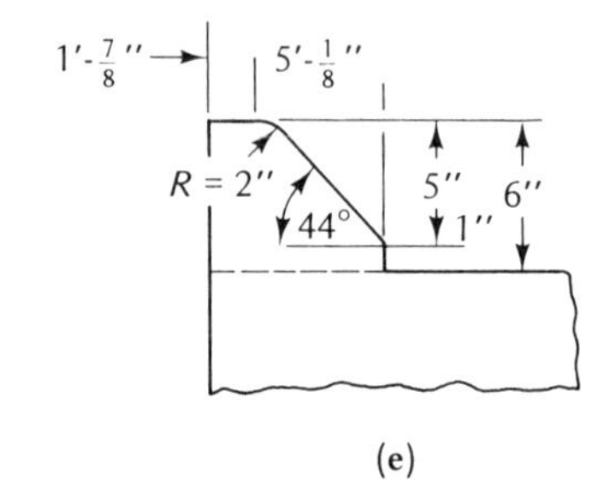

(e)

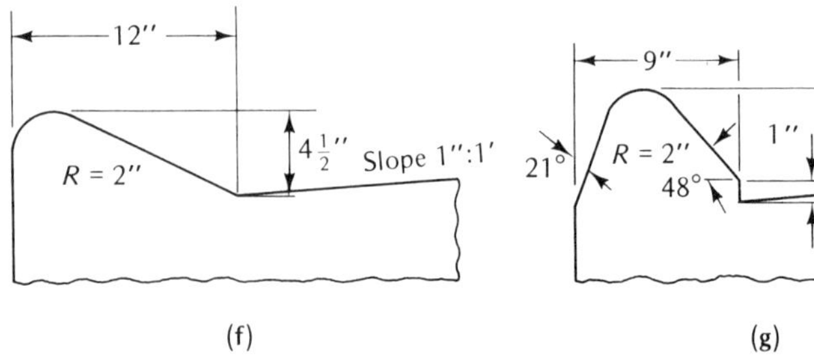

(f) (g)

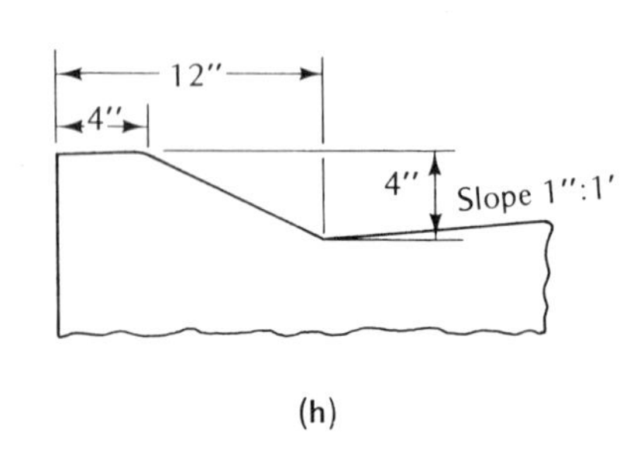

(h)

FIGURE 6–12
TYPICAL HIGHWAY CURBS

are to prevent. They are generally vertical or slightly sloped and should not be used where design speeds are above 50 mph (80 kmh). They are used with safety walks and in general to protect pedestrians. Barrier curbs have been used on bridges, and as protection against vehicle collision around piers and along walls. Curb height should not exceed six inches (0.15 m) adjacent to locations where vehicles are expected to stop.

Continuous barrier curbs should be offset at least 1 foot (0.3 m) from the edge of the through traffic lane. When curbs are introduced on sections of a roadway not having continuous curbs, they should be offset 2 or 3 feet (0.6 or 0.9 m) from the through traffic lane.

Where curbs are required in rural areas, they should be placed along the outer edge of the paved shoulder and should be mountable. Mountable curbs are also used at median edges, to outline channelization islands and at shoulder edges.

Gutter sections may be provided on the traveled-way side of either barrier or mountable curbs. They are usually 1 to 3 feet (0.3 to 0.9 m) wide.

High-visibility curbs are made by painting or use of reflective materials and are very advantageous in rain or fog.

Medians

A median is that portion of a divided highway separating the traveled ways for traffic in opposite directions. It serves to delineate the left extremity of the path of vehicle travel, decreasing vehicle encroachment into opposing lanes and providing space for vehicles running off the left edge of the pavement to regain control. The exact function that a given median is expected to perform depends on the degree of control of access on the street or highway facility. Thus, a median may serve to provide protection and control of left-turning or crossing vehicles, to provide refuge space for pedestrians, or to provide refuge space for disabled vehicles.

The importance of a median as space for drainage and snow storage depends on climate, topography, and the number of traffic lanes. Narrow medians in areas of heavy snowfall may require that most of the snow be pushed across several traffic lanes to the right for storage. Variations of roadway alignment, driving speed, landscaping, and the extent to which the roadway is lighted by other means influence the design of medians to reduce headlight glare.

The median concept also offers flexibility to the planner and the designer. Medians provide space for future addition of traffic lanes or the installation of other types of transportation facilities (exclusive bus lanes and transit). Depending on terrain and land values, medians can make separate roadway profiles possible, and provide space for structural appurtenances, lighting standards, etc.

Medians are classified as traversable, deterring, or barrier. A traversable median, usually consisting of paint stripes, buttons, and an area of pavement of contrasting color or texture, does not present a physical barrier to traffic movement. A deterring median is one that incorporates any of the features of a traversable median plus a minor physical barrier such as a mountable curb or corrugations. A barrier median consists of a guardrail, shrubbery or some type of wall that traffic cannot cross easily, if at all.

> In selecting the type of median barrier, it is most important to match the dynamic lateral deflection characteristics to the site. The maximum deflection should be less than one-half the median width. The median barrier should be designed to prevent penetration into the opposing lanes of traffic and should redirect the colliding vehicle in the same direction as the other traffic. In addition, the design should be aesthetically pleasing [Ref. (1)].

On heavily traveled facilities with narrow medians, a concrete barrier with a slope face has many advantages. This type will deflect a vehicle striking it at the likely slight impact angle. It has a pleasing appearance and requires little maintenance. The latter is a most important consideration, as on highways with narrow medians, maintenance operations on the barrier would encroach on the high-speed traffic lanes. Recently, particularly in urban areas, the "New Jersey" or "GM" type of deflecting median barrier has been used extensively.

Frontage Roads

A local road adjacent and parallel to a freeway or major arterial is called a frontage road. Its function is to permit the control of access to the main traveled way by providing access to the property adjoining the highway, and maintaining circulation of traffic on the street system on each side of the main highway. Access from frontage roads to the main traveled way is permitted only at specially designated locations. In urban areas, it is desirable to operate frontage roads as one-way pairs. However, in rural areas, the extreme distances between crossings of the main roadway often require two-way operation.

The frontage road system can add tremendous flexibility to the operation of a freeway when utilized as an auxiliary facility. It provides a means of utilizing a stage development program for the construction of urban freeways, as well as a means of handling traffic flow during the construction of the main freeway lanes. A continuous frontage road system provides maximum land service to properties abutting the freeway. It greatly increases the flexibility of the interchange and becomes an integral part of the overall street system. Some states require frontage roads on all their controlled access facilities.

However, for arterials with only partial control of access, but relatively high operating speed, and intersections at grade, continuous frontage roads may be undesirable. Much of the improvement in capacity and safety may be offset by the added hazard when the arterial and frontage road(s) intersect the cross street at grade. There are actually two or three intersections, resulting in problems in design and traffic control that are very complex. Reversed lot frontage roads (that is, frontage roads that have lots backing on the frontage road with major access from the parallel residential or other service street) can enhance the capacity, safety, and compatibility of the frontage road.

Right-of-Way Width

In the past, right-of-way widths were selected depending upon the class of highway to be built. This arbitrary width then became a controlling factor in design considerations. Although this practice still prevails to some extent, it is becoming more common to establish right-of-way requirements based on the final design of the cross-sectional elements of the facility.

Frequently, a minor roadway design may be determined to be adequate for a number of years, but future traffic is expected to require more lanes, grade separations, divided roadways, or other major changes. In these cases, economy can be realized through stage construction with the right-of-way to permit the ultimate development of the roadway procured initially.

Other Cross-Section Elements

Several other cross-sectional elements are important to geometric design and are discussed in the AASHTO design manuals [Refs. (1) and (2)]. They include

- Outer separations
- Fencing
- Borders and sidewalks
- Climbing lanes

OTHER ELEMENTS

There are several other elements of geometric design of lesser importance. These include horizontal and vertical clearance parameters, which are fairly well fixed and for which little or no analysis is required. Another element is the guardrail, which provides an opportunity to minimize construction costs by making steeper fill slopes possible. The geometric and structural design of guardrails provides opportunities for imaginative treatments to promote highway safety.

The following elements are listed, and their consideration in design is discussed in the AASHTO manuals [Refs. (1) and (2)]. Esthetics is briefly discussed in the next paragraph.

- Drainage
- Retaining walls
- Utilities
- Landscaping and erosion control
- Noise control
- Roadside control
- Esthetics
- Driveways
- Mail boxes
- Rest areas
- Traffic control devices
- Pedestrian crossings

Esthetics

Advanced geometric highway design practice requires that suitable attention be directed to the esthetics of the road and roadside from the viewpoint of the user and abutting and nearby property owners. Improving the visual appearance of the roadway requires attention to this aspect of design at the location stage as well as in final detailed design. Those roadways which have been judged to be particularly appealing visually usually are characterized by the use of generous rather than minimum design standards, particularly in right-of-way, structural treatment, and in horizontal and vertical alignment. These highways also have controlled the visual appearance presented by the road to the adjacent lands and vice versa. Final grading and planting are imaginatively done. These considerations necessitate the development of a sense of esthetic values by the engineer and often involve him with the planner and landscape architect.

LIGHTING

The highway transportation engineer frequently is responsible for lighting on highways, streets, and parking facilities. The facility design is not complete until lighting requirements are met. The development of a street lighting plan is based on scientific principles and requires engineering solutions, since street lighting design is quite complex.

The purpose of street lighting is to provide illumination adequate for the vehicle driver and pedestrian to see the necessary elements of the environment in order to accomplish the driving and walking task safely and confidently. The driver discerns an object in one of several ways: by silhouette, where the object to be seen is dark in contrast to its background; by reverse silhouette, where the object is brighter than the background; by detecting variations in brightness or color over the surface of the object; by specular or mirrorlike reflection from the object; and by the silhouette pattern of the object's background.

Thus, the ability to see an object is heavily dependent on the contrast between its brightness and that of the background. The size, shape, and texture of the object, as well as the time available for seeing, all affect visual perception.

A problem associated with illumination is glare. The two types of glare are discomfort glare and disability glare. Glare reduces the ability to see and is influenced by the following characteristics [Ref. (4)] :

1. Size of light source
2. Location of source relative to eye viewing direction
3. Adaptation level of eye
4. Time of exposure
5. Brightness of environment

The quality of lighting is determined by four characteristics:

1. The lighting level
2. The uniformity of this level
3. The restriction of glare
4. The degree to which it delineates the roadway alignment

Visual information needs of the driver fall into three levels:

1. Positional performance—routine steering, and speed
2. Situational performance—changes in speed, steering, etc.
3. Navigational performance—route guidance, etc.

A complete design is concerned with brightness, which characterizes the light striking the eye. Brightness, of course, depends on the characteristics of the light source as well as the reflectance capabilities of the object and background. Conventional design practice is to treat the reflectance of the object as essentially constant and to reduce the problem to one of providing a certain amount of illumination on the object.

The mechanisms used by the engineer to provide a desired level of illumination are (1) varying the energy of the light source, (2) controlling the direction of illumination by a lens system, and (3) varying the amount of light on the object by varying the mounting heights and orientation of the light source. Particularly on controlled access highways, there has been a rather extensive use of high-mast [over 100 feet (31 m)] clusters of luminaires. These arrangements provide safety advantages, since few obstructions are imposed. Widely varying lamp sizes are available with outputs from 2500 to more than 50,000 lumens, the measure of light energy. Lens distribution systems are available to provide a wide variety of light patterns, as shown in Fig. 6-13, in which the standard linear, cross, and circular patterns are shown. Dimensions are expressed in terms of the mounting height of luminaire (MH). Figure 6-14 shows some of the details of a Type II distribution.

Some of the benefits from lighting are (1) improved operations at intersections, interchanges, and weaving areas, (2) improved alignment delineation and evidence of decision points, (3) greater comfort in driving, particularly in inclement weather, (4) increased ease of policing, (5) increased visibility while clearing accidents, (6) possibly a slight increase in capacity, (7) possibly a greater use of the highway at night, (8) reduction of street crimes, and (9) enhancement of shopping districts. There is also some evidence that continuous fixed-source lighting reduces accidents, although studies to date have not produced conclusive results due to the numerous other factors that contribute to nighttime accidents, except for pedestrian accidents.

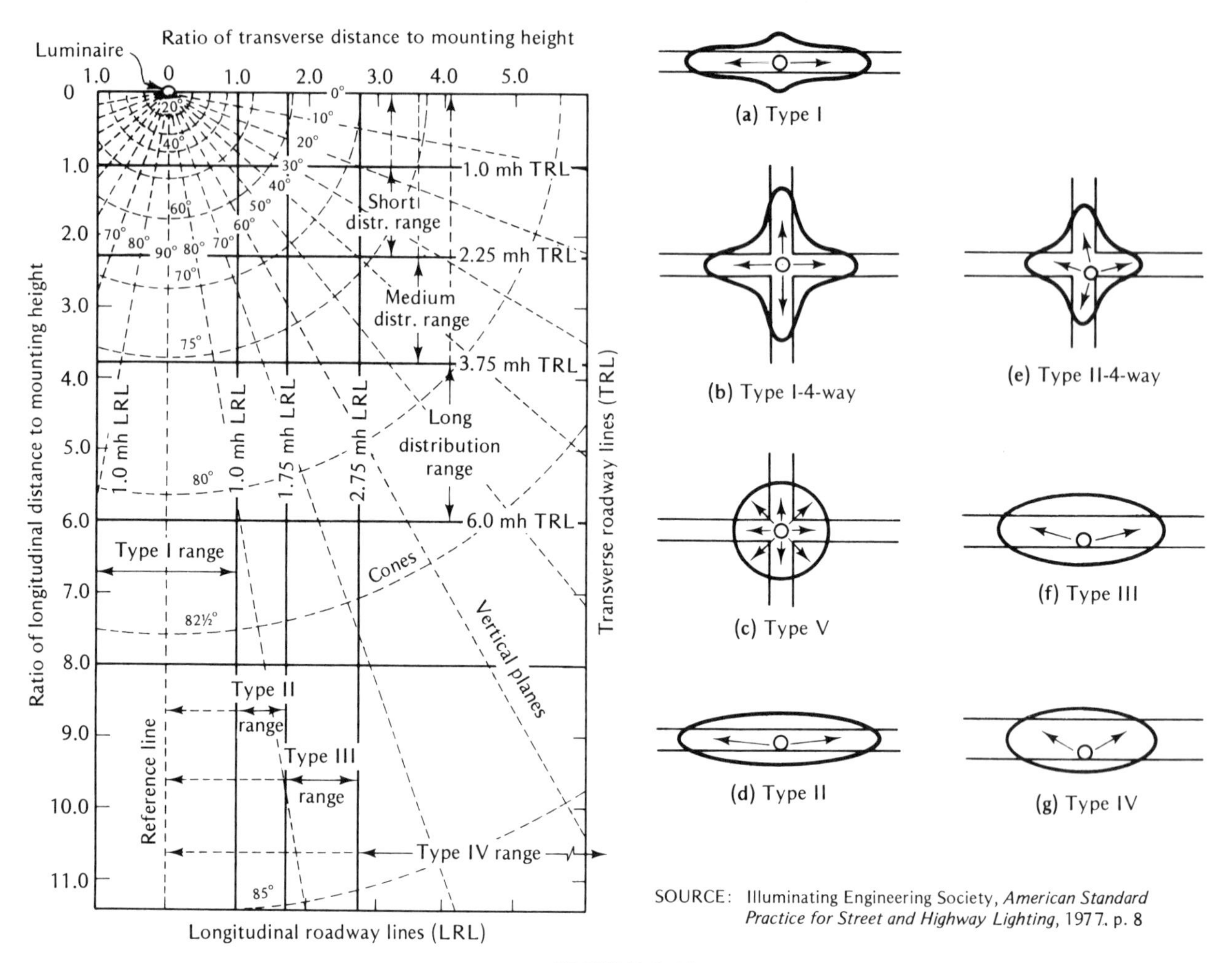

SOURCE: Illuminating Engineering Society, *American Standard Practice for Street and Highway Lighting*, 1977, p. 8

FIGURE 6-13
LUMINAIRE DISTRIBUTION PATTERNS

More detailed discussion of highway and street lighting can be found in the 1976 edition of the Institute of Traffic Engineers' *Transportation and Traffic Engineering Handbook* [Ref. (5)] and the 1976 AASHTO publication, "An Informational Guide for Roadway Lighting" [Ref. (6)].

SUMMARY

The geometric design of highways should be considered a dynamic process. An effective design for a highway requires both a correct concept of the overall plan and properly executed details of design [Ref. (7)]. *Conceptual design* and *detailed design* should be thoroughly considered and coordinated. Basic geometric design standards are intended as guides to design, not as rigid, inflexible dimensions and arrangements. The designer must "consider the interrelationships of all highway elements" in determining their effects on specific highway design. The reader is urged to review the two AASHTO design manuals [Refs. (1) and (2)], the ITE manual [Ref. (3)], and the Federal Highway Administration *Seminar Notes, Dynamic Design for Safety* [Ref. (7)].

REFERENCES

1. American Association of State Highway and Transportation Officials, *A Policy on Arterial Highways in Urban Areas.* American Association of State Highway and Transportation Officials, Room 341, National Press Building, Washington, D.C. 20045, 1973.

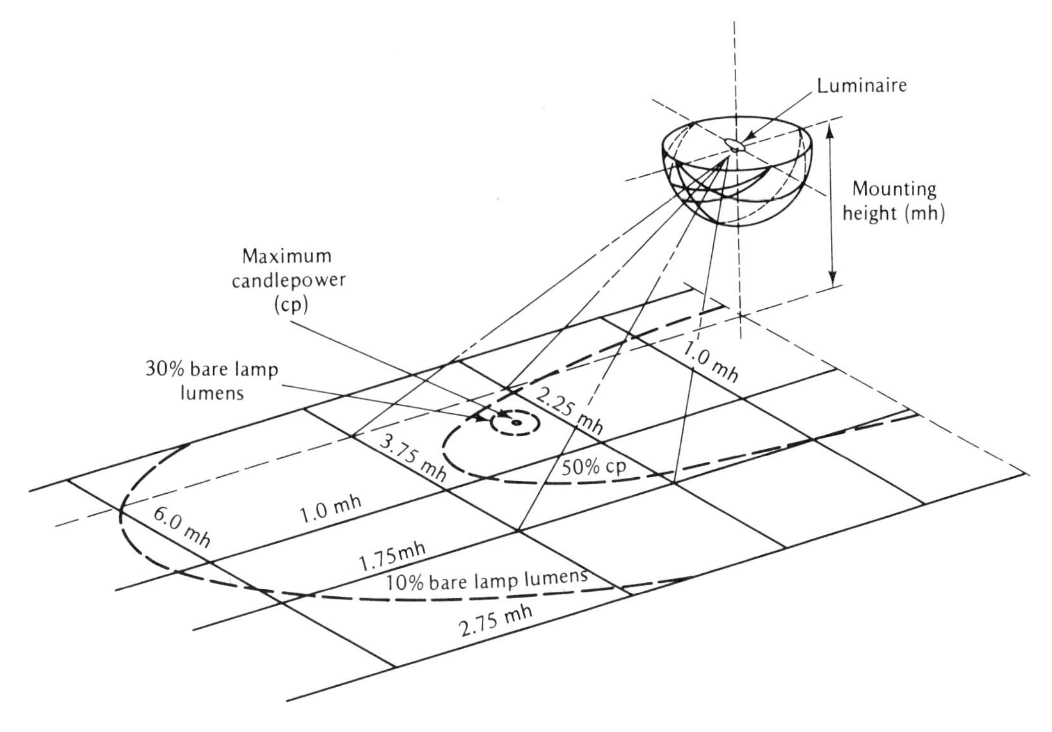

FIGURE 6-14
TYPICAL ILLUMINATION PATTERN

2. American Association of State Highway and Transportation Officials, *A Policy on Geometric Design of Rural Highways.* American Association of State Highway and Transportation Officials, Room 341, National Press Building, Washington, D.C. 20045, 1965.

3. Institute of Transportation Engineers, *Urban Street Design Standards.* Washington, D.C.: 1978.

4. Illuminating Engineering Society, *American National Standard Practice for Roadway Lighting.* Illuminating Engineering Society, 345 East 47th Street, New York, New York 10017, 1977.

5. Baerwald, John E., editor, *Transportation and Traffic Engineering Handbook,* Institute of Traffic Engineers. Englewood Cliffs, N.J.: Prentice-Hall, Inc., 1976.

6. American Association of State Highway and Transportation Officials, *An Informational Guide for Roadway Lighting.* Room 341, National Press Building, Washington, D.C. 20045, 1976.

7. U.S. Department of Transportation, Federal Highway Administration, *A Dynamic Design for Safety: Seminar Notes.* Institute of Transportation Engineers, 1975.

7

GEOMETRIC DESIGN OF INTERSECTIONS AND INTERCHANGES

An intersection is the area shared by the joining or crossing of two or more roads. The primary operational function of the intersection is to permit a change in travel route. Because of this, the intersection becomes a point of decision. The motorist must decide on one of the available alternative choices.

Thus an intersection presents the driver with added tasks not required at nonintersection points on the road. The highway designer must recognize the special problems of a driver passing through an intersection and make the driving task as simple as possible by the use of good geometric design.

INTERSECTIONAL CONFLICTS

Just as the intersection is a point of decision, it is also a point at which there are a number of possible conflicts. The movement of one vehicle can conflict with the movements of other vehicles in the same stream, the cross streams, the opposing stream, and pedestrians in crosswalks. An engineering analysis of an intersection is essentially a study of all aspects of the conflict problem. Good intersection design results from a minimization of the magnitude and characteristics of conflicts and a simplification of driver route selection decisions.

There are three types of conflicts: crossing, merging, and diverging. Figure 7-1 diagrammatically illustrates these conflicts at a four-approach intersection with two-way traffic.

An analysis of an intersection will show which of the possible conflicts are significant. The significance of a conflict depends on the type of conflict, the number of vehicles in each of the conflicting streams, the time spacings

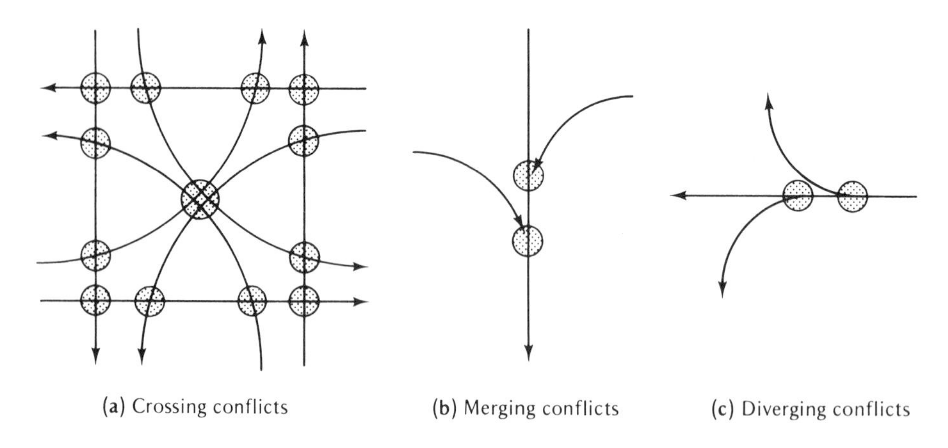

(a) Crossing conflicts (b) Merging conflicts (c) Diverging conflicts

FIGURE 7-1
EXAMPLES OF CONFLICTS

of the vehicular arrivals at the conflict point, and the speeds of the vehicles in the streams.

The relative speed of the conflicting vehicle streams is another factor affecting the significance of a conflict. Two vehicles approaching an intersection have a speed relationship to each other known as relative speed. This is defined as the difference of their velocity vectors.

The benefit of providing for low relative speed is twofold. First, events unfold more slowly, allowing more judgment time and, second, in case of an impact, the total relative energy to be absorbed is less and damage is less. In addition, when relative speed is low, the average motorist will accept a smaller time-gap space between successive vehicles to complete his move. This condition increases roadway capacity.

TYPES OF INTERSECTIONS

Intersections are classified as at grade or grade separated, depending upon the treatment of crossing conflicts. The grade-separated intersection, commonly called an interchange, is characterized by the separation of one or more crossing conflicts by overpasses or underpasses.

At-grade intersections may be subdivided according to the number of intersecting roadway legs and the angles and locations at which they intersect. Commonly used descriptions are (1) three-leg intersections (right angle or "T," skewed or "Y"); (2) four-leg intersections (right angle, skewed, off-set or jogged); (3) intersections with five or more legs, and (4) rotary intersections or traffic circles. Figure 7-2 shows some of these basic intersection forms for at-grade intersections [Ref. (1)]. The use of channelization in some of the intersections is also shown in Fig. 7-2.

PRINCIPLES OF INTERSECTION DESIGN

The smaller unit of intersection design is the individual maneuver area. Areas may be combined to produce alternative geometric designs for any intersection. To a considerable extent their arrangement is governed by traffic demands, topography, land use, and economic and environmental considerations; the proper compromise is a decision to be made by the individual designer. Intersection design should consider the ten following fundamental principles [Ref. (2)]:

Principle 1. *Reduce number of conflict points.* The number of conflict points among vehicular movements increases significantly as the number of intersection legs increases. For example, an intersection with four two-way legs has 32 total conflict points, but an intersection with six two-way legs has 172 conflict points. Intersections with more than four two-way legs should be avoided wherever possible.

Principle 2. *Control relative speed.* Relative speed is the rate of convergence or divergence of vehicles in intersection flow. A small difference (from 0 to 15 mph) in the speeds of intersecting vehicles and a small angle (less than 30 deg) between converging paths allow intersecting vehicular flows to operate continuously (uninterrupted flow). High relative speeds occur when there is either a large difference in vehicular speeds or a large angle of convergence. Since interrupted flow usually occurs under these conditions, traffic should be controlled by traffic

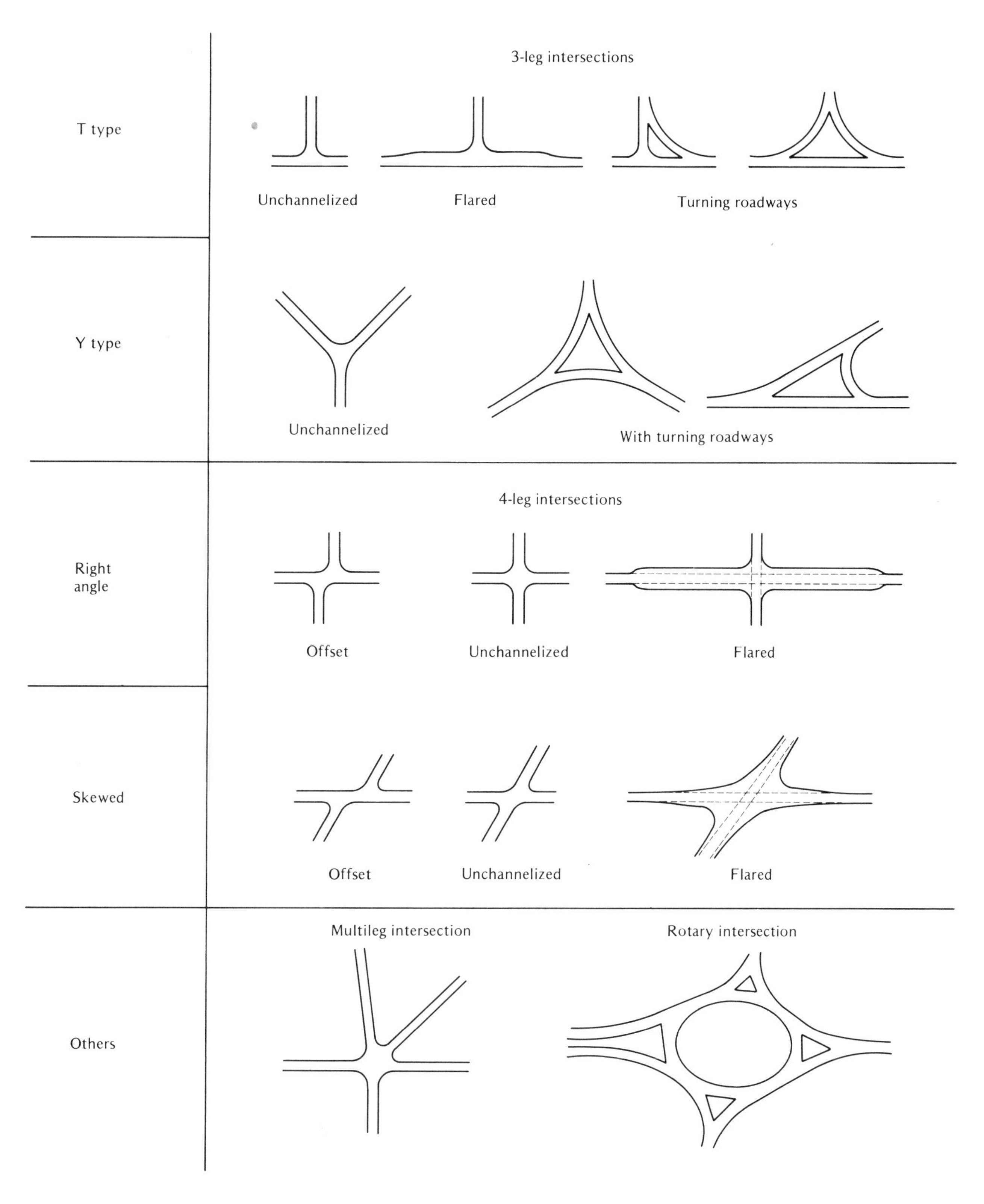

SOURCE: Adapted from American Association of State Highway Officials. *A Policy on Geometric Design of Rural Highways*, American Association of State Highway Officials, 1965, p. 388.

FIGURE 7-2
GENERAL TYPES OF AT-GRADE INTERSECTIONS

control devices. Any intersection can be designed for either condition. Low relative speeds require elimination of both speed differences and large angles between intersection flows through design. Intersection of flows at high relative speeds should be as close as possible to 90 deg.

Principle 3. *Coordinate design and traffic control.* Maneuvers at intersections accomplished at low relative speeds require a minimum of traffic control devices. Maneuvers accomplished at high relative speeds are unsafe unless traffic controls, such as stop signs and traffic signals, are provided. Designs should physically divert or block the path of vehicles making dangerous movements. Intersection design should be accomplished simultaneously with the development of traffic control plans.

Principle 4. *Use highest feasible crossing method.* Vehicle crossing manuevers can be accomplished in four ways: (1) uncontrolled crossing at grade, (2) traffic sign or signal-controlled crossing at grade, (3) weaving, and (4) grade separation. In general, both operational efficiency and construction cost increase in this order. The highest type should be used consistent with the numbers and types of vehicles using the intersection.

Principle 5. *Substitute turning path.* The method of providing turns can be changed. Separate lanes or roadways can be provided both for right- and left-turning vehicles, thereby reducing conflicts in the intersection area. For example, a direct connection can be provided to accommodate right turns at an intersection.

Principle 6. *Avoid multiple and compound merging and diverging maneuvers.* Multiple merging or diverging requires complex driver decisions and creates additional conflicts.

Principle 7. *Separate conflict points.* Intersection hazards and delays are increased when intersection maneuver areas are too close together or when they overlap. These conflicts may be separated to provide drivers with sufficient times (and distance) between successive maneuvers for them to cope with the traffic situation.

Principle 8. *Favor the heaviest and fastest flows.* The heaviest and fastest flows should be given preference in intersection design to minimize hazard and delay.

Principle 9. *Reduce area of conflict.* Excessive intersection area causes driver confusion and inefficient operations. Large areas are inherent in skewed and multiple-approach intersections. When intersections have excessive areas of conflict, channelization should be employed.

Principle 10. *Segregate nonhomogeneous flows.* Separate lanes should be provided at intersections when there are appreciable volumes of traffic traveling at different speeds. For example, separate turning lanes should be provided for high volumes of turning vehicles. When there are large numbers of pedestrians crossing wide streets, refuge islands should be provided so that more than three lanes do not have to be crossed at a time.

DRIVEWAYS

Typical driveways are really extensions of "T" intersections and should generally be designed according to the guidelines shown in Table 7-1. A minimum design for a driveway on a residential street is shown in Fig. 7-3. Table 7-1 also shows the guidelines for design of industrial driveways. Further information and details of driveway design, such as sight distance, maximum grades, maximum change in grade, and curb design and spacing can be found in References (3) and (4).

INTERCHANGES

A division of interchanges into geometric configuration is not as simple as that for at-grade intersections. Interchanges are described by the patterns of the various turning roadways or ramps. The common types include the diamond, cloverleaf, and trumpet. Figure 7-4 illustrates these three basic interchange forms. Actual interchange configurations are custom-designed to accommodate economically the traffic requirements of flow and direction of movements, type of controls and operations on the crossing facilities, and the physical requirements of topography, adjoining land use, and right-of-way. Another distinction made in type of interchange is between the directional and the nondirectional interchange. Directional interchanges are those having ramps that tend to follow the natural direction of movement. Nondirectional interchanges require a change in the natural path of traffic flow.

The design and operational characteristics of each of

TABLE 7-1
RECOMMENDED BASIC GUIDELINES FOR DRIVEWAYS

TYPE OF DEVELOPMENT SERVED	URBAN: High Pedestrian Activity[a] Major	URBAN: High Pedestrian Activity[a] Secondary	URBAN: All Other[b]	RURAL
RESIDENTIAL				
Width				
Minimum ft (m)	10 (3)	10 (3)	10 (3)	10 (3)
Maximum ft (m)	20 (6)	20 (6)	30 (9)	30 (9)
Right turn radius of flare[d]				
Minimum ft (m)	5 (1.5)	5 (1.5)	5 (1.5)	10 (3)
Maximum ft (m)	10 (3)	10 (3)	15 (5)	25 (8)
Angle[e]	75°	60°	45°	45°
INDUSTRIAL				
Width[c]				
Minimum ft (m)	20 (6)	20 (6)	25 (8)	35 (11)
Maximum ft (m)	35 (11)	35 (11)	40 (12)	40 (12)
Right turn radium of flare[d]				
Minimum ft (m)	10 (3)	15 (5)	15 (5)	25 (8)
Maximum ft (m)	15 (5)	20 (6)	25 (8)	50 (15)
Angle[e]	75°	60°	45°	45°

[a] As in central business areas, or in same block with auditoriums, schools, libraries.
[b] The remaining city streets, including neighborhood business, residential, industrial.
[c] Measured along right-of-way line, at inner limit of curbed radius sweep, or between radius and rear edge of curbed island at least $50\ ft^2$ in area.
[d] On side of driveway exposed to entry or exit by right-turning vehicles.
[e] Minimum acute angle measured from edge of pavement.

the major interchange types are discussed in the following sections.

Diamond Interchanges

The diamond interchange is the simplest form of grade-separated intersection between two roadways. The conflicts between through and crossing traffic are eliminated by a bridge structure. The left-turn crossing movement conflicts are reduced by eliminating the conflict with traffic in the opposing direction on one of the roadways, usually the more important of the two. The remaining left-turn conflicts and all merging and diverging movement conflicts take place at the terminal points of each ramp.

The diamond interchange requires a minimum amount of land, and it is economical to construct. In addition, a diamond interchange generally requires less out-of-the-way travel, and vehicle operating costs are less than those on most other types of interchanges. The single point of exit from the major roadway eases the problem of signing. With these advantages, diamonds appear to be the ideal solution to an intersection problem. However, the conflicts that occur where ramps meet the grade-separated cross street may require an alternative solution if ramp volumes are high.

Some signal timing problems at cross streets may result, and capacity may be inadequate for certain flows. A three-level design is sometimes used. This provides for interchanging at an intermediate level between the two through roadways.

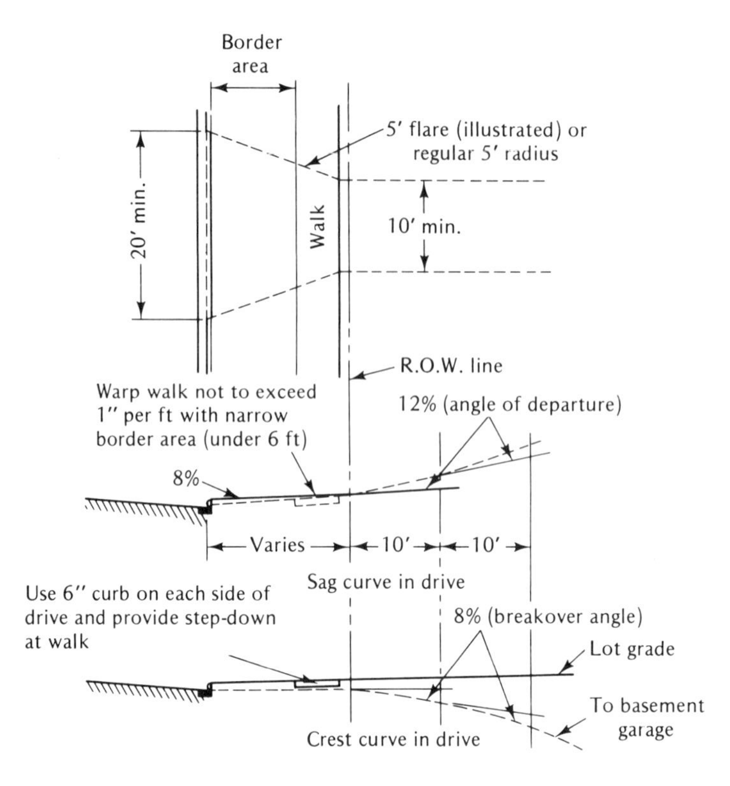

SOURCE: *Recommended Practices for Subdivision Streets* (1967), p. 8.

FIGURE 7-3
RESIDENTIAL DRIVEWAY DETAILS

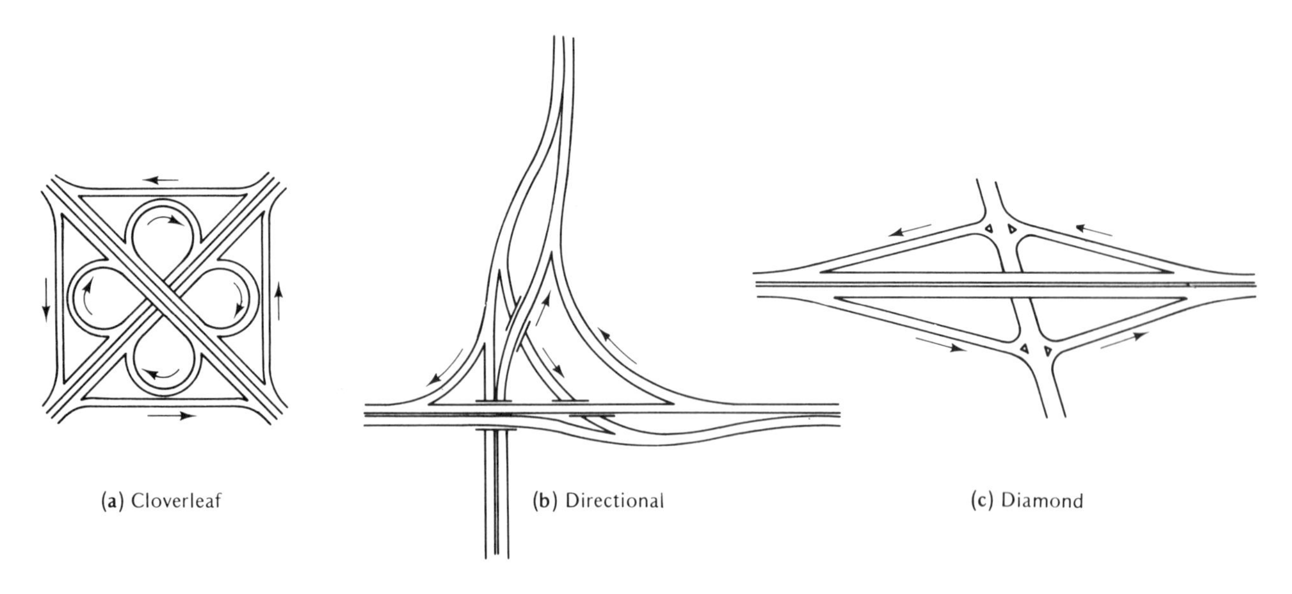

SOURCE: Adapted from American Association of State Highway Officials. *A Policy on Geometric Design of Rural Highways.* American Association of State Highway Officials, 1965, p. 494.

FIGURE 7-4
GENERAL TYPES OF INTERCHANGES

Cloverleaf Interchanges

The full cloverleaf interchange eliminates all crossing movement conflicts by the use of weaving sections. The weaving section replaces a crossing conflict with a merging, followed some distance farther by a diverging conflict. There are two points of entry and exit on each through roadway. The first exit, generally 1000 to 2000 feet (305 to 610 m) before the crossroad structure, provides for the right-turn movements. The second exit, immediately after the crossroad structure, provides for left-turn movements. The point of entry from the left-turn off the crossroad is immediately before the structure. A weaving section is created between the exit and entry points near the structure. This weaving section is a critical element of cloverleaf design. It must have length and capacity to allow for a smooth merging and diverging operation.

Although full cloverleaf interchanges eliminate the undesirable left-turn crossing movements of diamond interchanges, they have the disadvantages of greater travel distances, higher operating costs, difficult merging sections, and large rights-of-way occasioned by the radius requirements necessary for satisfactory speeds on the ramps.

Full cloverleaf development is not always required. A partial cloverleaf, or parclo, is a modification that combines some elements of a diamond interchange with one or more loops of a cloverleaf to eliminate only the more critical turning conflicts (see Fig. 7–5).

Another variation of the cloverleaf configuration is the cloverleaf with collector-distributor roads. This is also shown in Fig. 7–5. With the collector-distributor roadway, main roadway operations are much the same as in a diamond interchange. For each direction of travel, there is a single point for exits and a single point for entrances. Speed change, detailed exit directional signing, and the storage and weaving problems associated with a cloverleaf are transferred to the collector-distributor road, which can be designed to accommodate greater relative speed differences or encourage smaller ones. Although this configuration improves the operational characteristics of a cloverleaf interchange, the disadvantages of greater travel distances and the requirement of extra right-of-way are still present. The use of a cloverleaf with collector-distributor roads is appropriate at junctions between a freeway and an expressway or other divided-type roadway where a diamond interchange would not adequately serve traffic demands.

Directional Interchanges

Directional interchanges are those having ramps for one or more direct or semidirect left-turning movements. Interchanges of two freeways or interchanges with one or more very heavy turning movements usually warrant direct ramps, which have higher speeds of operation and higher capacities, compared to loop ramps. This type of design

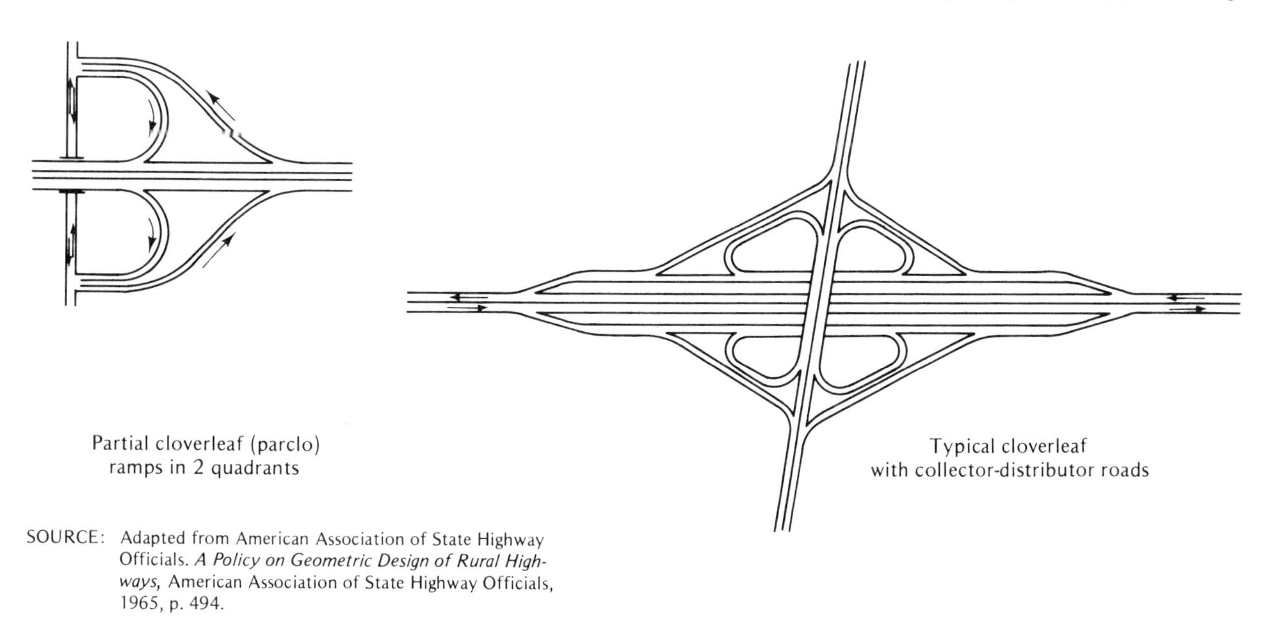

Partial cloverleaf (parclo) ramps in 2 quadrants

Typical cloverleaf with collector-distributor roads

SOURCE: Adapted from American Association of State Highway Officials. *A Policy on Geometric Design of Rural Highways,* American Association of State Highway Officials, 1965, p. 494.

FIGURE 7–5
CLOVERLEAF INTERCHANGES

requires more structures and more land than diamond interchanges and in some cases than cloverleafs, however, and requires left-turn entry or exit traffic, which can result in operational problems. Any one of several schemes using various combinations of directional, semidirectional, and loop ramps may be appropriate for a certain set of conditions, but only a limited number of patterns are utilized generally. These are the layouts that use the least space, have the fewest or least complex structures, minimize internal weaving, and will fit the common terrain and traffic conditions.

The basic patterns are as follows [Ref. (6)] :

1. With loops and weaving
2. With loops, no weaving
3. Fully directional, no weaving
4. Fully directional, multilevel structures

Adjacent Interchanges

Traffic operations at interchanges almost invariably introduce turbulent flow in the freeway traffic stream. This turbulence is dissipated "downstream" as vehicles adapt to the changed conditions. The distance required for this adjustment varies with the complexity of movements at the interchange. It is considered desirable that the next interchange not be located within this adjustment area, and distances between interchanges of one or more miles are often necessary.

To assist the driver in his behavior at interchanges, the operational character of successive interchanges should be similar. For example, consider the illustration of an inconsistent pattern of exits in Fig. 7-6, which shows a mixture of cloverleaf and diamond interchanges. In this situation, the cloverleaf and parclo interchanges have one or two exits. The diamond interchange has only one exit. The driver is never certain where his exit will be, causing hesi-

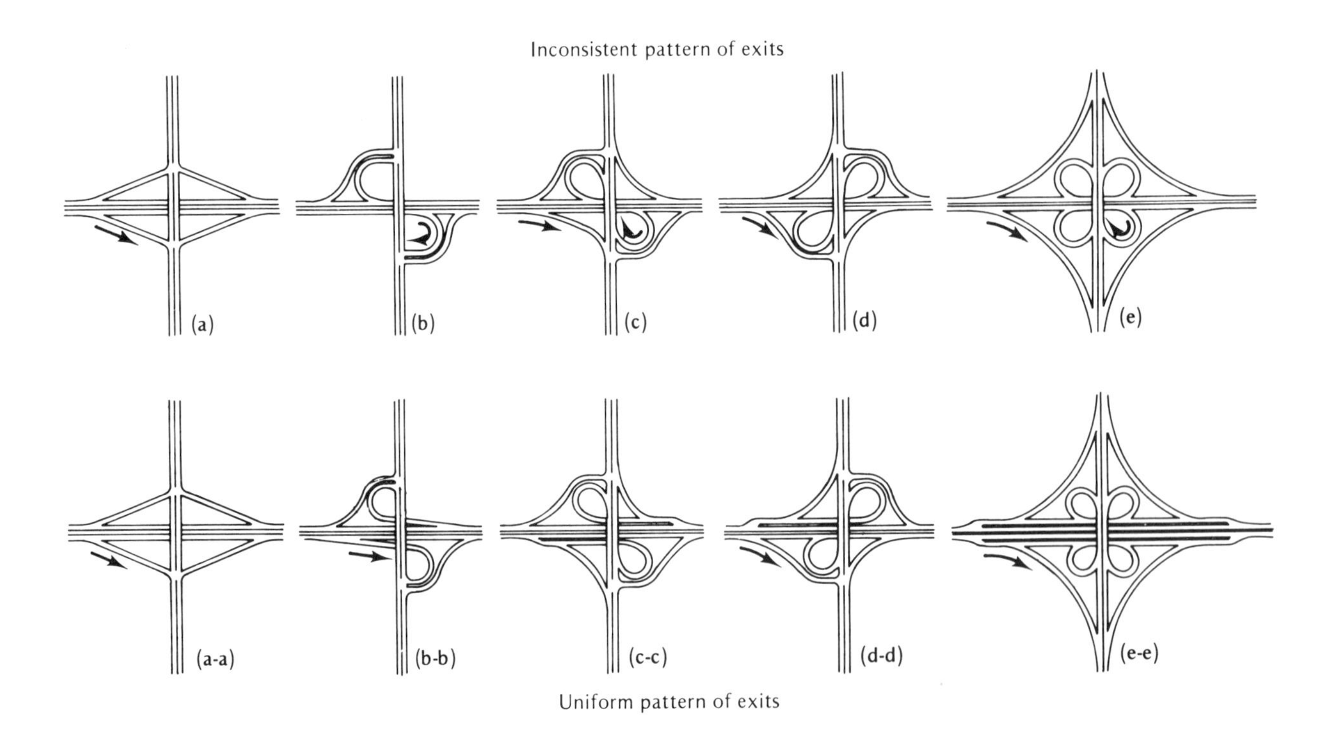

SOURCE: Jack E. Leisch, "Adaptability of Interchange Types on Interstate System," *Proceedings of the American Society of Civil Engineers,* Journal of the Highway Division, Vol. 84, No. HW-1, January 1958, pp. 1525-30.

FIGURE 7-6
ARRANGEMENTS OF EXITS BETWEEN SUCCESSIVE INTERCHANGES

tation at high freeway speeds or possible wrong-way entry into a one-way ramp. This should not be interpreted to mean that all interchanges must be diamonds or all cloverleafs. However, consistent design that will meet the needs of traffic volume and physical conditions is the designer's greatest challenge, requiring initiative and ingenuity, as is shown in the bottom portion of Fig. 7-6.

INTERSECTIONAL DESIGN ELEMENTS

Most of the roadway design elements of speed, capacity, horizontal and vertical curvature, sight distance, cross section, crown, and superelevation discussed in Chapter 6 are applicable in the design of intersections. However, the design problem is compounded, because an intersection requires the consideration of the simultaneous application of each element of two or more roadways and for two or more traffic streams. This section outlines the more important elements of design, with particular emphasis on their significance relative to intersections.

Sight Distance

A roadway designed for stopping sight distance as described in Chapter 6 has the sight distance required for intersections controlled by traffic signals. However, at uncontrolled intersections or those controlled by stop signs, each vehicle must have a clear view of all other vehicles approaching the point of intersection. Sight distance at these intersections can be limited by obstructions within the triangle formed by two approach legs.

The approach legs form two sides of the "sight triangle" that bounds the required area of unobstructed view. Figure 7-7 shows an example of a simple sight triangle. For intersection sight distance requirements, no object within the area of unobstructed view may protrude more than 3.75 feet (1.1 m) above the plane formed by the pavement elevations of the "sight triangle" A, B, and C.

Drivers approaching an uncontrolled intersection should have a clear view great enough to enable them to decide whether to stop, change speed, or continue through without speed change. The greatest sight distance is required for those conditions when it is necessary to decide whether or not to stop. Therefore, the sides d_a and d_b, of the sight triangle, Fig. 7-7, should be equal to the safe stopping sight distance as described in Chapter 6 for vehicle speeds U_a and U_b. For example, the National Committee on Urban Transportation recommends that the unobstructed

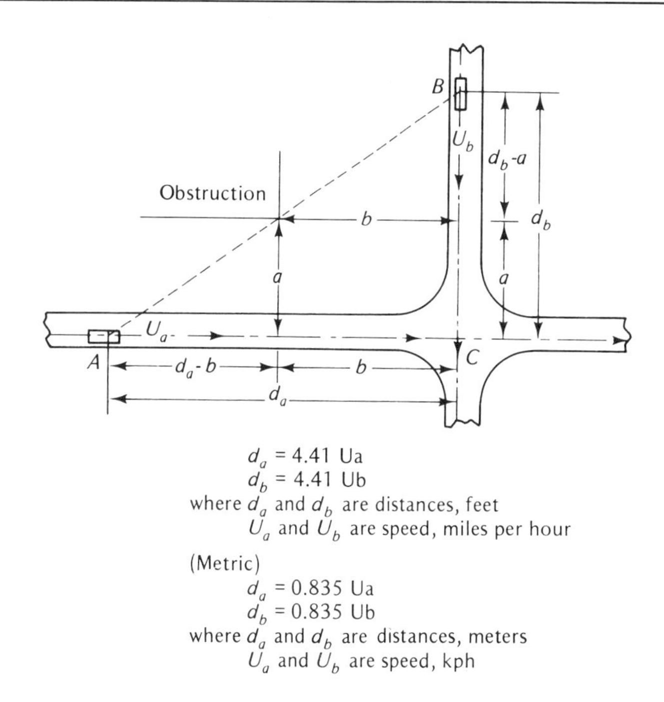

SOURCE: American Association of State Highway Officials, *A Policy on Geometric Design of Rural Highways,* American Association of State Highway Officials, 1965, p. 392.

FIGURE 7-7
SIGHT DISTANCE AT INTERSECTIONS

view across the corners of two local residential streets should be formed by a triangle whose sides are 110 feet measured along the centerlines of the approach streets from their point of intersection. This value of 110 feet is the approximate stopping sight distance for a speed of 25 miles per hour.

When one is accelerating from a stop sign of a major route, the sight distance required is the time to evaluate an acceptable gap and cross the intersection safely.

Turning Radii

Two factors important in the design of urban at-grade intersections are that (1) land area must be minimized, and (2) pedestrians must be able to cross the street safely and conveniently. These factors dictate that the minimum possible radius be used in the design of urban at-grade intersections. The minimum radius is governed by the assumed speed of operation and its resulting comfort requirements, as well as the extent to which the paths of design vehicles which will be making the turns encroach on other lanes. Figure 7-8 shows the relationship of curb

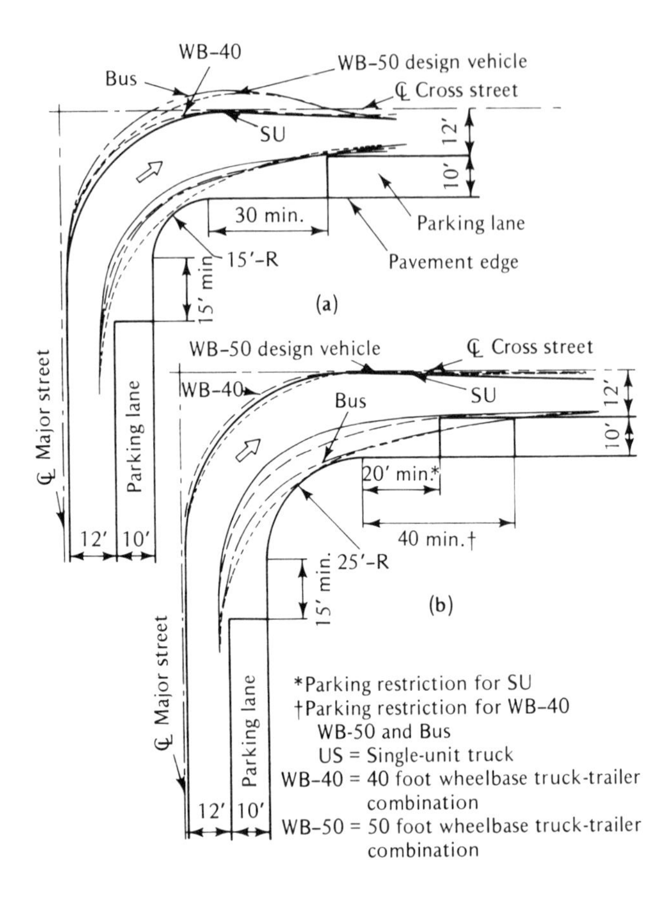

SOURCE: Adapted from American Association of State Highway Officials, *A Policy on Arterial Highways in Urban Areas,* American Association of State Highway Officials, 1973, p. 680.

FIGURE 7-8
EFFECT OF CURB RADII AND PARKING ON TURNING PATHS

radii and street lane width to the turning moves of several design vehicles [Ref. (6)]. A 15-foot (4.6-m) curb radius is adequate for passenger vehicles and single-unit trucks, but larger truck combinations encroach on opposing lanes of the cross street. Note in the figure that parking has been restricted in the intersection area to allow turning vehicles to encroach on the curb lanes. Residential streets that are infrequently used by large trucks can be designed with 15- to 25-foot (4.6- to 7.6-m) curb radii. Collector and arterial streets should have a minimum radius of 40 feet (12.2 m). This allows most types of vehicles to turn without encroaching on adjacent lanes. The effect of the 40-foot (12.2-m) radius on pedestrians is to increase exposure approximately five seconds more than a 15-foot (4.6-m) design radius.

In the actual design to accommodate turning vehicles, 3-centered compound curves are often used since this more closely fits the actual vehicle turning paths [see Ref. (3), Chapter VII or Ref. (6), Chapter L for a more detailed discussion].

In dealing with rural at-grade intersections, land conservation and pedestrian crossings are generally not significant problems. The magnitude of the turning movement and uniformity of treatment are guides for rural intersection design. As the importance of a turning movement increases or the route design standard improves, the turning radius is increased within the limit of economy. Although it would be desirable to have intersectional curves for the same speed as the approach speed, such design is generally not economical. Moreover, the design must accommodate the standard of the roadway that the turn enters so as not to create a new hazard. Commonly accepted design values for various turning speeds are presented below in Table 7-2.

TABLE 7-2
MINIMUM RADII FOR INTERSECTION CURVES

Design (turning) speed (V), mph (km/hr)	15 (24)	20 (32)	25 (40)	30 (48)	35 (56)	40 (64)
Side friction factor (f)	0.32	0.27	0.23	0.20	0.18	0.16
Assumed min. superelev. (e)	0.00	0.02	0.04	0.06	0.08	0.09
Total $e + f$	0.32	0.29	0.27	0.26	0.26	0.25
Calculated min. radius (R), ft (m)	47 (14)	92 (28)	154 (47)	231 (70)	314 (96)	426 (130)
Suggested curvature for design						
Radius—minimum, ft (m)	50 (15)	90 (27)	150 (46)	230 (70)	310 (94)	430 (131)
Degree of curve—maximum	—	64	38	25	18	13
Average running speed, mph (km/hr)	14 (22)	18 (29)	22 (35)	26 (42)	30 (48)	34 (55)

Note: For design speeds of 40 mph and over, values same as for open highway conditions.

SOURCE: *A Policy on Geometric Design of Rural Highways,* American Association of State Highway Officials, 1965, p. 325.

Auxiliary Lanes at Intersections

At intersections handling large trucks and allowing passenger cars to turn at speeds of 15 mph (24 kmh) or more, the operation, capacity, and safety can be improved by the use of auxiliary lanes. Auxiliary lanes include right-turn, left-turn, or through lanes which are added to the basic approach roadway cross section.

> The primary purpose of auxiliary lanes at intersections is to provide storage for turning vehicles, both left and right. A secondary purpose is to provide space for turning vehicles to decelerate from the normal speed of traffic to a stopped position in advance of the intersection or to a safe speed for the turn in case a stop is unnecessary. Additionally, auxiliary lanes may be provided for bus stops or for loading and unloading passengers from passenger cars [Ref. (6), p. 686].

Auxiliary lanes should be at least 10 feet (3.0 m) and preferably 12 feet (3.7 m) wide. The length of a turning lane consists of three components: (1) deceleration length, (2) storage length, and (3) entering taper. Although the total length should be the sum of all three components, on moderate-speed urban arterials the storage length plus taper is often used in practice.

The storage length should provide for the maximum number of vehicles to arrive in 1.5 to 2 cycles at a signalized intersection, or in 2 minutes at an unsignalized intersection.

The auxiliary lane allows the separation of the right-turning traffic from the through traffic stream. On city streets, the curb lane can sometimes be utilized as a right-turn lane by prohibiting parking. The more general case of a right-turn lane provides for an increase in the curb radius to allow for turning speeds at 15 miles per hour (24 kmh) or greater. In this latter instance, the lane is generally separated from the rest of the intersection by a traffic island. The island also serves as a space for pedestrian refuge.

Left-turn lanes reduce delays, rear-end collisions, and turning accidents, and they add to the roadway capacity at intersections. The separation of left-turning traffic into a distinct lane, clearly indicating the intent of those vehicles, eases the danger of the crossing and diverging conflicts.

The incorporation of left-turn lanes into roadways with medians can be accomplished by the construction of recesses in the medians. Median widths to accommodate left-turn lanes usually range from 14 to 20 feet (4.3 to 6.1 m) or more. Lengths of the recesses will vary, depending on distance between openings, demands for similar treatment at adjacent intersections, and the turning demand characteristics. On rural roadways without medians, a shift outward from the center in the alignment of the through lanes often can be made. Painted or raised islands are then provided in the center open portion to establish the left-turn lane. The transition of lanes prior to developing full added lane width is critical in a design of this type. The through-lane vehicle path must be smooth and natural if the rear ends of the left-turning vehicles are to be protected.

TAPERS

Although tapers of auxiliary lanes should follow the discussion of "Speed-Change Lanes," [Chapter J, Ref. (6)], shorter tapers are normally used on urban streets. A taper of 8:1 may be used for speeds up to 30 mph (48 kmh) but a 15:1 taper should be used for 50-mph (80 kmh) operations. Both straight-line taper and tapers with reverse curves (one at either end) are used on urban highways. The length of a taper depends on the rate of change in lateral position (sometimes assumed as ⅓ second/ft of displacement).

PEDESTRIANS

Adequate protection for pedestrians is an important intersection design factor in urban areas. Since pedestrian crossings are supposed to take place at intersections, design must recognize their presence and provide as safe a crossing as possible.

Experience with the usage of specially provided bridges and tunnels is disappointing, since pedestrians have shown a reluctance to use these facilities when a direct grade crossing is available. They will use tunnels and bridges, however, if the curb is fenced for a considerable distance or if moving traffic constitutes an extreme hazard.

The general solution to the pedestrian problem is to provide convenient, well-marked crosswalks at grade. Intersections that include right- or left-turning lanes or skewed and multilane approaches should be analyzed for the effect of width on the pedestrian crossing. Pedestrians, walking at the average rate of four feet per second, are exposed to traffic an additional three seconds for each added lane. A design possibility to aid the pedestrian movement is the inclusion of pedestrian safety islands.

CHANNELIZATION

Channelization is the separation or regulation of conflicting traffic movements into definite paths of travel by means of traffic islands or pavement markings to facilitate the safe and orderly movements of both vehicles and pedestrians. Proper channelization increases capacity, improves safety, provides maximum convenience and instills driver confidence. Improper channelization has the opposite effect and may be worse than none at all. Over-channelization should be avoided as it could create confusion and worsen operations [Ref. (6)].

Several rules govern the application of channelization [Ref. (1)]:

1. Islands should be arranged so that the driving path seems natural and convenient.
2. There should be only one path for the same intersectional movement (separate conflicts).
3. A few well-placed, large islands are better than a confusion of small islands.
4. Islands should be offset 2 ft (0.6 m) or more from the edge of normal traveled way.
5. Adequate approach-end treatment is required to warn drivers and to permit gradual changes in speed and path.
6. Curving roadways should have radii and width adequate for the governing design vehicle.
7. Adequate visibility should be provided drivers approaching the intersection. There should be no hidden obstructions, and islands should be well-defined. In many cases some form of illumination will be necessary, depending on the type of curb used to outline the island.

Utilizing these rules should result in the following:

1. Reduce area of conflict.
2. Design crossing at 90-deg angles.
3. Merge streams at small angles (10–15 deg).
4. Bend and funnel traffic streams.

To accomplish its purpose, channelization must simplify the intersection driving task by guiding the motorist along natural paths of movement. Since human behavior is not as easily translated into geometric form as are vehicle and physical design factors, it is sometimes advisable to install temporary channelization (traffic cones, sand bags, etc.) and to determine its effect on traffic operations, before installing permanent curbing and markings.

Conditions at two intersections are seldom identical. Although certain classical forms of channelization are particularly suited to specific types of intersections (see Fig. 7-9), final design is a compromise of human, vehicle, and physical factors.

REFERENCES

1. Baerwald, John E., *Transportation and Traffic Engineering Handbook.* Englewood Cliffs, N.J.: Prentice-Hall, Inc., 1976.
2. Matson, T.M., W.S. Smith, and F.W. Hurd, *Traffic Engineering.* New York: McGraw-Hill Book Co., 1955.
3. American Association of State Highway Officials, *A Policy on Geometric Design of Rural Highways.* Washington, D.C.: 1965.
4. American Association of State Highway Officials, "An Informational Guide for Preparing Private Driveway Regulations for Major Highways." Washington, D.C.: 1960.
5. Institute of Traffic Engineers, "Guidelines for Driveway Design and Location: Recommended Practice." Washington, D.C.: 1973.
6. American Association of State Highway Officials, *A Policy on Design of Urban Highways and Arterial Streets.* Washington, D.C.: 1973.

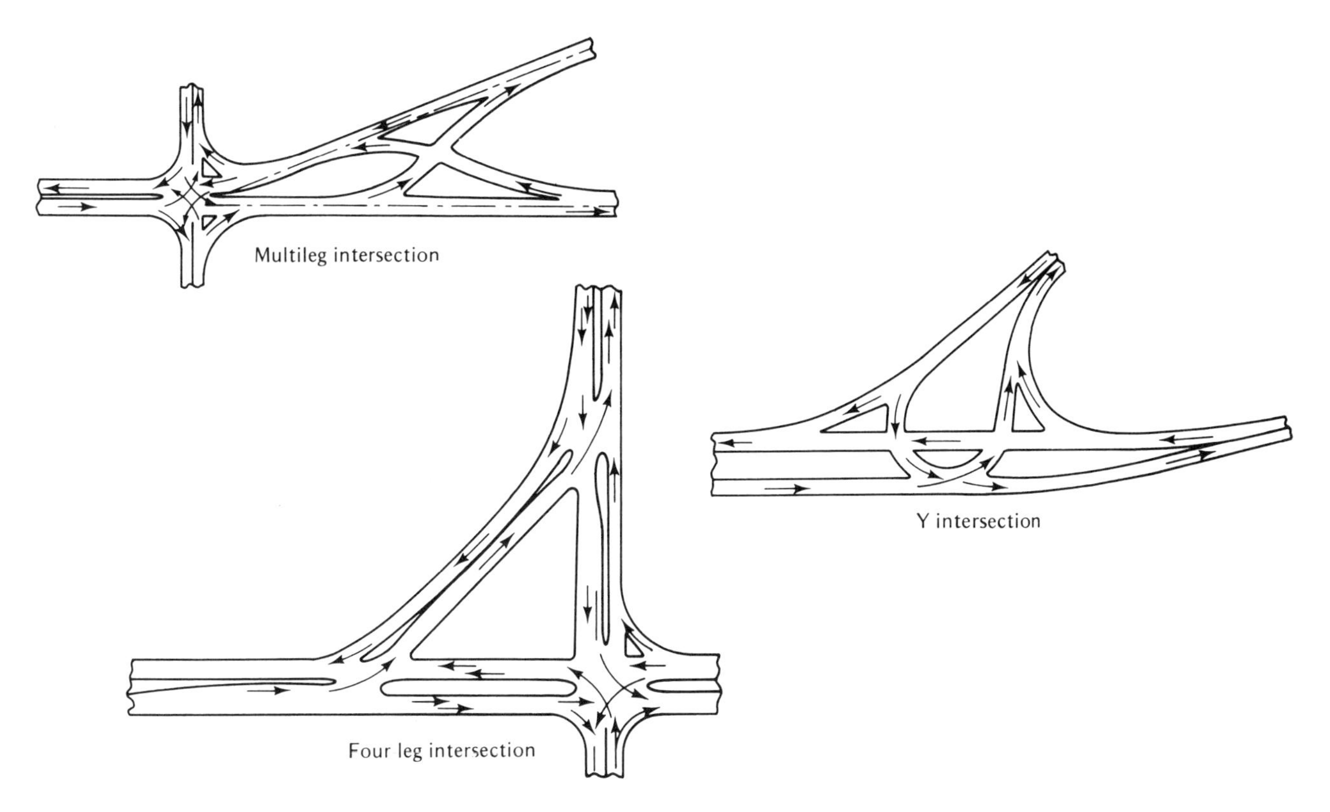

SOURCE: American Association of State Highway Officials, *A Policy on Geometric Design of Rural Highways*, American Association of State Highway Officials, 1965, pp. 445, 465, 477.

FIGURE 7-9
CHANNELIZED INTERSECTIONS

8

GEOMETRIC DESIGN OF PARKING AND LOADING FACILITIES

The physical plant of any transportation system has three basic elements: the vehicle, the roadway, and the terminal. For highway transportation the terminal is a parking space, whether at the curb or off-street in a lot, garage, shopping center, or private driveway; a bus stop or terminal; a taxicab stand, or a truck loading zone or specially designed terminal. Regardless of location or type, the terminal facility is an essential part of the highway transportation system.

Ideally, road users would like to park immediately adjacent to their destination at a minimum cost—preferably nothing. On-street curb parking can serve this need adequately in areas of low density of trip generation. However, as the intensity of land use and trip destination increases, the few feet of curb in front of a particular building cannot accommodate all those whose destinations lie within. Furthermore, as street space is increasingly needed for moving traffic, the supply of curb parking spaces is correspondingly reduced. Table 2-7 in Chapter 2 indicates the decreasing role of curb parking in central business districts as the size of the urban area increases. Therefore, in areas of concentrated parking demand, increased dependence must be placed on off-street facilities (lots, garages, and special terminals).

If these off-street facilities are to function effectively and efficiently, they must be an integral part of the highway system and must serve the needs and desires of the drivers who use the system. The supply of parking must be adequate for the concentrated demand generated at locations such as the central business district (CBD), regional shopping centers, educational campuses, airports, transit stations, or major industrial plants, unless land use or environmental policies call for deliberate shortages to force changes in travel patterns. At the same time, the location, layout, and operation of this parking must be acceptable to the drivers and adequate for their vehicles.

Parking facility planning and design require a determination of the number of spaces needed, the proper location for these spaces, and a workable layout with acceptable operating controls. In these determinations an understanding is essential of the characteristics of both the individual parker and the total parking operation in the district to be served as well as the parking generating characteristics of the major types of land use.

This chapter is concerned primarily with the geometric design of terminal facilities. The reader should consult Chapter 2 for a discussion of parking characteristics. Chapter 12 discusses the role of bus terminals within a transit network. Chapter 15 deals with the finance and administration of parking programs.

PARKING DESIGN FOR AUTOMOBILES

The design of automobile parking facilities must take into account the concerns of drivers as well as the viewpoint of the owners of the facilities. Parkers may wish to park their own vehicle, may wish an attendant to park it, or may be indifferent. If they park their own vehicle, they may be willing to squeeze a little when getting in or out of their vehicle or may insist on a full unobstructed door opening. They may be skilled in moving into a small space or require much maneuvering area. They may require their vehicle immediately or be willing to wait up to 15 or more minutes. The purpose of the trip and the duration of the stay vary widely and affect the parkers' willingness to pay and their service demands. The problems of the parking facility operator are those of meeting the requirements of the drivers served, yet obtaining the maximum return from the facility in terms of usage or profit.

The engineer involved in the geometric design of parking places transforms the above requirements into a set of design standards and operating criteria. These design and operational standards vary substantially and offer an opportunity for the engineer to evaluate a number of alternatives to determine which best meets the overall community and operator objectives.

Curb Parking Design

The determination of appropriate curb use requires a study of the requirements of both moving traffic and abutting properties. The control of curb use is described in Chapter 11. Having determined that curb parking will be permitted at a location, it becomes necessary to specify the geometric configuration. This design depends basically on low-speed driver-vehicle performing characteristics, although parking duration, turnover, and space occupancy may affect design somewhat.

Figure 8-1 shows the geometric requirements for the parking stalls themselves. It can be seen that more spaces per lineal foot of curb can be obtained with angle parking. However, angle parking restricts the moving traffic use of a street much more than parallel parking, as the street manuevering and parked-vehicle space requirements are greater for angle than parallel parking. In addition, angle parking results in a higher accident rate than parallel parking at the same location. Because of these adverse operational characteristics, the use of angle parking is generally limited to wide minor streets carrying slow-moving traffic. The tandem parallel parking arrangement reduces parking maneuver time and is advantageous on major streets.

The geometry of bus, truck, and taxi loading zones is based on the size of vehicle using the facilities. Spaces at the end of blocks and adjacent to driveways and alleys reduce parking maneuver time for these vehicles and expedite traffic flow. Curb bus loading zones are discussed in detail in Chapter 12.

Off-Street Facility Entrances and Exits

For larger off-street parking facilities the movements in and out generally must be related to the traffic operations on the adjacent streets. The capacity and spacing of access points must be adequate to accept the incoming vehicles without storage on the street and to discharge exiting vehicles smoothly. The maximum useful size of many facilities, especially of those used for short-term parking, may be determined by the ability of adjacent streets to deliver and receive cars (access and egress capacity) rather than by the parking demand (Fig. 8-2). For this reason, sites with frontage on at least two streets are preferred.

Entrances and exits are often paired. They should be located as far away from intersections as possible so that the driveways do not interfere with the intersection operation and so that the chance of driveways being blocked by traffic backed up from an intersection is minimized. Considering also the fact that corner sites in many districts command premium rents, the best location is a mid-block parcel running through the block and fronting on two parallel streets.

The entrance must be large and obvious enough to at-

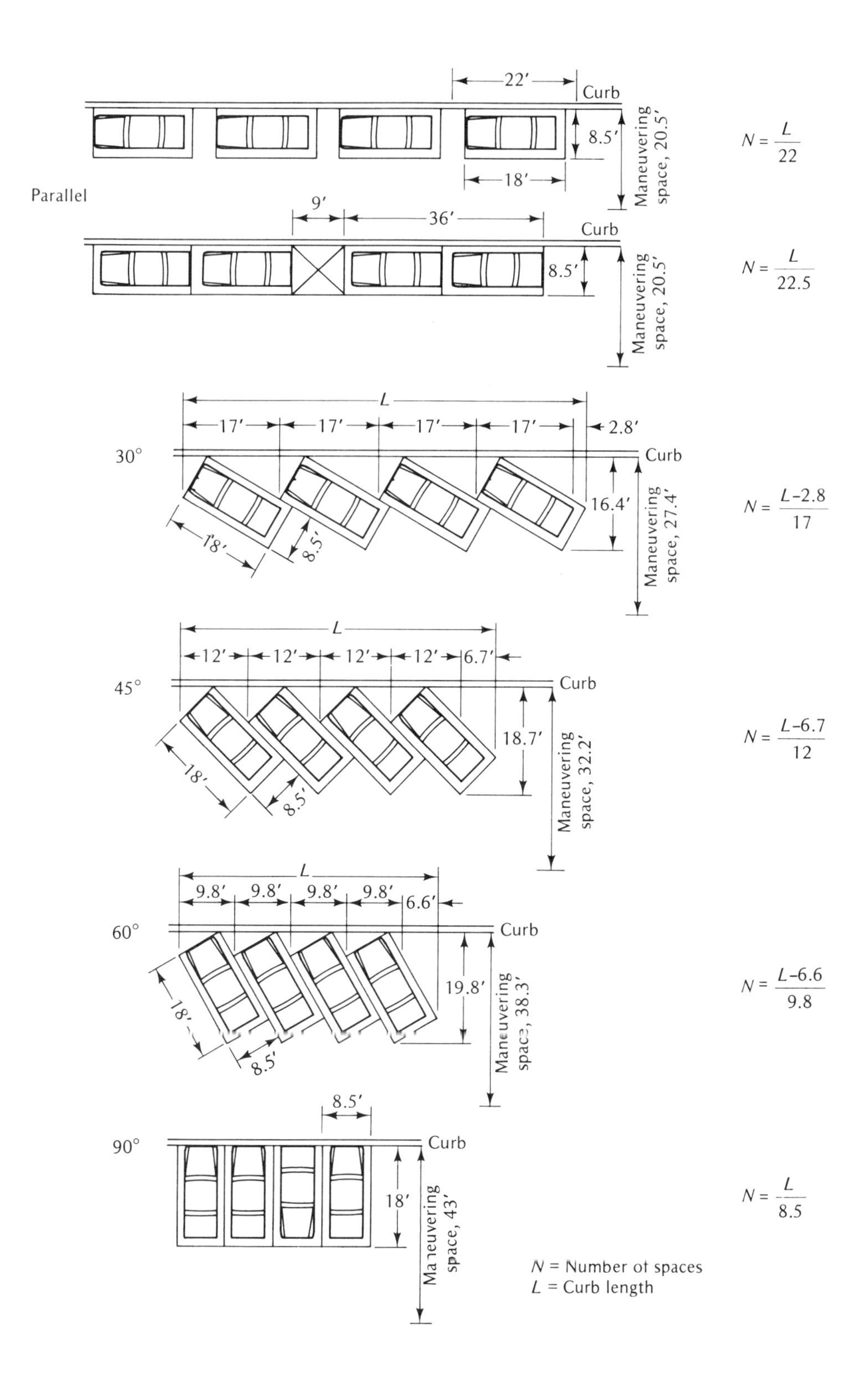

SOURCE: Adapted from R. H., Burrage, D. A. Gorman, S. T. Hitchcock, and D. R. Levin, *Parking Guide for Cities,* U.S. Government Printing Office, 1956, p. 125.

FIGURE 8-1
CURB PARKING GEOMETRY

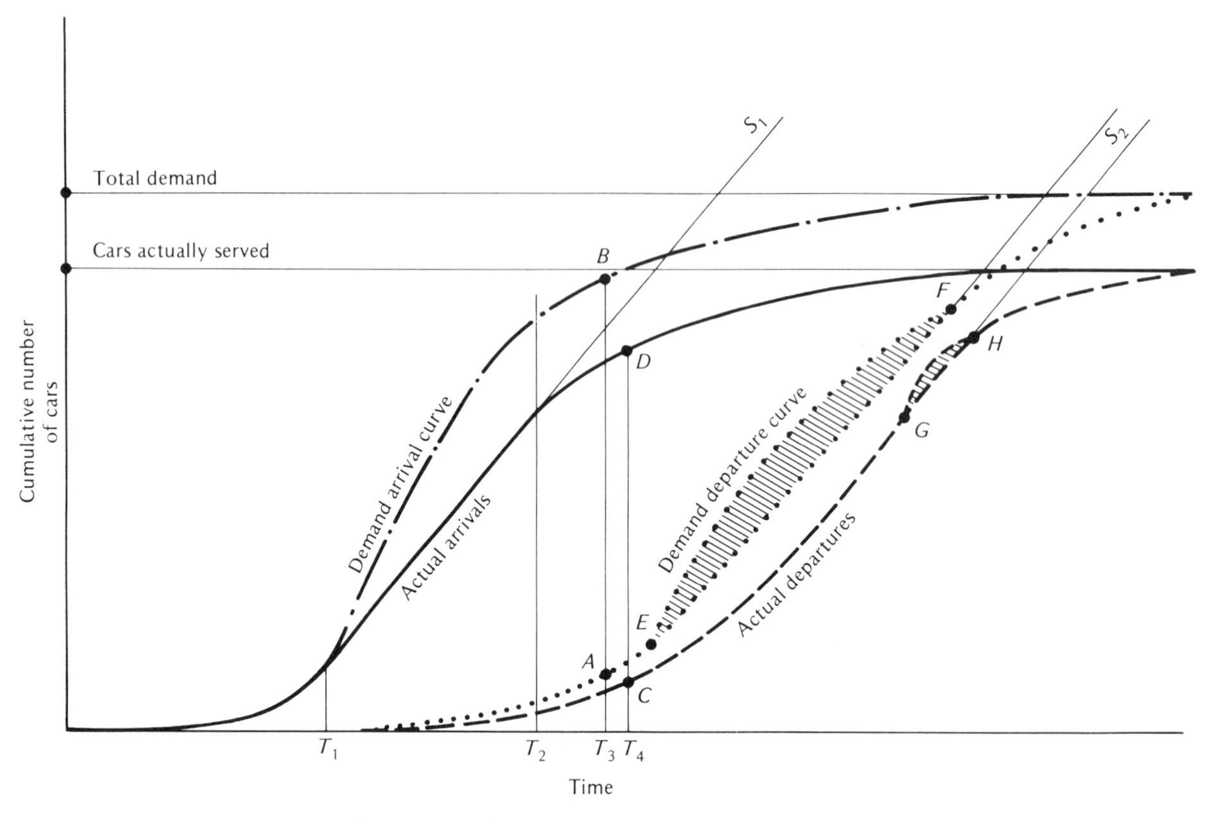

S_1 = Slope representing access capacity
S_2 = Slope representing egress capacity
T_1 = Time when demand rate begins to exceed S_1
T_2 = Time when demand rate drops below S_1, and after which the demand and actual arrival curves become parallel
T_3 = Time of theoretical maximum accumulation, when demand arrival and departure curves are parallel
T_4 = Time of actual maximum accumulation, when actual arrival and departure curves are parallel
AB = Maximum demand accumulation (computed facility size)
CD = Maximum actual accumulation (required facility size)
EF = Queueing diagram if demand arrivals could be accommodated while egress capacity remained unchanged
GH = Queueing diagram if desired departure rate of cars actually parked exceeds S_2

FIGURE 8-2
EFFECT OF ACCESS AND EGRESS CAPACITY ON MAXIMUM ACCUMULATION IN PARKING FACILITIES

tract patrons and make it easy for drivers to move into the facility. In the case of a garage, ramps should not be immediately inside the entrance or exit, since they cause incoming vehicles to slow down, and exiting vehicles to wait on a slope. Columns and walls should not be placed where they block the exiting drivers' view of pedestrians on sidewalks. Where volumes of traffic generated by the parking facility and of conflicting pedestrian streams are both large, a grade separation may be desirable; this can often be provided for by starting the ramps within the sidewalk area and moving the sidewalk toward or inside the building line to pass over or under the ramps.

Operation

The internal operations of an off-street parking facility may be classified in two different categories: self-parking by the vehicle driver and parking by attendants. The differences in operational characteristics affect almost every aspect of design and are so great that the decision as to which will be utilized must be made early in the design of a particular facility.

Self-parking is universally used in shopping centers. It is popular with most car owners, who prefer to handle their own car, lock it, and have access to it at any time. It has the obvious advantage of minimizing the number of employees, thus greatly reducing operating costs. However, as it requires more area per car space in both aisles and stalls, the initial development cost per space is higher than for attendant-operated facilities.

Attendant parking is attractive to those motorists who do not wish to park their own cars. Since attendants are experienced parkers, they can place cars in smaller stalls. In addition, double parking is often possible. These measures improve space utilization from 30 to 50 percent. However, at peak unparking times considerable delays are encountered by parkers awaiting delivery of their cars. Also, rising labor costs have reduced interest in attendant parking facilities except in areas of very high land values.

Layout

In the layout of parking stalls the engineer should try to obtain the maximum amount of storage capacity from the given workable area. However, it must be remembered that overcrowding that results in the restriction of necessary vehicle movements will decrease the handling rate and may result in less efficient use of the lot or garage.

Two main points must be kept in mind during the layout design procedure: (1) the layout of parking must be flexible enough to adapt to future changes in automobile dimensions; (2) ramp, stall, and aisle dimensions must be compatible with the type of operation planned for the facility.

The critical dimensions in laying out storage areas are the width and length of stalls, width of aisles, angle of parking, and radius of turns. These are related to car dimensions and performance characteristics, as shown in Fig. 3-1. Values of each of these dimensions for models in current production are published by the automobile companies for each model year and can be used to select a design vehicle for any group of cars [Ref. (3)]. Other vehicle characteristics that affect design include the height of hood, top, and rear deck, the underclearance, and the width required to open the door sufficiently for the driver to get out comfortably.

Recommended stall dimensions and the sizes of layout resulting therefrom are shown in Fig. 8-3 for four typical parking angles. For easy maneuvering, stall widths of 9.0 ft (2.75 m) are often used; in some facilities that anticipate extensive use by shoppers with packages the stall width may be 9.5 ft (2.90 m).

The key design dimension is the "module," including two rows of stalls and the aisle between them. In Fig. 8-3 this module is shown as bounded by a wall on the left side and by an interlocking row of stalls on the right. Except at the 90-deg angle, interlocking saves space, as shown by the lower dimensions for the width of the rows on the right of the aisle.

Stalls are often laid out at right angles, since this is the most efficient use of space and provides for two-way movement in the aisles. Two-way movement with angle parking is sometimes used. This type of design requires wider aisles than those shown in the figure. When orderly traffic movement is needed and can be self-enforcing, one-way traffic aisles and angle parking can be used. Many lot dimensions are more adaptable to one-way traffic and angle parking.

Any parking angle between 30 and 75 deg can be used to optimize the use of the site, and different angles can be used in the same facility where they do not interlock. Below 30 deg, excessive amounts of wasted space occur at the end of rows. Above 75 deg the layout is usually no more efficient than the 90-deg (right angle) pattern, since the modules are of almost equal width. The 45-deg angle is unique in that it permits interlocking in a herringbone pattern if the layout requires adjacent aisles to move in the same direction (Fig. 8-3).

Cars may be parked in double rows in an attended parking facility. With this, a maximum use may be made of odd-dimensioned land parcels. However, the moving and replacement of a front-row car to unpark a back-row car materially increases car delivery time. Double parking is not feasible in self-parking lots and garages.

The width of the aisles should permit a vehicle to be parked or unparked in a single maneuver. Where traffic must turn, the minimum turning radius of the outside front bumper of the largest vehicles expected to use the facility must be accommodated. Generally this requires an outside aisle radius of 30 ft (9.2 m) and—to provide adequate width—an inside radius of 18 ft (5.5 m).

PARKING GARAGES

Multistory parking garages are usually constructed in areas of high land cost because of their economic use of ground space. To determine the optimum facility, an analysis of

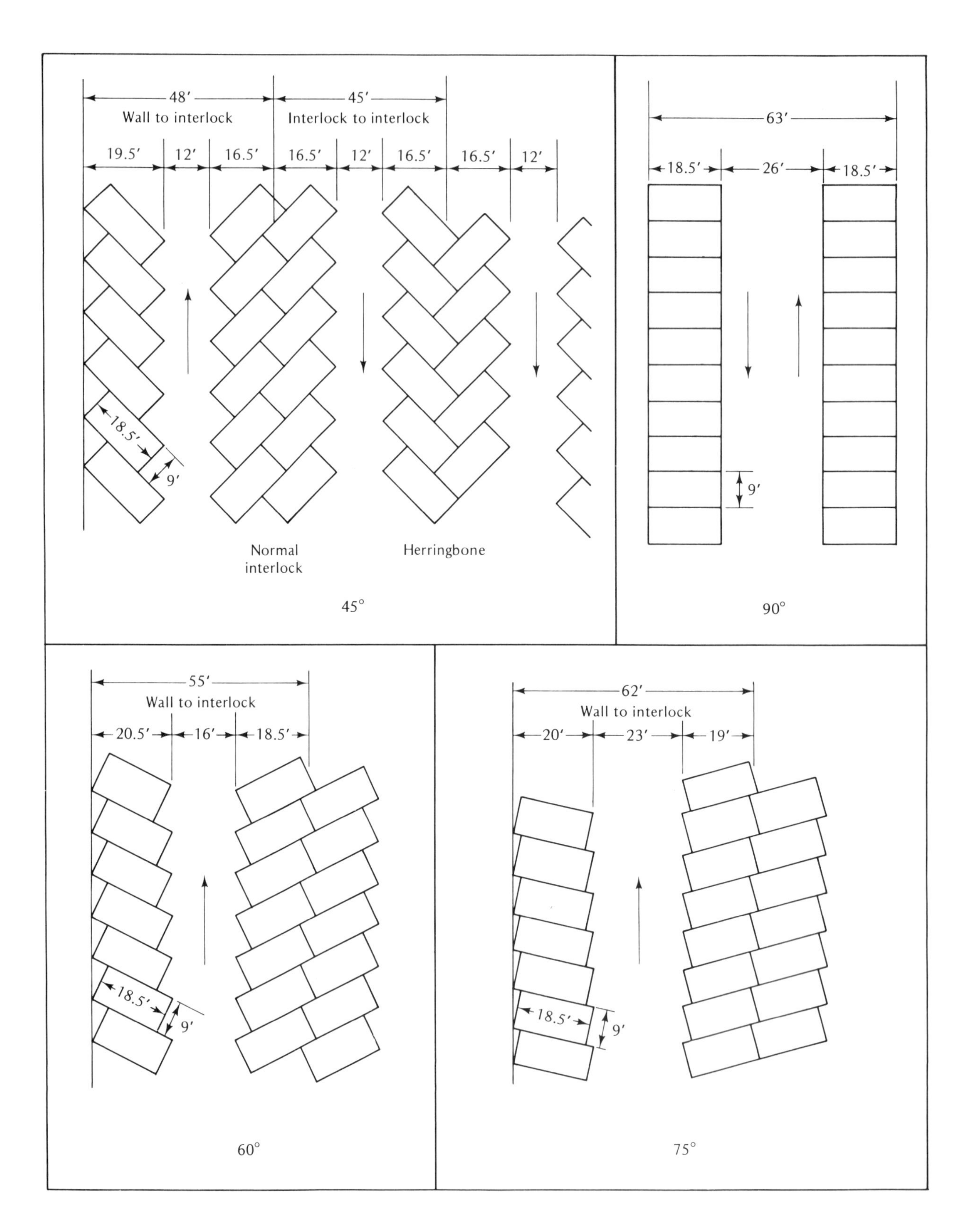

FIGURE 8-3
TYPICAL SELF-PARKING DIMENSIONS

the type sketched in Fig. 8-4 can be used. Values of the construction cost per space for lots and for structures of different heights must be obtained. The gross area per space (total area of layout divided by the number of spaces) will vary somewhat with the shape of the site and the stall dimension standards adopted. It also increases with structure height, as ramps, stairs, elevators, and columns must be accommodated. For any unit cost of land the minimum-cost facility is easily found. The total amount of land can then be determined by dividing the maximum accumulation to be served by the number of levels obtained in the calculation.

Garages may be built above or below the level of the adjacent land. Underground garages are considerably more expensive than above-grade structures because of excavation costs and the requirement for elaborate ventilation equipment. However, since they are usually placed under high-rise office or residential buildings or under a park, part of the land cost can be charged to the other uses of the site. Above-grade structures are usually designed to achieve maximum economy, with open walls, low ceiling heights, and without heating or ventilation.

Operationally, garages may be classified by their vertical vehicular connection design as ramp, sloping floor, or mechanical facilities. The choice here also relates to the cost of land. Ramp garages require somewhat more gross area per space, while sloping floor garages economize by combining the ramp function with the aisles. Mechanical garages have advantages only on very small sites in very high-rent districts.

Because of site and access conditions, no two garages will be designed identically. Figure 8-5, however, shows typical layouts for ramp and sloping floor garages.

Ramps

There are many possible ramp systems. For a specific design, the shape of the land parcel and the operating system

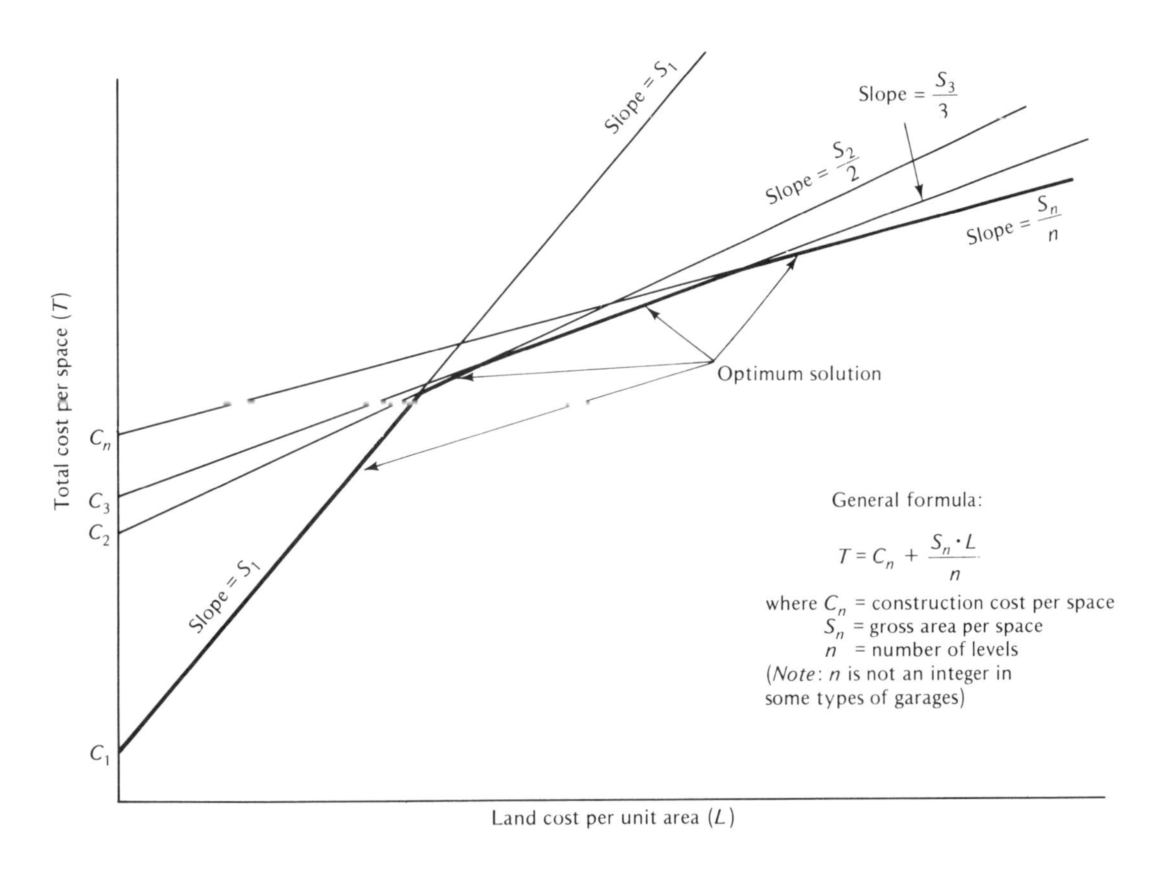

FIGURE 8-4
DETERMINATION OF OPTIMUM HEIGHT OF PARKING FACILITIES AS A FUNCTION OF THE COST OF LAND

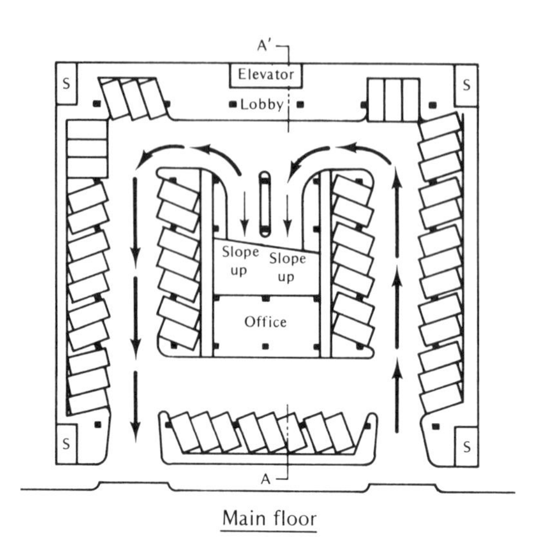

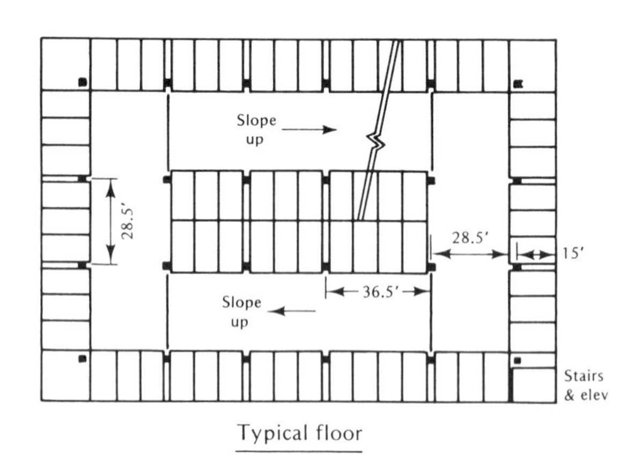

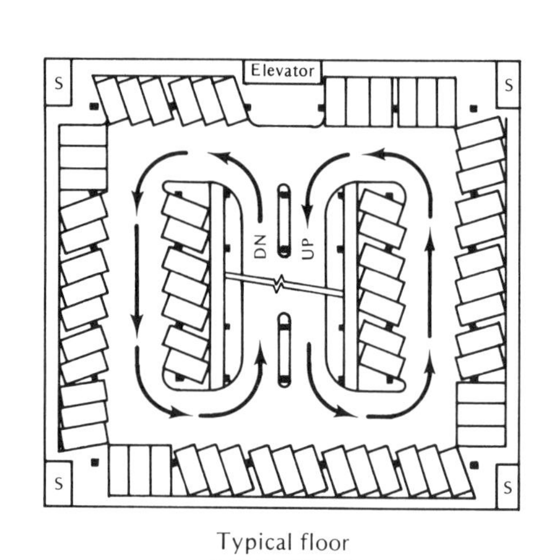

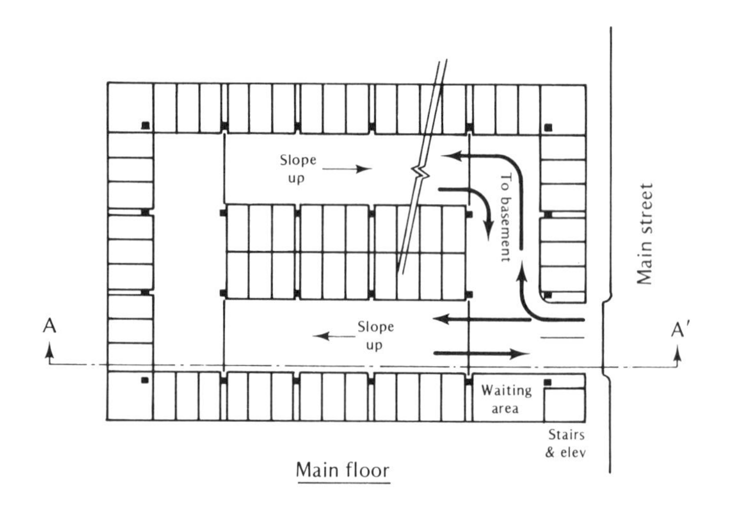

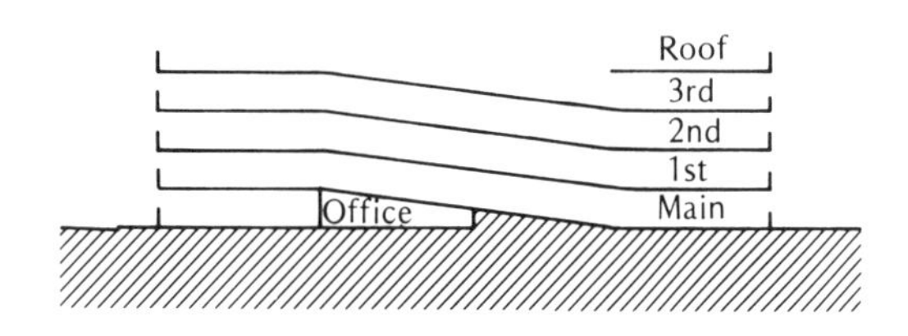

Section A–A′

Elev. +45.0′
+35.0′
+25.0′
+15.0′
+5.0′
−5.0′
Roof
3rd level
2nd level
1st level
Main level
Basement
Main st.

Section A–A′

(a) Functional plan—straight ramp garage

(b) Functional plan—sloping floor garage

FIGURE 8-5
TYPICAL GARAGE LAYOUTS

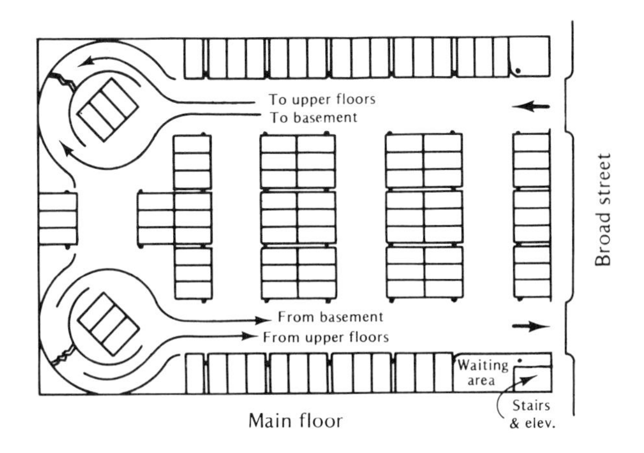

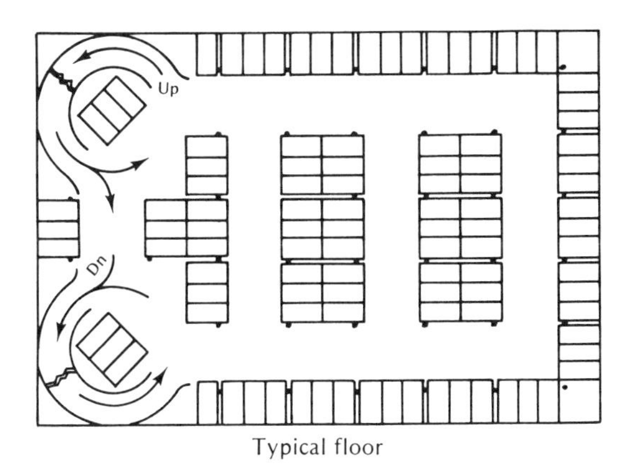

(c) Functional plan—twin-spiral garage

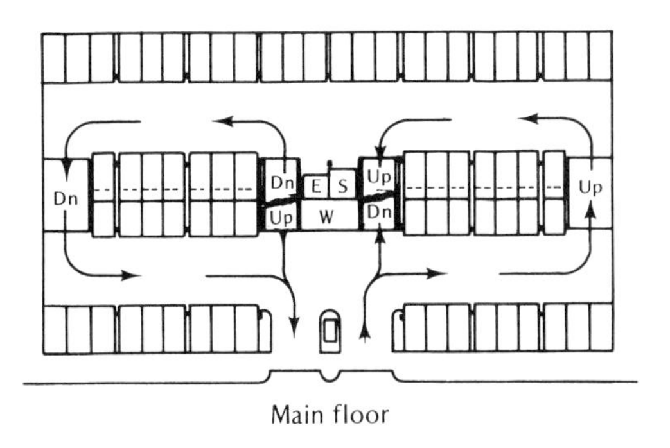

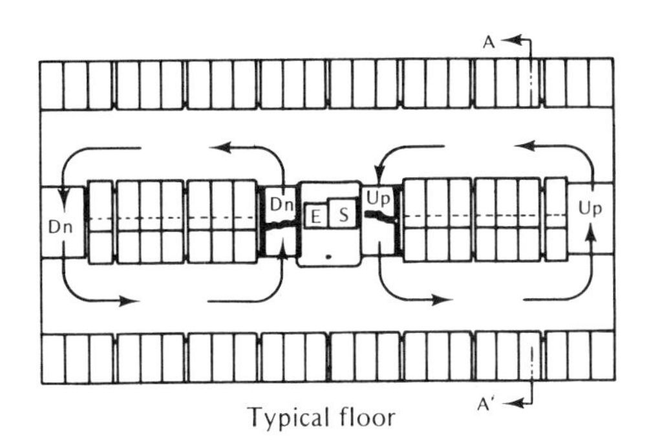

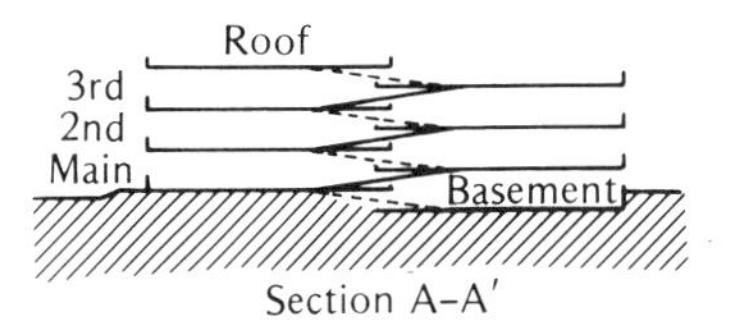

(d) Functional plan—staggered-floor garage

SOURCE: Robert E. Whiteside, *Parking Garage Operation.* Saugatuck, Conn.: Eno Foundation, 1961.

FIGURE 8-5
(CONTINUED)

planned for the garage will determine the system used. Several different characteristics can be identified, which are categorized as follows:

- *One-way* or *two-way,* according to the direction of travel.
- *Straight* or *curved.* Curved ramps may be circular or semicircular.
- *Concentric* or *tandem,* distinguishing one-way ramp systems in which the travel paths revolve about the same or different centers.
- *Opposed* or *parallel,* as to whether the ramp surfaces are sloped so that travel paths revolve in the same or opposite directions for up and down movements.

Ramp grades, particularly in self-parking garages, are limited more by characteristics of drivers than of vehicles. The average vehicle is capable of ascending and descending grades of at least 20 percent, but the average driver does not feel at ease driving on ramps much over 12 percent. This is particularly true on circular ramps. In attendant garages, where drivers have greater skill and more operating experience, ramp grades can be as high as 20 percent.

A minimum ramp width of 10 ft (3 m) for straight ramps and 16 ft (5 m) for curved ramps is recommended. In the design of ramp curvature, the radius of the outer edge should be in the range of 32–38 ft (10–12 m). The maximum superelevation of curved ramps should be limited to between 5 and 10 percent.

In special situations a staggered floor garage may be advantageous. The floors on one side of the structure are located vertically halfway between those on the other side. Relatively short ramps connect these half-stories. A small amount of valuable space may be saved in the center by placing the front 3 feet (1 m) of one row of stalls beneath the row one-half floor above, since this space will be occupied by the hood area of cars only. This detail can be seen in Fig. 8–5.

Sloping Floors

Whereas in ramp garages the floors are parallel to each other and roughly horizontal, in sloping floor facilities the entire parking area becomes a sloping ramp. Cars enter a single ramp and drive up or down until they find a space. Since the parking spaces are also on the slope, the grade should be limited to 5½ percent, and the parking angle should equal or exceed 60 deg. In structures with no more than about three floors, two-way ramps for both entry and exit are used in conjunction with 90-deg stall layouts. For taller structures, or where 60–70-deg parking provides major advantages, a separate exit ramp is desirable so that one-way flow can be maintained. This exit ramp should be accessible at several levels of the main parking ramp.

Receiving Area

The choice of attendant or self-parking operation greatly affects the design of the receiving area, which is generally located just inside the entrance. If cars are to be self-parked, they need pause only long enough to be issued a ticket, or to pick up one from a machine, and little or no reservoir space is needed.

In an attendant parking operation reservoir space must be provided. Here the driver and other vehicle occupants get out, and the attendant moves the car to the storage area. A number of spaces must be available for temporary storage of entering cars. The rate at which they are moved away from the reservoir depends on passenger unloading time, the number of attendants, and the time required for storage of each vehicle. If the capacity of the reservoir space is not adequate to accept all vehicles wishing to enter at peak periods, they accumulate in the street, or move on to some other facility.

The supply of reservoir spaces must be adequate to allow for the necessary transfer time of vehicles from their drivers to the parking attendants and the temporary storage of vehicles whenever the arrival rate exceeds the average. The Poisson theory of probability can be applied to determine the excess number of vehicles accumulating in short increments of time.

Based on overloading less than one percent of the time, Fig. 8–6 shows the relationship between storage rates slightly greater or less than the arrival rate and the size of reservoir necessary to accommodate the fluctuations in traffic flow. Cars arriving at an average rate of 120 per hour would require a reservoir capacity of 27 vehicles if the storage rate were to equal the arrival rate. If more attendants were added, approximating a storage rate 1.10 times the arrival rate, a reservoir capacity of 15 vehicles would be sufficient.

Vertical Person Movement

Passenger elevators are needed in multifloor self-parking garages. Attendants often make use of manlifts. Stairways are required by fire regulations and are useful for travel between two or three floors. The location of these vertical travel facilities must be determined from studies of both the main floor and the storage floors. On the main floor they should be located near the cashier's cage with convenient access to the street. On the storage floors they should be near the centroid of the parking area and placed so as to minimize walking along the aisles, particularly those that carry movements between ramps. It is sometimes possible to connect upper floors of garages to adjacent buildings by pedestrian bridges, thus reducing the capacity for vertical movement needed in the garage itself while supplying a valued short cut to users of the facility.

Mechanical Garages

Several types of mechanical devices for lifting and storing cars have been developed. They were designed to obtain maximum utilization of space on restricted sites, particularly small lots, by eliminating the need for ramps and aisles, although the lifting devices require some space. Each car can be handled rapidly, but the rate of storage and retrieval is limited by the number of lifts or elevators. A considerable storage reservoir at the street level may be needed.

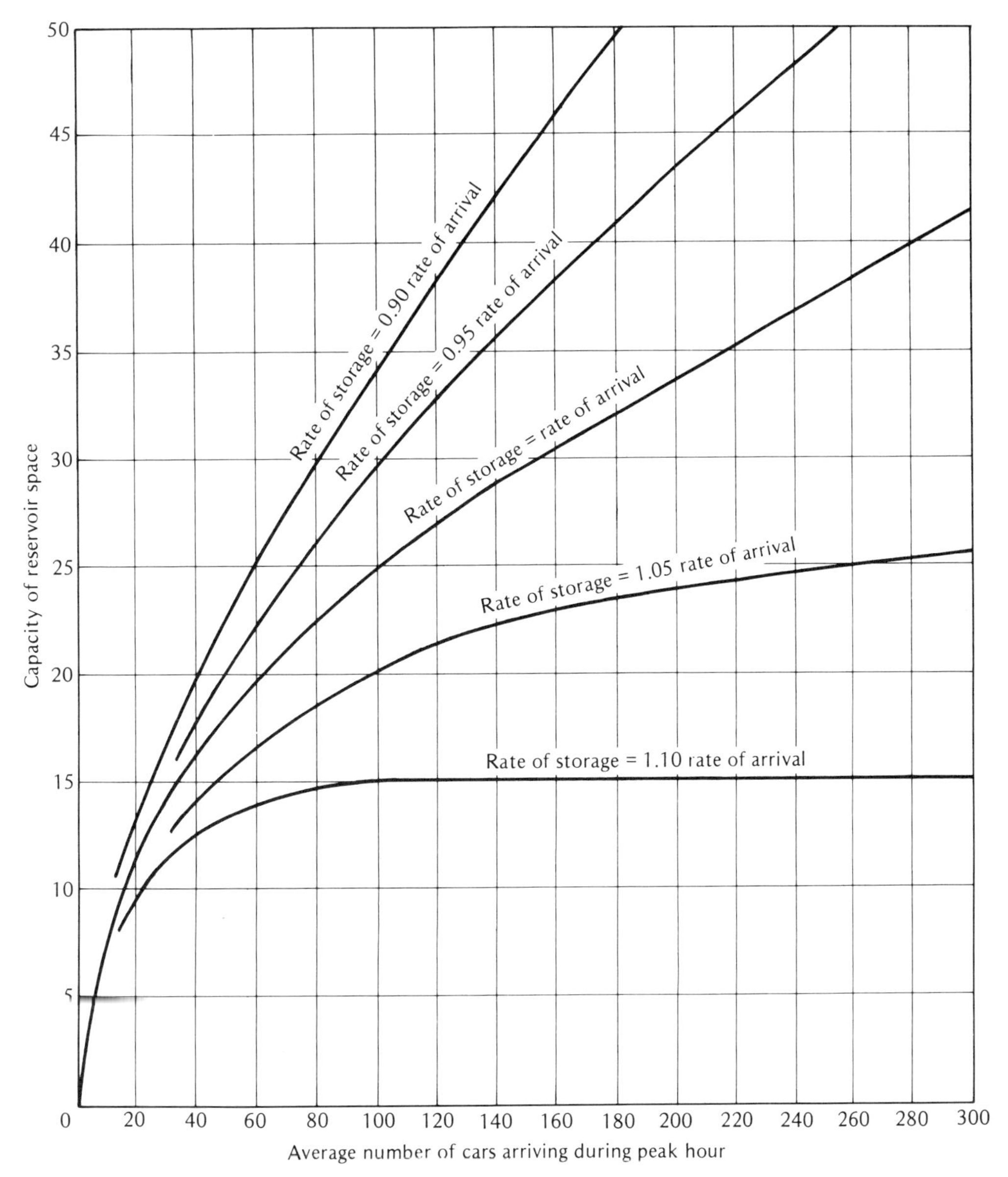

SOURCE: Edmund R. Ricker, *Traffic Design of Parking Garages.* Saugatuck, Conn.: The Eno Foundation for Highway Traffic Control, 1957, p. 51;

FIGURE 8-6
RESERVOIR SPACE REQUIRED IF OVERLOADED LESS THAN 1 PERCENT OF TIME

Most of these installations belong to one of the two following types: traveling elevator or fixed elevator. With the traveling-elevator type (see Fig. 8-7), the car is mechanically lifted into the elevator, transported horizontally and vertically, and driven to the stall. With the fixed-elevator type a dolly lifts the car on or off an elevator that is restricted to vertical movement and is operated in conjunction with car platforms that move on two parallel rows of tracks on each floor. Each platform can move forward, backward, or sideways under its own power.

A third type of system resembles a Ferris wheel. Cars are loaded at the street level into an empty stall, which

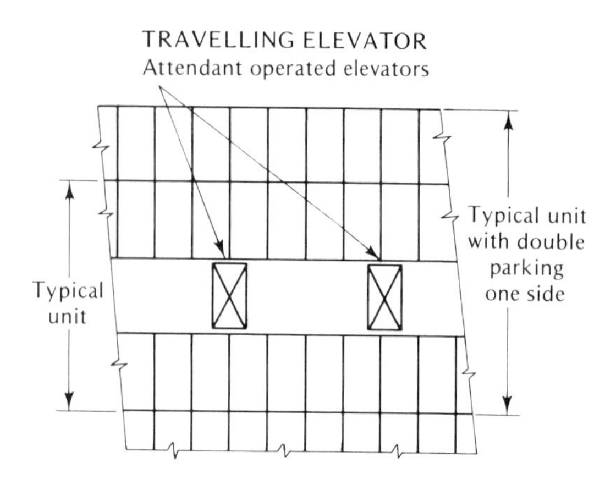

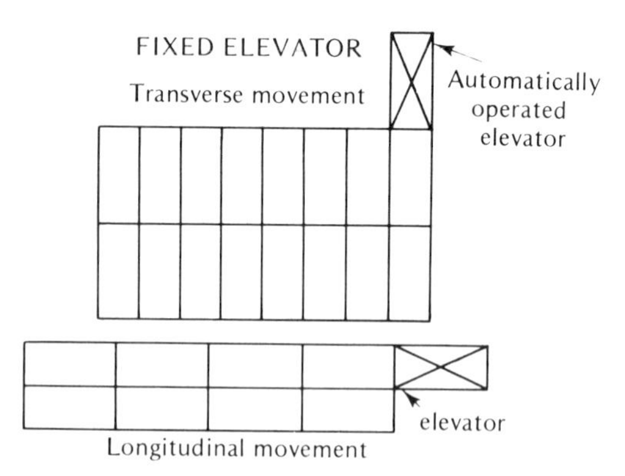

Note: In both movements one stall is always vacant.

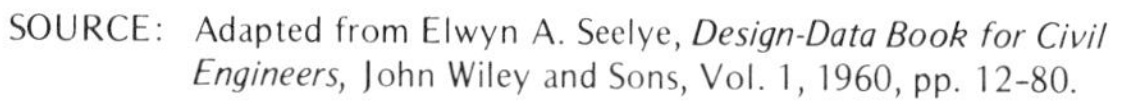
SOURCE: Adapted from Elwyn A. Seelye, *Design-Data Book for Civil Engineers*, John Wiley and Sons, Vol. 1, 1960, pp. 12-80.

FIGURE 8-7
MECHANICAL GARAGES

then rotates vertically in a rectangular path as other stalls are loaded or unloaded. Because this device is open, it presents environmental problems that limit its usefulness.

Although this type of garage may be operated with a minimum number of employees, both the capital and operating costs are high. For this reason, mechanical parking has lost its popularity in recent years.

BUS TERMINALS

The need for bus terminals as well as methods for calculating the number of loading positions required are discussed in Chapter 12. The typical layout of a terminal includes entry and exit roadways, a network of roads to, from, and bypassing the loading positions, pedestrian walks, and often stairs and escalators leading to a grade-separated circulation area and to adjacent streets and facilities. Also included may be comfort and concessionaire facilities for the travelers, ticket offices, parking facilities for buses, and rest rooms for drivers. Long-distance bus terminals also require installations for processing baggage and package shipments.

Many of these features are shown in Fig. 8-8, which depicts a terminal of modest size designed primarily for suburban commuter buses, but also serving some long-distance traffic. This modern facility is connected directly to a freeway as well as to city streets and to a nearby bus maintenance and storage yard.

Large terminals, such as the Port Authority terminal shown in Fig. 12-4, involve elaborate design procedures. Ramps connect directly to a freeway, bridge, or tunnel to remove buses from the city streets. Long-distance buses, because of their unique needs, are separated from commuter buses and are served by stops with freight loading space; these berths are usually of the sawtooth pattern. Commuter buses unload and load at straight platforms. Each route is assigned a separate departure location to make it easy for passengers to locate the correct bus on a daily basis. Arrivals may be handled at the same locations or, especially if the bus is not scheduled for departure within a few minutes, drivers may be instructed to discharge their patrons at any convenient position. A central controller often supervises operations.

TRUCK TERMINALS

Highway freight transportation is conducted in two types of vehicles. Small trucks, generally single-unit vehicles 30 feet (9 m) or less in length, pick up and deliver many small shipments and, if the destination is in the same general area as the origin, also transport the consignments. Larger vehicles, including truck-trailer and tractor-trailer combinations, perform "over the road" services, carrying shipments of all sizes between regions. Where an individual consignment is too large for pickup and delivery by small truck, the larger vehicle also performs these functions.

Three types of freight-handling situations must, therefore, be considered by the traffic engineer and transportation facilities planner. Loading and unloading of smaller

FIGURE 8-8
BUS TERMINAL (COURTESY OF THE CALIFORNIA DEPARTMENT OF TRANSPORTATION)

trucks is often performed at curb truck loading zones, because shippers and receivers of small amounts of freight neither need nor can afford off-street facilities. However, where large quantities of freight trips are generated, such as at warehouses and retail shopping centers, off-street arrangements must be made. Because of their dimensions, large trucks and combinations must be processed off the street. Transshipment between fleets of small trucks making local pickups and deliveries and large over-the-road vehicles is best carried out at specially designed freight terminals.

The geometric layout of truck loading docks is based on the maneuvering capabilities of the vehicles to be served. Table 8-1 indicates the "apron space" (maneuvering space perpendicular to the dock face) required for different lengths of vehicles and varying widths of stalls provided. Where only one truck is accommodated at a time, the apron is measured from the edge of the dock (left diagram in Fig. 8-9). Where physical obstructions, such as posts supporting a canopy or other vehicles parked in the vicinity, are present, the apron is measured from the outer edge of such points (remainder of Fig. 8-9).

The height of the dock itself is about 44 inches (1.10 m) for pickup and delivery trucks and 48-52 inches (1.20-

TABLE 8-1
APRON SPACE REQUIRED AS A FUNCTION OF TRUCK LENGTH AND STALL WIDTH

OVERALL LENGTH OF TRUCK COMBINATION		WIDTH OF STALL		WIDTH OF APRON	
ft	m	ft	m	ft	m
40	12.2	10	3.05	46	14.0
		12	3.65	43	13.1
		14	4.25	39	11.9
50	15.2	10	3.05	60	18.3
		12	3.65	57	17.4
		14	4.25	54	16.5
60	18.3	10	3.05	72	22.0
		12	3.65	63	19.2
		14	4.25	60	18.3

SOURCE: "Recommended Yard and Dock Standards," *Transportation and Distribution Management* (October, 1966) p. 27.

1.32 m) for larger units. Some difference between the height of the dock and the truck floor can be accommodated by dock boards, wedges, or ramps. However, if the dock is higher than the truck bed, opening of the truck doors becomes difficult or impossible. The depth of the dock is determined by the type of equipment (pallets, fork lifts, carts) that is to move over this space. The vertical clearance to the roof of interior docking space or to the canopy is based on the height limit of trucks, or about 14 ft (4.25 m).

In truck terminals designed for transshipment of freight, the layout provides for segregation of local trucks and over-the-road combinations usually on opposite sides of a long building. The smaller trucks are best located nearest the gates to the street system. The central building serves to sort and route freight between the two fleets of trucks and to store it for a limited amount of time if necessary. Freight may be moved by fork lifts, electric trucks pulling carts, or overhead hoists. Allowance in the layout must be made for reservoir space for trucks awaiting a dock space, for some truck and employee automobile parking, and perhaps also for truck refueling and light maintenance. In very large terminals there may also be crew dormitories and showers and a restaurant.

REFERENCES

1. Highway Research Board, *Parking Principles.* Washington, D.C.: Transportation Research Board, Special Report 125, 1971, 217 pp.
2. Levinson, Herbert S., *Bus Use of Highways: Planning and Design Guidelines.* Washington, D.C.: Transportation Research Board, National Cooperative Highway Research Project, Report 155, 1975, Chapter 6.
3. Motor Vehicle Manufacturers Association, *Parking Dimensions.* Detroit, Michigan: Annual.
4. Whiteside, Robert E., *Parking Garage Operation.* Saugatuck, Conn.: The Eno Foundation, 1961.

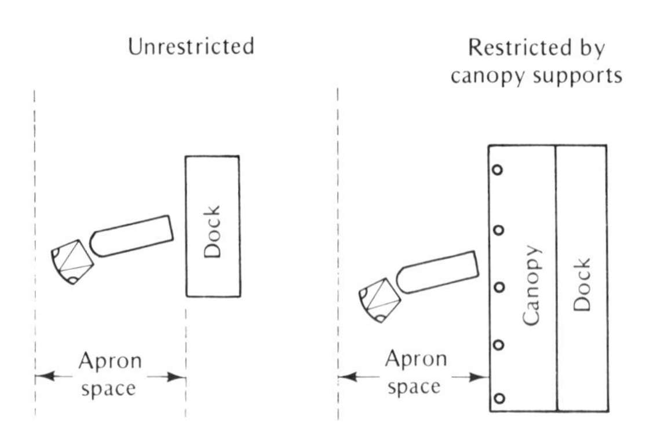

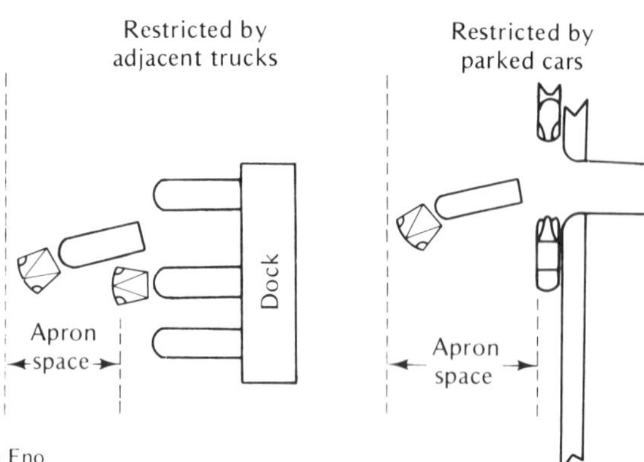

FIGURE 8-9
LOADING DOCK LAYOUTS

FIGURE 8-10
TRUCK TERMINAL (COURTESY OF THE PORT AUTHORITY OF NEW YORK AND NEW JERSEY)

9

REGULATION OF VEHICLES AND DRIVERS

With more than 131 million registered motor vehicles and even more licensed drivers in the United States today (1975) there is an evident need for regulation of both vehicles and drivers. Widespread ownership of motor vehicles requires regulations to ensure the operating ability and financial responsibility of the vehicle operator. Steps must be taken to keep mechanically unsafe vehicles off the road. These regulations have been developed primarily in the interest of property protection and highway safety.

There are certain fundamental "rules of the road," which all drivers, pedestrians, cyclists, and other road users must know and follow for orderly traffic movement. Vehicles must be safe to operate and be equipped with a variety of devices for safety and environmental compatibility. Their dimensions must not be such as to impair traffic safety or cause undue congestion. Weight limits have been established to meet the design limitations of bridges and pavements. These regulations, then, provide a set of standards with which drivers and vehicles must comply before they can operate on the highways.

Regulation of the public is always a difficult problem, particularly with regard to the use of the highways. Roads are built and maintained by government with public financing. Consequently, people expect the highway system to be available for whatever uses they wish to make of it. For this reason, it must be shown that traffic regulations and controls that limit this freedom are necessary for the general welfare.

Unreasonable restrictions or regulations are not likely to be obeyed. Regulations are effective only when they have compliance. They must be generally accepted by the public as reasonable, and self-enforceable, since surveillance problems preclude close, continuing observation of the road network. The highway transportation engineer must recognize that regulations are only one element in a highway transportation system. Good regulations are essential, but they cannot completely compensate for deficiencies in highway planning and design.

The authority to regulate traffic stems from the general powers of government to promote general welfare and safety, as indicated in state statutes or constitutions or, by enabling legislation, from local governments. Since the 1966 Motor Vehicle Safety and Highway Traffic Safety Acts, the federal government has been actively engaged in safety regulation of vehicles and drivers. Primary responsibility for enforcement and implementation of safety regulations still rests with the state.

In addition to equipment standards for commercial vehicles engaged in interstate commerce, established by the Interstate Commerce Commission (ICC), the 1966 acts extended the federal activity to safety design features of motor vehicles and their equipment. This seems to have occurred because inadequate safety features in automobiles are a national problem, difficult to cope with on a state, local, or industry level. Results of the 1969 National Environmental Policy Act, the Clean Air Act (with subsequent amendments), and the national policy toward energy conservation have led to national standards for antipollution devices and for gasoline conservation (mileage).

Although local regulation of traffic is logical and proper, the rapid growth of highway transportation in the United States has resulted in a mass of nonuniform local statutes and ordinances. In this era of high-speed travel over a broad network of good roads, drivers may not realize that they have passed from one jurisdiction to another and have become subject to a new set of traffic regulations. In extreme cases this can result in confusion and a greater accident potential for the driver in unfamiliar surroundings. One definite consequence of nonuniformity is a lessened respect for regulations, with concomitant burdens on law-enforcement agencies. Each locality has traffic problems peculiar to the area, but this should not prevent local authorities from using uniform regulations to the fullest practicable extent. Standardization of traffic regulations makes them more effective, easier to enforce, and more respected by the driving public.

In 1924, at the First National Conference on Street and Highway Safety in Washington, recognition of the need for uniformity in traffic regulations prompted the organization of a national committee. This group, now known as the National Committee on Uniform Traffic Laws and Ordinances, was organized to provide the leadership necessary to attain uniformity in traffic regulations in all states, counties, and muncipalities.

The committee is composed of specialists representing federal, state, and local government agencies and private organizations. These representatives are in daily contact with the complex problems of highway transportation. They meet periodically to study and evaluate research findings, which often lead to traffic law revisions required by new concepts in motor vehicle transportation.

The committee prepared the original text for the *Uniform Vehicle Code* and the *Model Traffic Ordinance* (both of which cover basic traffic regulations). These publications were combined in the 1968 edition [Ref. (3)]. The *Uniform Vehicle Code* has done much to bring about the enactment of modern, comprehensive, and uniform motor vehicle laws in the United States. It combines the best features of existing traffic laws in the various states and guides states in the continuing study and evaluation of their traffic laws. It is examined periodically and revised in accordance with current thinking and experience in the field of traffic law.

The *Uniform Vehicle Code* provides for a separate motor vehicle administration in the state government under the control of a "Commissioner of Motor Vehicles." The code outlines standards covering motor vehicle registration; certificates of title; registration and license fees; liens on the vehicle; special registration for dealers, manufacturers, and transporters; licensing of dealers and wreckers, and antitheft laws. It also deals with the minimum age and other requirements for operators' and chauffeurs' licenses; cancellation, suspension, or revocation of licenses, and procedures covering violation of license provisions. There is a section on financial responsibility covering responsibility and security following traffic accidents, the mechanics of future proof of financial responsibility, and penalties for the violation of the above provisions. A section on civil liabilities discusses those of states, counties, municipalities, and other corporations. It also includes the assignment of negligence or willful misconduct of operator to owner, liability for injury or death of a guest occupant, service of process on nonresidents, and the control of owners of foreign vehicles. Finally, the Code provides for the regulation of traffic on the highways, the authority for setting up traffic devices and measures, procedures for initiating and processing accident reports, the fundamental rules of the road, pedestrian rights and responsibilities, vehicle equipment and inspection, and size, weight, and load regulation. The *Uniform Vehicle Code* is the backbone of modern vehicle and traffic regulation. The remainder of this chapter will discuss driver controls and vehicles controls in some detail.

The *Model Traffic Ordinance,* which contains many of the provisions for regulating highway traffic found in the *Uniform Vehicle Code,* serves as a guide for municipal traffic control. Traffic administration within the framework of the city government is discussed in Chapter 15.

Finally, the federal government has issued 18 highway safety program standards as authorized by the 1966 act [Ref. (4)].

1. Periodic motor vehicle inspection
2. Motor vehicle registration
3. Motorcycle safety
4. Driver education
5. Driver licensing
6. Codes and laws
7. Traffic courts
8. Alcohol in relation to highway safety
9. Identification and surveillance of accident locations
10. Traffic records
11. Emergency medical services
12. Highway design, construction, and maintenance
13. Traffic engineering services
14. Pedestrian safety
15. Police traffic services
16. Debris hazard control and cleanup
17. Pupil transportation safety
18. Accident investigation and reporting

Forty-six vehicle safety standards have also been issued by the U.S. Department of Transportation, and include: (a) Precrash items, such as brakes, lights, and mirrors; (b) Crash items such as restraint systems, steering controls, etc; and (c) Postcrash items such as fuel tanks, flammability, etc.

DRIVER CONTROLS

To maintain a safe and efficient highway transportation system all motor vehicle operators must have sufficient driving skill and knowledge of the fundamental rules of the road. Traditionally it has been through driver licensing that control over the driving population has been exerted. Driver licensing was originally conceived as a means to identify drivers and provide revenue to the state. Today the licensing of drivers is a means for insuring that drivers have the necessary qualifications for satisfactory driving. Ideally those who are incompetent are weeded out through some selection process. However, people vary widely in physical characteristics. Research has shown that even drivers with characteristics markedly inferior to the average can be good drivers. The ideal selection process must admit as qualified drivers a great variety of people while weeding out those who are potential hazards to the safety of others. Since driver education is a continuing process, its impact is still being evaluated.

Driver Licensing

The issuance of licenses for drivers dates back to 1899 in New York City. As the number of vehicle accidents increased, cities and towns began to require licenses for every driver. States quickly took over the responsibility, however. All states now require qualification by examination before the issuance of a driver's license.

Most states consider the issuance of a license a privilege allowing the holder to operate a motor vehicle on public highways. By not considering driving a right, these states can justifiably impose whatever regulations they deem appropriate. There is a trend in other states now to consider driving a right that can be denied only through due process of the law where the general welfare is balanced against individual rights. In any event, court decisions through the years have demonstrated considerable reluctance in denying or revoking a driver license from application of administrative regulations alone, tending instead to apply due process of individual rights. Due process is characterized by consideration and objectivity which override unconditional application of regulations on the books.

Most states permit unaccompanied driving at age 16, complying with the recommendation of the *Uniform Vehicle Code.* There is a tendency toward requiring driver education for applicants under 18, and many states consider a license in this age group to be probationary. The code recommends that bus and truck drivers be licensed at age 21. ICC restrictions on drivers engaged in interstate commerce require a minimum age of 21 plus a year's driving experience. It has frequently been demonstrated that drivers under 25 years of age, particularly male drivers, are involved in more than their share of traffic accidents. Most trucking companies require that drivers be 26 years of age, have two years of driving experience on similar equipment, and pass the Department of Transportation's physical examination, as well as driver aptitude and knowledge tests [Ref. (5)]. Despite this, there is no evidence that raising the age of first licensing would significantly decrease the accident rate per licensed driver. Studies in California found that the accident rate does not increase between the ages 16 and 19, but the miles driven per year increase considerably, indicating that several years of driving experience before age 18 and high school graduation might be beneficial. Table 3-8 also shows a very high victim rate in the 18-24 year age bracket. For drivers only, this is illustrated by California experience, as shown in Table 9-1.

The *Uniform Vehicle Code* recommends the following tests for drivers' licenses: eyesight, ability to read and understand traffic signs and control devices, knowledge of state traffic laws, a demonstration of driving ability, and additional physical and mental examinations where appropriate.

All states require a driving test of applicants for a first license. These tests cover only the basic knowledge for motor vehicle operation and do not necessarily test the applicant's ability to handle difficult traffic situations. All states require a vision test and a knowledge of traffic laws, although a few still do not require literacy. Some states require physical or mental examinations for first licenses, but such examinations do not seem to have noticeably improved driving in those states. In general, there has been very little evaluation of the impact of these tests on driver quality. Other methods to measure reaction time, speed estimation, and glare and depth correlation have been developed over the years, but researchers have found it difficult to correlate their results with driving records.

In recent years there has been increasing interest in the detection of medical conditions that might impair driving ability. The following medical conditions are often cited: epilepsy, cardiovascular diseases, diabetes, alcoholism, drug usage, mental illness, and other disorders primarily involving coordination or mobility. A California study concluded that drivers with these medical impariments have twice the accident rate of normal drivers. More recently, the National Highway Traffic Safety Administration (NHTSA) has research underway to determine whether certain drugs contribute to a disproportionate share of highway accidents and if so, what measure might counteract the situation. Thirty-five Alcohol Safety Action Projects (ASAP's) were established throughout the country, using the latest research to identify drinking drivers and to reduce the problem by better enforcement, better court procedures, and close scrutiny of subsequent driving records [Ref. (6)].

All states have authority to deny a license to a person judged physically incompetent to drive. Detection of impairment is the biggest problem for motor vehicle administrators. In California, applicants for license renewal must appear in person so that obvious impairments can be observed by officials. Applicants are asked to state if they have a current physical or mental impairment that might limit their driving ability and, if so, to grant permission for access to medical files. The licensing department also has available police and accident records, which are studied

TABLE 9-1
AGE DISTRIBUTION OF DRIVERS INVOLVED IN FATAL AND INJURY ACCIDENTS (CALIFORNIA, 1975)

AGE GROUP	TOTAL	PER 100,000 LICENSED DRIVERS	"HAD BEEN DRINKING"	PER 100,000 LICENSED DRIVERS	PERCENTAGE OF ALL INVOLVED DRIVERS "HBD"
0-14 years	558	a	23	a	4.1
15-19 years	47,812	4,351	5,554	506	11.6
20-24 years	55,760	3,068	9,684	533	17.4
25-34 years	69,961	2,140	10,808	331	15.4
35-44 years	36,841	1,607	5,359	234	14.5
45-54 years	30,202	1,325	3,873	170	12.8
55-64 years	20,513	1,240	1,834	111	8.9
65 and over	13,774	1,195	600	52	4.3
Total—age stated	275,421	2,031	37,735	278	13.7
Age not stated	12,300	...	598	...	...

[a] Ratios are ∞ since there are no licensed drivers in this age group.

with attention to frequency of violations and accident involvement. A section of the state's health and safety code also requires physicians to report the names of persons suffering disorders that are characterized by lapses of consciousness through local health officers to the state licensing department. Each year the state evaluates over 100,000 drivers out of a total population of thirteen million licensed drivers in following up on medical and mental problems, including alcoholism and drug abuse, as well as over 200,000 drivers with high violation point counts. Pennsylvania attempts to accomplish this detection by means of compulsory general medical examinations.

In all states, driver licenses have a limited duration so that there is an opportunity to review the qualifications of the driving population. Durations range from one to five years, with two or three years most common. Such a review is especially important for elderly drivers. License fees in all states are nominal. The revenue derived from license fees is often employed to finance the licensing program and improvements in driver selection and training. Thirty-six states reexamine for knowledge or vision, or both, at least every four years [Ref. (6)].

Enforcement and the Court

The driver control function extends much further than the issuance of a driver's license. When drivers obtain their licenses they accept the responsibility to obey traffic laws and regulations. When they fail in this responsibility they can be subjected to disciplinary action, which may include the suspension or withdrawal of their license.

All states have authority to suspend or withdraw the driving privilege when a driver's record warrants it. Grounds for suspension or withdrawal differ considerably among the states. Most states base their decision on a review of the offender's past record as well as the latest offense. If the record is considered bad enough, drivers may be sent a written notice that their license is in jeopardy, or they may be required to appear for a personal review of their record.

The question of drivers' retaining their driving privilege is only a small part of the enforcement problem. Adequate enforcement is necessary to ensure respect for traffic regulations on the part of the driver. Every year approximately 35 million traffic citations are issued in the United States [Ref. (7)]. These range from parking tickets to serious offenses such as manslaughter. Most citations are processed by mail, and the offenders pay fines according to an established fee schedule. If they neglect to settle the citation, they may be brought to trial. Unfortunately, in many cities today this sytem has hopelessly bogged down. Violations agencies are anywhere from several months to years behind in processing unheeded citations, leading to driver disrespect for the system. So-called "scofflaws" flaunt the law with their accumulations of overdue tickets. Honest drivers resent this and also lose respect for traffic authorities. The solution in some cities has been an overhaul of the system by the processing of citations by computers.

It is estimated that about nine million cases are heard in the nation's traffic courts each year [Ref. (7)]. The courts are so overloaded with cases that they are in continual need of more efficient operation. As a result of this load on the courts, the violations bureaus currently handle many cases administratively. State and local procedures for determining which cases should go to the courts also need to be reviewed continually.

The Traffic Court Program of the American Bar Association has exerted a major influence in steps to improve traffic court organization and procedure. The objectives of the program are to increase the efficiency and effectiveness of traffic courts by upgrading the status of the traffic court judge and bench to correspond with the salaries and benefits of the rest of the American judicial system, removing the judge's salary from dependency upon the fines collected, enabling professional members of the bar to act as prosecutors independent of the judge, and incorporating the traffic courts into the state judicial structure.

This organization has promulgated a set of national standards for traffic courts which promotes the above objectives, particularly with regard to establishing uniform traffic court procedures and the professional independence and competence of traffic court judges. At this time no state fully complies with the standards, although half the states have liberated court financing from the fee system. Most states have subscribed to canons of ethics for traffic courts and use the system of uniform traffic complaints. It is the contention of the Traffic Court Program that the judge should be knowledgeable in traffic matters, and the courtroom procedure should be such as to impress offenders with the seriousness of their acts without putting them on trial like common criminals.

Since the court procedure is considered an educational experience for the offender, the Traffic Court Program considers that appearance in court should be mandatory for violations that are hazardous to life and property. These violations would include reckless driving and driving while under the influence of alcohol or drugs. There is some room for contention as to what constitutes a hazardous violation, whether moving or nonmoving. Since nonhazardous violations are presumably handled administratively by the violations bureau, the border between judicial and administrative enforcement is a little hazy.

Whether judicial or administrative, enforcement must necessarily entail a penalty. Penalties can take the form of warnings, fines, or jail sentences. Licenses may be suspended for a definite period or revoked altogether. Fines and jail terms have been criticized for their ineffectiveness and their discrimination against the poor.

There has been some research into the effectiveness of taking away a driver's license. A 1965 study in California of drivers who lost their licenses during the previous six years disclosed that 33 percent of the suspended drivers and 68 percent of the revoked drivers drove during the period they were without a license. These discouraging results have been confirmed by other studies. In fact, a 1977 NHTSA report by McGuire and Peck indicated that the situation found in California may have worsened slightly and that even this may be a conservative statement [Ref. (8)].

Such drivers are most often young males. The situation is unfortunate, not only in terms of the ineffectiveness of the law, but because the law has failed to keep many dangerous drivers off the road. Increased surveillance of traffic and stiffer penalties have been suggested to rectify the situation.

The *Uniform Vehicle Code* gives state motor vehicle departments authority to suspend or revoke licenses. Half the states now allow traffic courts to suspend licenses. One method of providing for administrative action is the "point system." Many states have adopted it for their driver improvement program. With this system each violation is assigned a point value scaled in accordance with the seriousness of the offense. Points are accumulated for each driver, and each addition is kept on his record for a predetermined length of time, usually several years. If at any time the total number of points exceeds a certain level, then the driver's record is examined and appropriate action is taken. In a sense, the point system is a convenient method for determining which cases come up for review. It is simple to administer and easily understood by the driving public. Some states print their point system on the back of their driver licenses. Table 9-2 shows the point system used in Maryland.

If a Maryland driver accumulates eight points within two years, the license is suspended. Twelve points within two years results in license revocation.

Most programs to restrain or discourage the errant driver are termed "driver improvement actions." There is considerable interest on the part of motor vehicle agencies in improving driver performance by nonpunitive efforts to educate and reform. In modern driver improvement programs the identification of problem drivers by means of point systems and the like is only the first step. The American Association of Motor Vehicle Administrators recommends that local agencies attempt to rehabilitate these drivers through a program of systematic retraining.

Mixed reactions have resulted from the establishment of traffic violator schools, driver improvement clinics, and group hearings. Many drivers attending these sessions improve their driving performance. However, some researchers have found that a long personal interview and discussion of a driver's record and traffic laws is just as effective as enrollment in a school or clinic.

A special problem of enforcement is that of alcohol. Alcohol has been identified as an important factor in severe automobile accidents. Data from various parts of the United States indicate that over half of all fatally injured drivers had been drinking to an extent that their driving was probably impaired.

The *Uniform Vehicle Code* requires the revocation of the driving license of anyone convicted of driving while intoxicated. The code has provided for the implied consent law under which the license of a driver suspected of being under the influence may be suspended or revoked for refusal to submit to a chemical test for intoxication. All of the states now have adopted such a policy and, when individuals are accused and refuse to submit to a chemical test, their license is revoked in most states. Establishing a definition of intoxication is difficult because of individual driver differences. The *Uniform Vehicle Code* recommends that a blood-alcohol level of 0.10 percent be the level to define intoxication.

Driver Education

Many people believe that regulatory and enforcement problems can be reduced if new drivers undergo a comprehensive training program of driving skills and knowledge of traffic rules and regulations under the supervision of competent instructors. New drivers would also benefit from authoritative instruction on the mechanics of motor vehicles, first aid and accident procedures, and the use of seat belts and other safety items along with an understanding of their significance. Although the high school enrollments are slightly less than a decade ago, the number of students taking driver training is over one million, and there is difficulty in meeting all of the demand.

TABLE 9-2
POINT SYSTEM FOR DRIVER'S LICENSES IN MARYLAND

Violations	Point Value
Any moving violation not listed below and not contributing to an accident	1
Violations contributing to an accident	3
Exceeding posted speed limit by	
(a) 10 miles per hour or more	2
(b) 30 miles per hour or more	5
Reckless driving	4
Driving without a license or out of classification	5
Failure to report an accident	5
Driving on an instruction and examination license unaccompanied	5
Participating in a racing or speed contest on the public highways	5
Driving while ability impaired by consumption of alcohol	6
Turning off lights of a vehicle to avoid identification	8
Failing to stop after accident resulting in damage to attended or unattended vehicle or property	8
Failing to stop after an accident resulting in injury or death of any person	12
Driving after refusal, suspension, cancellation, or revocation	12
Any violation of Driver's License Provisions	12
Homicide or assault committed by means of a vehicle	12
Driving in an intoxicated condition or under the influence of narcotic drugs or any other drug	12
Any felony involving use of a vehicle	12
Fleeing or attempting to elude a police officer	12
The making of a false affidavit or statement under oath, or falsely certifying to the truth of any fact or information to the administration under this article or under any law relating to the ownership or operation of motor vehicles	12

There are about 2500 commercial driving schools in the United States training between two and three million persons a year, many of them young people. About half the states regulate these schools in some manner.

There is undoubtedly much room for improvement in driver training programs. Possibilities for more efficient instruction include the use of driving simulators, which can subject students to a variety of situations and test their reactions. All of these educational programs attract a great many young people sincerely interested in learning to become good drivers. It may be for this reason that trained drivers are more accident-free.

Financial Responsibility

Financial responsibility laws were first promulgated in the 1920's to protect injured parties from financially irresponsible motorists. All states have adopted some type of financial responsibility law. The most common law requires that drivers involved in accidents show evidence that they have adequate insurance or other resources to settle damage claims of an established amount. Inability to do so results in suspension of one's driver's license and vehicle registration.

For those drivers convicted of certain offenses that result in suspension or revocation of their driving privileges and for those who have defaulted in the payment of judgments for previous accidents, proof of financial responsibility for the future is required.

Some states attempt to protect the public from uninsured drivers by establishing a special fund with fees collected from uninsured drivers at the time of vehicle registration. Drivers unable to collect from the uninsured motorist can recover from this fund. A few states have adopted compulsory public liability insurance laws.

Civil Liability and Motor Vehicles

The *Uniform Vehicle Code* provides that states, counties, and muncipalities be liable for civil damages resulting from the operation of a motor vehicle by public officials while acting within the scope of their official duties. Most states have laws that provide that a vehicle owner or driver is not responsible for hitchhikers and gratuitous guests unless driver negligence can be proven. Owners are liable for the negligent driving of an agent, servant, or member of the family.

VEHICLE CONTROLS

Controls over motor vehicles date back to the beginning of the century. These controls govern vehicle registration, equipment requirements, sizes and weights of commercial vehicles, and vehicle inspection.

Vehicle Registration

Vehicle registration records generally include vehicle classification and identification, the name of the owner, a unique chassis number (federal standard) and a registration number. This information has proved valuable for the identification of stolen vehicles and other enforcement matters, for taxation, as a measure of the vehicle population and distribution for use in highway planning, and for other statistical uses.

Registration fees are an important source of revenue for highway improvement and administration. Registration fees for passenger vehicles are usually based on vehicle weight, horsepower or value, and generally range from $5 to $40. Commercial vehicles must pay registration fees to the state motor vehicle department, to the Interstate Commerce Commission, if they are engaged in interstate commerce, and sometimes to other jurisdictions with special regulations. The various states determine fees from gross weight, empty weight, rated capacity, or a combination of these.

Vehicle Equipment

States have the authority to regulate motor vehicle equipment. All states have requirements for braking and lighting and require turn signals, rearview mirrors, and windshield wipers. The National Traffic and Motor Vehicle Safety Act of 1966 directs the Secretary of Transportation to "establish by order Federal motor vehicle safety standards." Standards have been promulgated on safety performance criteria for vehicles, their equipment, and tires, as indicated previously and include windshield washers, defrosters, and seatbelts.

In addition, under the Highway Safety Act of 1966, the Secretary of Transportation is responsible for administration of a coordinated national highway safety program through matching grants to the states to assist them in developing and improving traffic safety programs in conformity with the highway safety program standards listed previously.

In recent years considerable attention has been paid to vehicular safety and pollution control equipment. The state of California led the way prior to federal requirements for antipollution control on crankcase and exhaust emissions. See Chapter 3 for further discussion.

Administrative control of both the vehicle and highway safety standards is vested in the National Highway Traffic Safety Administration of the U.S. Department of Transportation.

Compulsory Vehicle Inspection

Vehicle inspection programs have been lauded for improving the general standard of automobile conditions, informing drivers of the condition of their vehicle, developing better garage workmanship in repairs, and educating the driver of the necessity for periodic mechanical inspection. Inspection laws were first adopted before World War II in the interests of safety. All states forbid the operation of unsafe vehicles, and compulsory inspection is used by some states as the most direct means of ensuring that all vehicles are safe. Thirty-two states and the District of Columbia require periodic motor vehicle inspection, and 12 states have substitute inspection programs approved on a trial basis to meet the Federal Safety Standard. Only six states have no motor vehicle inspection, based on the argument that compulsory inspection is unusually expensive, both to the state and the driver, who supposedly already realizes that the vehicle must be taken care of mechanically. The states with compulsory inspection use one of two systems: state-owned and -operated inspection stations, or authorized private garages that do the inspection required in the state law.

Most officials connected with state inspection programs feel that inspection makes a worthwhile contribution to traffic safety. Rejection rates vary considerably between states, and the reasons for rejection also vary, apparently a result of the variety in inspection programs and policies. Rejections range between 25 and 50 percent of the vehicles inspected, and even higher if inspection is especially rigorous. As might be expected, the rejection rate is higher for older vehicles. The most common reasons for rejection are the lights or brakes.

On the basis of the rejection rates during inspections, it is apparent that many vehicles involved in accidents do have mechanical defects. A comparison of mileage fatality rates in the various states indicates that states which have some form of vehicle inspection do have lower fatality rates. It is difficult to separate the many factors that may contribute to traffic safety, but federal officials believe there is a definite correlation between compulsory inspection and safety and have issued the safety program standard for periodic motor vehicle inspection.

Size and Weight Regulations

Weight restrictions on commercial vehicles were first imposed in the 1920's, when the seriousness of pavement damage from overweight vehicles was first appreciated. All states now have restrictions on weight, height, width, and length. These restrictions meet considerable opposition from commercial vehicle interests. From the latter's point of view, the use of bigger trucks with more capacity is more economical. Pavement can be designed to withstand heavy loads, and the question becomes one of highway cost allocation. Highways must be built with heavier structures and gentler grades or truck climbing lanes to properly accommodate large trucks, but these expensive modifications are not needed for passenger cars. Because of the effects of large, slow-moving vehicles on the traffic stream, it becomes necessary to provide additional facilities if satisfactory passenger vehicle levels of service are to be maintained.

No area of traffic enforcement may cause more difficulties than the enforcement of size and weight regulations. Enforcement can be accomplished by allowing officers to weigh or measure any vehicle that they suspect of being in violation. Penalties include fines and short jail sentences. Many states have established permanent weighing stations at strategic points and have portable scales for use in spot checks.

REFERENCES

1. Arthur D. Little, Inc., *The State of the Art of Traffic Safety: A Critical Review and Analysis of the Technical Information on Factors Affecting Traffic Safety.* Acorn Park, Cambridge, Mass.: for the Automobile Manufacturers Association, 1966.
2. Matson, Theodore M., Wilbur S. Smith, and Fred W. Hurd, *Traffic Engineering.* New York: McGraw Hill Book Co., 1955.
3. National Committee on Uniform Traffic Laws and Ordinances, *Uniform Vehicle Code and Model Traffic Ordinance.* Washington, D.C.: 1968 (with 1972 Supplement and 1975 Revisions). Printed by the Michie Co., Charlottesville, Va.
4. U.S. Congress, Public Law 89-564, Highway Safety Act of 1966; and Public Law 89-563, National Traffic and Motor Vehicle Safety Act of 1966.
5. Taff, Charles A., *Commercial Motor Transportation.* Cambridge, Md.: Cornell Maritime Press, 1975, p. 225.
6. National Highway Traffic Safety Administration, *Traffic Safety '75.* Digest of Activities of NHTSA, 1976.
7. Estimated from statistics available for the State of Maryland and from discussions with NHTSA personnel.
8. McGuire, John P. and Raymond C. Peck, "Traffic Offense Sentencing Processes and Highway Safety," Final Report, DOT, HS-802-326, April 1977, Washington, D.C.

NO
PASSING
ZONE

10

TRAFFIC CONTROL DEVICES

A highway system is not complete, even though it may contain properly designed roadways, competent drivers, and well-designed vehicles. Traffic control is the missing essential. Vehicles on the road must be regulated to minimize conflicts with other vehicles and to protect them from environmental conditions and to protect "environmental conditions" (e.g., pedestrians) from vehicles.

A moving traffic stream under normal flow conditions must be kept separate from parallel opposing and concurrent streams. This same stream must be advised continually of conditions ahead that necessitate a change in the flow pattern and warrant close attention by the individual driver. Engineers use traffic control devices to warn of potentially hazardous conditions and to foster efficient traffic flow.

Minimization of vehicular conflicts is most critical at intersections where the conflict is one of right-of-way determination. Various means are available for intersection control, and this topic is treated in detail in the next chapter.

Modern traffic control extends to more than the minimization of vehicular conflicts. The roadway is not an entity existing apart from its environment. Vehicular conflicts with abutting land uses are regulated by speed controls, parking restrictions, and the like.

As the network of usable motor roads spread early in the twentieth century, it became evident that driver disorientation with his surroundings could be a severe problem. This led to privately initiated directional and informational markers and later to government-installed route numbering systems, street signs, geographical markers, and informational signs. Traffic signs of this type transmit to the driver information unrelated to the immediate traffic situation. Such informational signs can be considered as traffic control devices that promote more efficient operation of the highway system. However, *uniformity* (of design, application, and location) is almost more important than actual use of control devices.

Traffic control devices include all traffic signs, signals, pavement markings, and other devices placed on or adjacent to a street or highway by a public authority.

The general criteria for traffic control devices include:

1. Standards—as shown in the *Manual on Uniform Traffic Control Devices (MUTCD)* (1971 edition—also see the new 1978 edition [Ref. (1)]) and revisions by the National Joint Committee on Uniform Traffic Control Devices. However, each state has its own standards manual, which might differ from or go beyond the MUTCD.

2. Basic requirements
   - *a.* Fulfill a need.
   - *b.* Command attention.
   - *c.* Convey a clear, simple meaning.
   - *d.* Command respect of road users.
   - *e.* Be located to give adequate time for response.
   - *f.* Be sanctioned by law if it controls or regulates traffic.

3. Function—to facilitate traffic by:
   - *a.* Regulating.
   - *b.* Warning.
   - *c.* Guiding.

State laws usually require that all traffic control devices conform to state standards, and unauthorized control devices are not permitted. All states and many local jurisdictions have adopted all or part of the *MUTCD.* Control devices supplement the driver's knowledge of driving conditions and of certain basic rules that all competent drivers must know. These "rules of the road" affect pedestrians, cyclists, and other road users as well as motor vehicle operators. They are the foundation of traffic control. It is appropriate here to discuss them before entering upon the specifics of traffic signs, markings, markers, and signals.

RULES OF THE ROAD

The "rules of the road" are general guides to traffic behavior intended for statewide application. The "rules of the road" include such basic regulations as obedience to traffic laws, requirements for driving on the right, procedures for overtaking and passing, meaning of and required obedience to traffic control devices, right-of-way at intersections, procedures for starting, stopping, and turning vehicles, and speed limits. Many are so fundamental that in the interest of safety and uniformity, authority for such controls is retained by the state. Local jurisdictions, however, have been granted broad powers over parking, bicycles, stop signs on city streets other than state highways, and other traffic rules covering problems found mainly in municipalities.

As indicated in Chapter 9, the *Uniform Vehicle Code* provides that all vehicles and persons on public ways be governed by the motor vehicle codes and those persons delegated to administer and enforce them [Ref. (2)].

All states have speed controls. The general maxim is that drivers maintain a speed reasonable and proper for prevailing roadway conditions. The relationship of speed and accidents is a commonly accepted warrant for speed controls, and research shows that speed affects the severity of accidents. State authorities can set speed limits and speed zones on state highways based on engineering or traffic considerations. Local authorities can modify limits in areas under their jurisdiction somewhat, but not so as to impose unreasonably low limits outside urban areas or change the limit too frequently along a given road.

There are two types of speed limits, absolute and prima facie. Exceeding the absolute speed limit is sufficient to warrant arrest, while exceeding the prima facie limit places the burden of proof on the driver that the vehicle was being operated safely. Law enforcement officials quite naturally prefer the former. Most of the states have absolute speed limits. Speed limits prior to the energy crisis were 50 to 70 miles per hour (80 to 112 km per hour). Congress adopted a new national speed limit of 55 mph (88 kmph) in 1974. There is a general agreement that the results are both energy saving and improved safety with about 10,000 fewer highway fatalities in 1974 compared to 1973. A similar reduction occurred in 1975 compared to 1973. With increased travel, the reduction in 1976 was somewhat less. The states are monitoring the new speed limit, accidents, and enforcement, and numerous studies have been initiated [Refs. (3) and (4)]. Another study indicates that the initial effect of the reduced speed limit may be diminishing [Ref. (5)]. Still another study indicates that one significant impact of the reduced speed limit is to cause the upper and lower vehicle speeds to be closer together, resulting in increased safety [Ref. (6)]. Some objection to the national speed limit has been voiced, as indicated in a 1977 publication by an ITE local section committee, citing a lack of conclusive evidence of the effectiveness of the speed limit in reducing accidents [Ref. (7)].

The *Uniform Vehicle Code* covers the rights and duties of pedestrians. According to the code, both pedestrians and motorists must obey traffic signals [Ref. (2)]. Drivers shall yield the right-of-way to pedestrians in crosswalks at intersections or elsewhere, but where no crosswalk provisions exist the pedestrian must yield to the vehicle. However, drivers must exercise due care to avoid colliding with pedestrians, no matter what controls exist. Pedestrians are expected to use sidewalks, but where they are not provided, pedestrians should walk on the roadway side facing traffic.

Modern traffic ordinances incorporate regulations for bicycle and motorized cycle operation. Regulations cover licensing, registration, bicycle lights, number of riders per bicycle, and brakes, as well as rules of the road applicable to bicycles.

TRAFFIC SIGNS

There are six basic principles to be followed in the design, installation, and maintenance of traffic signs:

1. *Interpretation.* Traffic signs should convey a simple, clear meaning to the driver, and words and symbols on signs should be designed to minimize the possibility of misinterpretation.
2. *Continuity.* Signs should be designed and installed in context with the other signs on the highway so that continuity is achieved over long stretches of highway. This is one means of commanding the respect of drivers for traffic signs.
3. *Advance notice.* The placement of a sign should allow adequate time for easy response by the driver. Placement would also depend on the response required of the driver. If the sign is an advanced warning of a traffic signal, the driver may be required to stop and hence would need a greater distance within which to respond than for a sign that cautions one to slow down for a hazard. Too much advance notice is also bad, for the sign will become less effective.
4. *Relatability.* Directional sign messages should be in the same terms as information available to the driver from tourist maps and other sources. This is a critical problem in freeway signing, where such signs must indicate the proper exit for drivers who may be disoriented.
5. *Prominence.* The sign should compel attention. This is a function of the size and position of the sign, as well as the number of times the sign or its message is repeated. In urban areas a traffic sign frequently has to compete with advertising signs and other roadside distractions.
6. *Unusual maneuvers.* Specially designed signs may be warranted at points where drivers must make unexpected or unusual movements. Thus, when freeway lanes are closed for repair or construction, an elaborate system of warning signs, barricades, and flashing beacons is needed to guide traffic.

Control devices develop and maintain their effectiveness through proper design, placement with respect to the driver's line of vision, adequate maintenance and application, and uniformity of design, location, and application. These will be considered in order below, along with those aspects that are unique to signs, pavement markings, and signals.

TRAFFIC SIGN DESIGN

A good signing program should aim for consistency in application and reasonableness in restrictions and warnings. Signs should not be viewed as a means for covering up defects in roadway design, but they should be an integral part of highway design.

The characteristics of a sign that command the attention of the driver even before he can read its message are the size, shape, and color. Shape and color are almost as important as the message itself in conveying a meaning to the driver. The combination of color and shape alert the driver long before he can read the sign message.

A considerable degree of standardization in traffic signs has been achieved in the United States through the application of the *Manual on Uniform Traffic Control Devices* [Ref. (1)]. The latest edition shows an increased usage of symbols and international standards. The following standard shapes are employed (see Fig. 10-1):

1. The octagon, for stop signs.
2. The circular, for advance warning of railway grade crossings and for Civil Defense evacuation route markers.
3. The pennant, an isosceles triangle with its longest axis horizontal, for warning of no-passing zones.

30″ x 30″ octagon

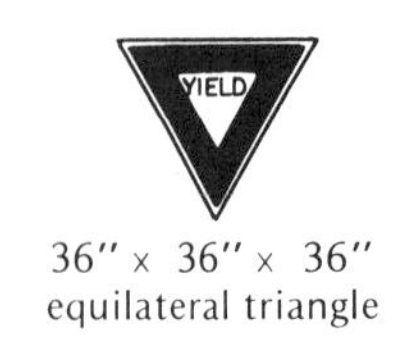

36″ x 36″ x 36″
equilateral triangle

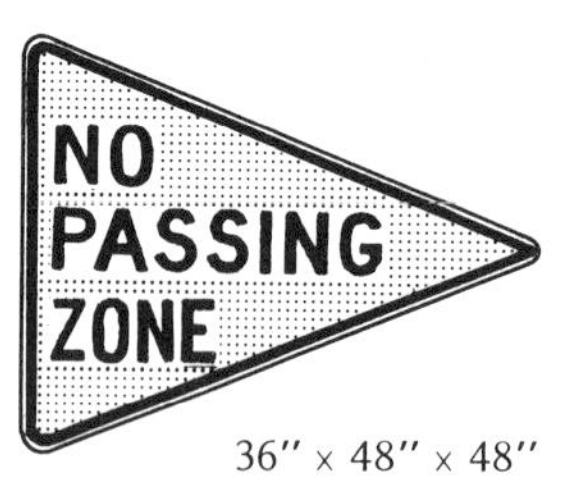

36″ x 48″ x 48″

24″ x 24″
24″ x 18″

36″ diameter
circle

36″ x 24″

30″ x 30″ Square 30″ x 30″
(Mounted vertically along diagonal)

36″ x 36″

48″ x 9″
(drilled for 90-degree mounting)

Variable size
rectangle

48″ x 24″
rectangle

Interstate shield
24″ x 24″

U.S. route marker
24″ x 24″

30″ x 30″

24″ x 24″
24″ x 18″

24″ x 30″
rectangular

State route marker
24″ x 24″

Variable size

Variable size

SOURCE: Adapted from Federal Highway Administration, *Manual on Uniform Traffic Control Devices.* Washington, D.C.: U.S. Government Printing Office, 1971, various pages.

FIGURE 10-1
UNITED STATES STANDARD SIGN SHAPES

4. The diamond, for advance warning of potential road hazards and pedestrians.

5. The rectangle with the major axis horizontal, for guide and directional signs, and also for regulatory one-way signs.

6. The rectangle with the major axis mounted vertically, to convey regulatory messages such as speed restrictions.

7. The equilateral triangle with the one point downward, for the yield right-of-way sign at intersections.

8. Characteristic route marker shields for interstate, Federal, state, county, and special highways.

9. The trapezoid for recreational area guide signs.

10. The pentagon, point up, for school advance and school crossing signs.

11. Other shapes for special purposes.

Standard international road signs employ the triangular shape for warning signs, a circular shape for regulatory messages, and rectangular shapes for guide and informational signs.

Shape effectively draws attention to signs because of the contrast of simple geometric shapes against irregular natural backgrounds. For this reason, irregular shapes such as cut-out arrows should be avoided. The shape of a sign should be efficient for the purpose designated. Rectangular signs are most efficient for word legends.

Unique sign shapes such as the octagonal stop sign or the trapezoid are useful for indicating a specific hazard or regulation. The octagon shape for stop signs evolved from large eight-sided coffin-shaped stop signs first used in Detroit in 1919 to intimidate the driver. The novelty of such signs quickly wears off. But unique shapes are a recognition aid.

The color a sign is just as important as shape in commanding the attention of the driver. Visual impact depends on the color brightness of the sign, the contrast between colors used in the sign, and the contrast between the sign and its background. Another consideration in color impact is the number of color-blind drivers. As many as eight percent of United States males have difficulty in distinguishing red from green. The green and red colors used in traffic signals have been shifted toward blue and yellow, respectively, as an aid to these drivers.

The *Manual on Uniform Traffic Control Devices* recognizes eight different colors for use on standard traffic control devices; black, yellow, white, green, red, blue, brown, and orange [Ref. (1)]. The normal viewer can distinguish far more colors than these, of course, and four other colors have been identified for highway use, adding flexibility for future use.

The colors used on standard signs are [Ref. (1)]:

1. Red. Used only as the background color for STOP signs, multiway supplemental plates, DO-NOT-ENTER messages, WRONG WAY signs and on Interstate route markers; and as a legend color for YIELD signs, parking prohibition signs, and the circular outline and diagonal bar prohibitory symbol.

2. Orange. Used as the background color for construction and maintenance signs and shall not be used for any other purpose.

3. Yellow. Used as the background color for warning signs except where orange is specified and for school signs.

4. Black. Used as the background on ONE WAY signs, certain weight station signs, and night speed limit signs as specified. Black is used as a message on white, yellow, and orange signs.

5. White. Used as the background for route markers, guide signs, the fallout shelter directional sign, and regulatory signs, except STOP signs, and for the legend on brown, green, blue, black, and red signs.

6. Brown. Used as the background color for guide and information signs related to points of recreational or cultural interest.

7. Green. Used as the background color for guide signs (other than those using brown or white), mileposts, and as a legend color with a white background for permissive parking regulations.

8. Blue. Used as the background color for information signs related to motorist services (including police services and rest areas) and the evacuation route marker.

Sign shape and color effectively convey a general meaning to the driver and influence a sign's target value and contrast with natural backgrounds. This is quite desirable for alerting the driver before he or she is close enough to discern the message contained on the sign. But to respond correctly and effectively, the driver must comprehend this specific message and have sufficient time to react properly. The sign must be large enough to carry a sufficiently legible message that can be read and understood within approach distances related to vehicular speeds. Finally, the placement of the sign with regard to the "hazard" must be such as to give the driver adequate response time at the given velocity.

Sign messages are conveyed either by printed words or symbols. In the United States words were widely used to convey sign messages until the 1971 edition of the *MUTCD* [Ref. (1)], which adopted many symbol signs. Such signs have been used in Europe and other parts of the world for decades. Examples of standard international road signs used in Europe and elsewhere are given in Fig. 10–2 [Refs. (8), (9), and (10)].

These signs avoid written words altogether. A well-chosen pictorial symbol is naturally preferable to an abstract symbol. But many situations requiring signing do not lend themselves to a clear pictorial symbol. As a result, an all-symbol signing system is a code that must be learned by all prospective drivers, and the large number of situations requires an extensive set of abstract symbols difficult to memorize.

As early as 1949, a United Nations study group met and recommended a worldwide signing standard. This group reconvened in Vienna and developed a 1968 draft convention on road signs. The warning signs from these deliberations are shown in Fig. 10–3 [Ref. (11)]. Some features of both the American and the international systems were used. Americans seem to have some difficulty with some of the symbol signs, but are gradually adjusting to more symbols since the 1971 *MUTCD* [Ref. (1)].

It is difficult to use symbols effectively for some regulatory and directional signs. Such signs are trying to convey a message of many words through a symbol in order to maximize legibility. Complicated parking restrictions are particularly difficult to present effectively. Color coding is a help. Traffic engineers for years have experimented with a variety of designs to make these signs easier to read and more effective.

For directional signs the written messages should be incorporated into the sign design on the basis of priority values. People tend to start reading a sign from the left as they would a page of writing, so important information should start from the upper left of the sign. For multiple destinations each destination should be separated rather than listed on one sign.

Message Legibility

Under daylight conditions the legibility of a sign depends on the contrast of the message with the background color, the forms of the elements of the message, the physical size of the message, the spacing of elements in the message, the visual acuity of the observer, the motion of the observer, and changing viewing angles, weather conditions, and traffic conditions.

The distance at which a message can be read or recognized is called the "pure" legibility distance. The distance at which a message can be read at a glance is called the "glance" legibility. The latter depends on eye movement and eye focusing. There is also a "dynamic" legibility distance, which takes into consideration the motion of the viewer. Dynamic visual acuity is not as great as static acuity.

Word messages should be kept as short as practicable in consideration of glance legibility. A driver has a glance area of about five feet in diameter at a distance of 100 feet. In moving traffic there may be no time to study the message, so the message should be legible at a glance and contain no unfamiliar words or symbols.

Alphabet design should be as simple as possible. Rounded block letters are somewhat more legible than pure block letters, and the use of lowercase letters also improves legibility. The four most commonly used capital letter designs and the accepted lowercase letter design are illustrated in Fig. 10–4 [Ref. (12)]. The different series of letters are characterized by the height-width ratio and stroke width. The wider letters are favored for better visibility but require more space. The daytime legibility of a Series D letter is approximately 50 feet per inch of letter height, as compared with 42 feet per inch for Series C.

The spacing between letters also greatly affects legibility. The trend in highway letter design is toward greater stroke width and increased spacing between letters. There should be ample spacing between words and lines, and messages should not be cramped by the size of the sign. Providing a margin around the sign serves to outline the shape and provide a frame of reference for reading the message. Symbols used on signs, such as arrows, should also have simple and bold design.

Sign sizes are also standardized in the *Manual on Uniform Traffic Control Devices* [Ref. (1)]. Some of these dimensions were given in Fig. 10–1. Oversized signs are warranted on high-speed routes, at places where the degree of hazard is high, or where the sign must compete with other signs and distractions for attention.

On freeways and other controlled-access facilities oversized signs are required because of the lateral distance between the signs and the vehicles from which they must be read. In fact, overhead signs, which are somewhat larger than side-mounted signs, are used on most freeways (Fig. 10–5 is an example). Traffic demand on urban freeways has created congestion during peak hours and has led to the use of variable message signs to inform the motorist of unusual conditions ahead [Ref. (13)].

Filling station

Telephone

Breakdown service

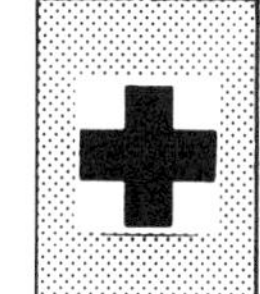
First aid station

Hotel or motel

Camping site

No entry for motorcycles

No entry for cycles

No entry for mopeds

No entry for goods vehicles

No entry for any power-driven vehicle drawing a trailer other than a semi-trailer or a single axle trailer

No entry for pedestrians

No entry for animal-drawn vehicles

No entry for handcarts

Give way

No entry for vehicles carrying more than a certain quantity of substances liable to cause water pollution

No entry for power-driven vehicles

No entry for power-driven vehicles or animal-drawn vehicles

No entry for vehicles having an overall width exceeding . . . . . . metres (. . . . . . feet)

No entry for vehicles having an overall height exceeding . . . . . . metres (. . . . . . feet)

No entry for vehicles exceeding . . . . tons laden weight

No entry for vehicles having a weight exceeding . . . . . . tons on one axle

No entry for vehicles or combination of vehicles exceeding . . . . metres (. . . . . . feet) in length

Driving of vehicles less than . . . . . . metres (. . . . . . yards) apart prohibited

No left turn

No right turn

No U-turns

Overtaking prohibited

Overtaking prohibited by goods vehicles

Maximum speed limited to the figure indicated

Use of audible warning devices prohibited

Closed to all vehicles in both directions

FIGURE 10-2
INTERNATIONAL REGULATORY AND INFORMATION SIGNS

No entry for any power-driven vehicle except two-wheeled motorcycles without side cars

End of speed limit

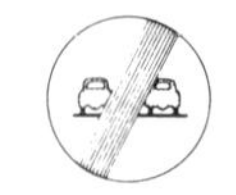

End of prohibition of overtaking

Parking prohibited

Standing and parking prohibited

Alternate parking; prohibited on odd number dates

Alternate parking; prohibited on even number dates

Limited duration parking zone

Exit from the limited duration parking zone

SOURCE: "European International Road Sign System." Washington, D.C.: International Road Federation, May 1973.

FIGURE 10-2
(CONTINUED)

For more data on sign design details, the reader is referred to the *Uniform Manual* [Ref. (1)] and the *Transportation and Traffic Engineering Handbook* [Ref. (14)].

Sign Illumination and Materials

Sign reflectorization or artificial illumination is required for all highway signs that must be legible at night. This includes all regulatory signs except parking signs in urban areas, all warning signs, and all guide signs on important state and local roads. Even the best means of illumination are inferior to natural lighting, so sign dimensions should take this into account. Also at night a driver's vision may be hindered by other headlights and distracting lights. Some recent work by the Virginia Highway and Transportation Research Council was reported by Robertson and Shelor [Ref. (15)], and indicated that in the majority of locations modern high-intensity reflectorization (e.g., reflectorized sheeting) is sufficient for night-time visibility requirements. Care must be exercised to ensure that the artificial lighting does not appreciably alter sign color.

Considerable progress has been made in recent years in sign reflectorization. Financial considerations may dictate the use of reflectorized signs instead of illuminated signs. However, if a given sign needs illumination in order to function it should be provided.

There are two basic types of reflecting elements in use: reflector buttons and reflective coatings. The buttons are set in or on the sign to outline letters, symbols, and the border of the sign. They have the property of reflex or retrodirective reflection, where incident light is reflected almost directly back toward its source. These buttons are made of glass or plastic and have lenses molded into them as reflectors.

Reflecting coatings can be used for the entire sign, enabling color and shape to be reflectorized as well as letters and symbols. Greatly improved coatings have been developed in recent years. Most such coatings consist of tiny glass spheres embedded in a painted or plastic surface or completely enclosed in a transparent binder. The latter technique prevents water and dirt from getting onto the glass beads and impairing reflectivity. Each sphere focuses the incident light entering the sphere, by refraction onto a reflecting surface in the back.

Figure 10-6 diagrams the reflective property of a sphere. Colored light reflection is realized by use of suitable dyes or transparent pigments in the binder.

The divergence angle, or angle between the incident and reflected beams, must be large enough to include the driver's field of vision. Reflective materials must also be effective at wide entrance angles of incident light, so that the sign can be read as the driver is passing it.

The most commonly used materials for highway signs are aluminum, steel, plywood, and tempered masonite. Substantial development work in recent years has led to safer break-away design and some new designs to reduce wind loads and structural requirements. A more detailed description of these various aspects can be found in the

Dangerous bend: left bend

Dangerous bend: double bend

Steep descent or ascent: dangerous descent

Carriageway narrows

Carriageway narrows

Swing bridge

Road leads onto quay or river bank

Uneven road: bad condition

Uneven road: hump bridge or ridge

Uneven road: dip

Slippery road

Loose gravel

Falling rocks

Pedestrian crossing

Children

Cyclists

Animals crossing

Road works

Light signals

Airfield

Cross wind

Two-way traffic

Other dangers

Approach to intersection: general priority rule

Approach to intersection

Approach to intersection: side road

Approach to intersection: merging traffic

Approach to intersection: roundabout

Level crossing: with gates or staggered half gates

Level crossing: without gates

Intersection with tramway line

Count-down posts

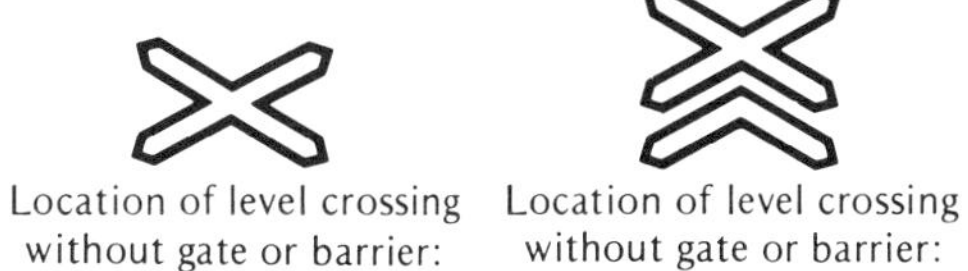

Location of level crossing without gate or barrier: one track

Location of level crossing without gate or barrier: at least two tracks

SOURCE: "European International Road Sign System," Washington, D.C.: International Road Federation, May 1973.

FIGURE 10-3
INTERNATIONAL WARNING SIGNS

ABDEGMORS3568
Series "C"
ABDEGQRS389
Series "D"
ABDEGOS389
Series "E"
AEGORSU368
Series "F"
abefgiklorstuw
Lower case

SOURCE: Federal Highway Administration, "Standard Alphabets for Highway Signs" Washington, D.C.: U.S. Government Printing Office, 1966, pp. 5-24. (Reprinted 1971)

FIGURE 10-4
STANDARD LETTERING FOR HIGHWAY SIGNS

Transportation and Traffic Engineering Handbook [Ref. (14)].

Installation and Maintenance

The placement of a sign depends on its function. Regulatory signs are quite naturally located where the regulation is pertinent, or at the locations where the regulation begins and ends. The latter would include speed zone signs and restricted parking signs. "Reminder" signs are desirable where the regulation extends for a considerable distance. Warning signs should always be placed well in advance of the hazard point or the beginning of the danger area. Guide signs are commonly placed at intersections and between intersections for confirmation.

By convention, signs are usually placed along the side of road nearest the road user for whom the sign is intended. Overhead signs, as in Fig. 10-5, are placed directly over the lane or lanes to which they apply, as are lane control signs. Overhead signs are also used where roadside de-

FIGURE 10-5
OVERHEAD FREEWAY SIGNS (COURTESY OF THE MISSOURI STATE HIGHWAY COMMISSION)

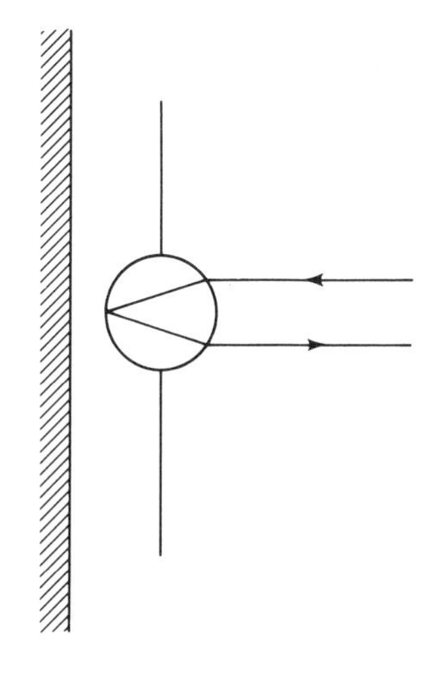

(a)

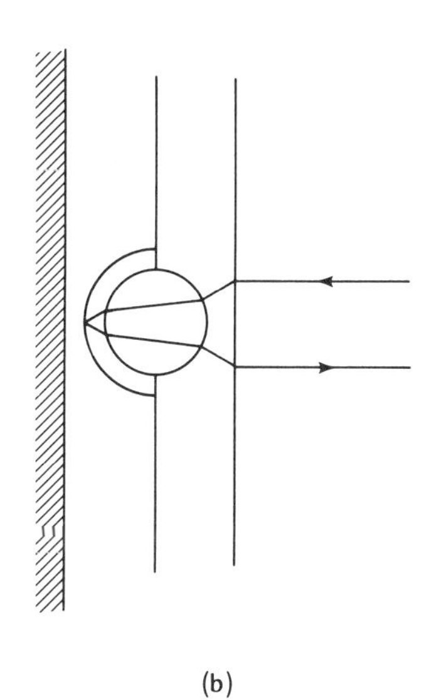

(b)

FIGURE 10-6
LIGHT REFLECTIONS WITH A TRANSPARENT GLASS SPHERE

velopment or parking hinders the visibility of roadside signs.

As sign reflectorization has improved, there has been a trend toward moving signs upward and outward away from the roadway to reduce the accumulation of dirt and splash from passing traffic, which seriously reduces sign reflectivity. Lateral clearances between the sign and the pavement surface are generally 6 to 12 feet (1.8 to 3.7 m) in rural areas, and the elevation of the lower edge of the sign at least 5 feet (1.5 m) above the ground. Overhead signs must provide clearance for all vehicles. Urban roadside signs must be high enough that they can be read over parked automobiles and not present a hazard for pedestrians. Signs should not be so located that reflectivity is impaired or that the driver must take his eyes too far off the roadway.

Whenever possible, each traffic sign should be mounted on an individual post to prevent competition between sign messages. On high-speed roadways, breakaway sign supports are being used in the interests of traffic safety. In the case of massive overhead structures, as in Fig. 10-5, a guardrail should be used at the base of the supports to deflect vehicles out of danger.

To serve the long-distance traveler, the placement of guide signs is a special problem. On freeways confusion is reduced and disruption of traffic flow lessened when adequate advance warning of interchanges is given. Many cities install name signs with large letters on major urban streets in advance of their intersections with other arterials as an aid to drivers. Other signs provide the driver with clues to assist in orientation to unfamiliar surroundings.

Research is continuing in various areas of communicating with the driver, including better sign design and legibility, changeable messages, etc., as well as in-vehicle radio or other audible techniques.

To be effective, deteriorating signs must be replaced, and all signs should be cleaned at intervals.

PAVEMENT MARKINGS AND MARKERS

Pavement markings and markers guide and regulate lateral vehicle placement and supplement some traffic signs. They channelize traffic into the proper positions on roadways and separate opposing streams of traffic. They delineate no-passing zones and turn lanes at intersections. Markings are used to outline pedestrian ways across intersections. As is the case with other traffic controls, markings should be uniform in design and function. National standards are incorporated in the *Manual on Uniform Traffic Control Devices.*

Traffic markings and markers can be classified into the following types:

1. Longitudinal markings (typically 3, 4, or 6 inches wide). These include pavement centerlines, lane separation lines, pavement edge lines, no-passing zone markings, special turn lane markings between intersections, channelizing lines in advance of traffic islands, transition markings where lanes terminate and two traffic streams must merge, approach markings for obstructions in the roadway, such as a bridge pier, and turning lane lines at intersections. The colors (white or yellow) are as indicated in the *Uniform Manual* [Ref. (1)].

2. Transverse markings (usually 12 to 18 inches wide). These include crosswalks and vehicle stop lines (see Fig. 10-7).
3. Message markings. These include both words and symbols, including arrows.
4. Miscellaneous markings. These include curb painting, parking stall markings, road grade crossings, etc.
5. Object markings. These include markings for highlighting obstacles near the roadway, such as bridge piers, temporary barricades or culvert headwalls. The obstacle itself may be marked, or separate markers may be installed.
6. Markers, reflectors, and delineators. These include devices other than traffic signs that are erected to guide, warn, or regulate traffic. A common example is the reflectorized road delineator positioned along the shoulder of the road to indicate the course of the road ahead at night. Raised traffic buttons (reflectors) are employed as substitutes for painted markings. Ribbed pavements and raised concrete bars are used as rumble strips to warn drivers away from traffic islands and other nontraffic areas. Traffic cones made of rubber or similar material are useful for the temporary guidance of traffic.

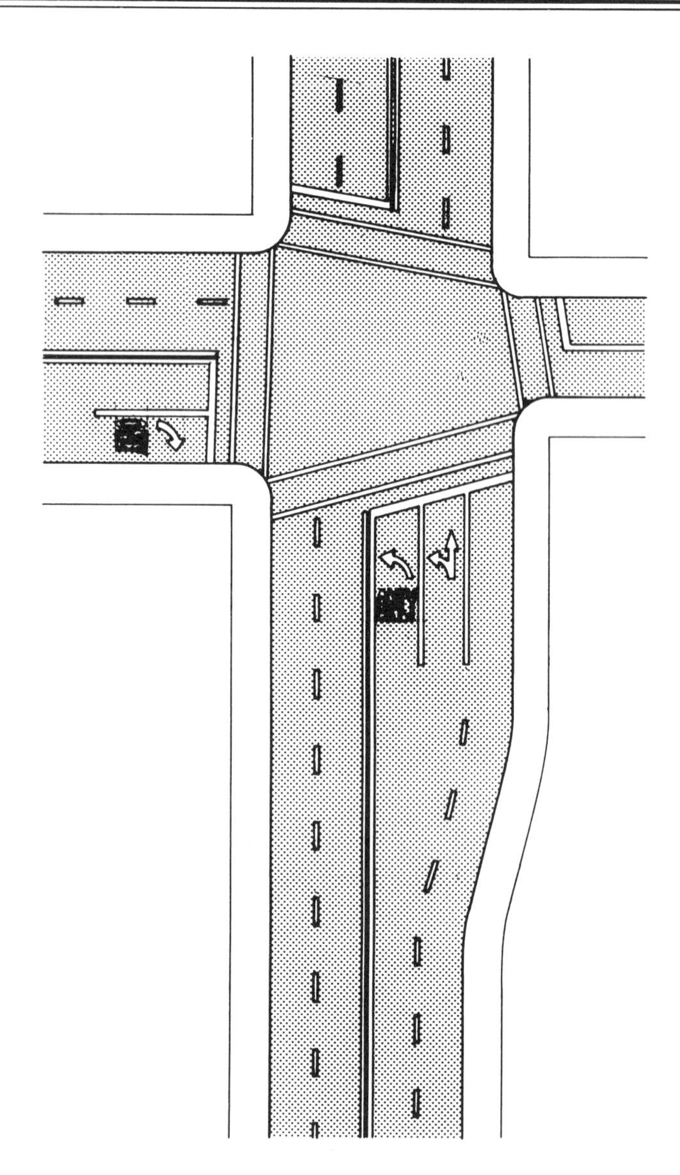

SOURCE: Federal Highway Administration, *Manual On Uniform Traffic Control Devices*, Washington, D.C.: U.S. Government Printing Office, 1971, p. 208.

FIGURE 10-7
INTERSECTION MARKING PLAN

Today there is quite a variety in pavement marking materials. This situation has been created by competition among suppliers and the search for durable and quick-drying materials. The more important materials are listed and described below:

1. Paint. Paint was the first marking material used extensively for pavement surfaces, and it remains the most commonly used today. A modern traffic paint is extremely durable and rapid setting. Since most traffic paints are reflectorized with glass beads, the paint surface is either nonglossy or semiglossy. The paint must also not be impervious to water vapor, or else vapor from the pavement will push the paint off the surface.
2. Thermoplastic striping. Both hot-placed and cold-rolled thermoplastic markings have been used successfully. Technological advances in the past few years have made this material very cost effective in areas with severe winters, requiring salt, sand, and snow plowing. The hot-placed materials are usually at temperatures of 130 deg Centigrade and have much greater durability than paints sprayed. Extruded and cold-rolled thermoplastic materials have also been used.

Both types allow traffic to cross almost as soon as they are placed and despite the increased cost, the durability of this material makes it competitive with paint.

3. Prefabricated tape markings. These markings, generally with an adhesive backing that is designed to conform closely to the texture of the roadway surface, have found wide use for temporary markings for routing traffic during construction and for semipermanent markings such as pavement symbols, parking stalls, and parking lot markings. More recent materials give a measure of wet night reflection as well. When longer service is needed for this kind of marking, a primer should be applied to the road surface prior to the placing of the tape [Ref. (14)].
4. Raised markings. These are usually used in areas where there is little or no snow plowing or use of studded tires and are very good at night or in wet weather. They are usually from ¼ in. (0.62 cm) to 1 in. (2.5 cm) high [Ref. (14)].
5. Reflectorized striping powder. Recently developed reflectorized striping powder provides "instant" track-free marking, used in crosswalks, school zones, etc. [Ref. (14)].
6. Other materials. Stainless steel studs, other insets, etc. have been used, as have glass beads, the last primarily with paint.

Standard pavement markings are either white or yellow. White is the most commonly used color (e.g., delineation of traffic flows in the same direction) whereas yellow is used to delineate traffic flows in opposite direction, both prohibited (solid yellow) and permissive (broken yellow). The solid yellow includes double centerlines on multilane pavements, no-passing zones, pavement-width transitions, approaches to obstructions that must be passed on the right, and approaches to railroad crossings. Yellow is also used for curb markings to indicate where parking is prohibited, and for the curbs or traffic islands. Obstacles near the roadway, such as bridge piers, are marked with reflective yellow or black and white striping.

To ensure attention, longitudinal pavement markings, except edge lines, are at least four to six inches wide. Stop lines range from 12 to 24 inches in width, and crosswalk lines are at least six inches wide. Words to be applied on pavements are elongated to compensate for driver perspective, as illustrated in Fig. 10-8. The design of pavement markings used for channelization around obstacles takes into account prevailing vehicular speeds and the width of the obstacle, as illustrated in Fig. 10-9.

Reflectorization of pavement markings is generally accepted as necessary to maintain effective contrast between the marking and the pavement at night. Reflectorized paint is imperative for marking obstacles on or near the roadway. Glass beads are almost universally used as the reflecting agent mixed in with or applied on paint or embedded in the plastic tape mentioned previously.

A drawback to glass-bead reflectorization is that when the pavement is wet, pavement markings lose their effectiveness. The film of water acts like a mirror in bouncing most of the incident light away from the driver. To circumvent this, traffic wedges and mushroom buttons have been developed. Two types commonly used are glass-beaded polyester wedges cemented to the pavement with epoxy adhesive, and similar plastic wedges with molded lens reflectors. "Pancake" marker buttons with small incandescent light bulbs contained within each one and protected from traffic have been developed for use in lighted guidelines in heavy fog. The visual range of the driver can be doubled by use of reflective or lighted buttons instead of ordinary reflective markings. But these devices also have a major failing. Since snow can cover pavement markings, traffic buttons are often plowed up by snow removal operations. This limits their use to areas that have no snow or to seasonal applications. Where they can be used, however, they have the added advantage over conventional markings of providing a rumble strip effect when drivers cross a line of buttons.

OTHER TRAFFIC CONTROL DEVICES

Other traffic control devices include:

1. Barricades. Temporary devices used to warn drivers of construction or maintenance activities.
2. Traffic cones. Portable, temporary devices used to guide traffic through or around a hazard.
3. Barricade warning lights. Portable lens-directed, enclosed lights, mounted on top of barricades.
4. Rumble strips. A device used to alert drivers of changing conditions ahead.
5. Mileposts. Markers indicating the exact location with reference to a common point (e.g., eastern State border).

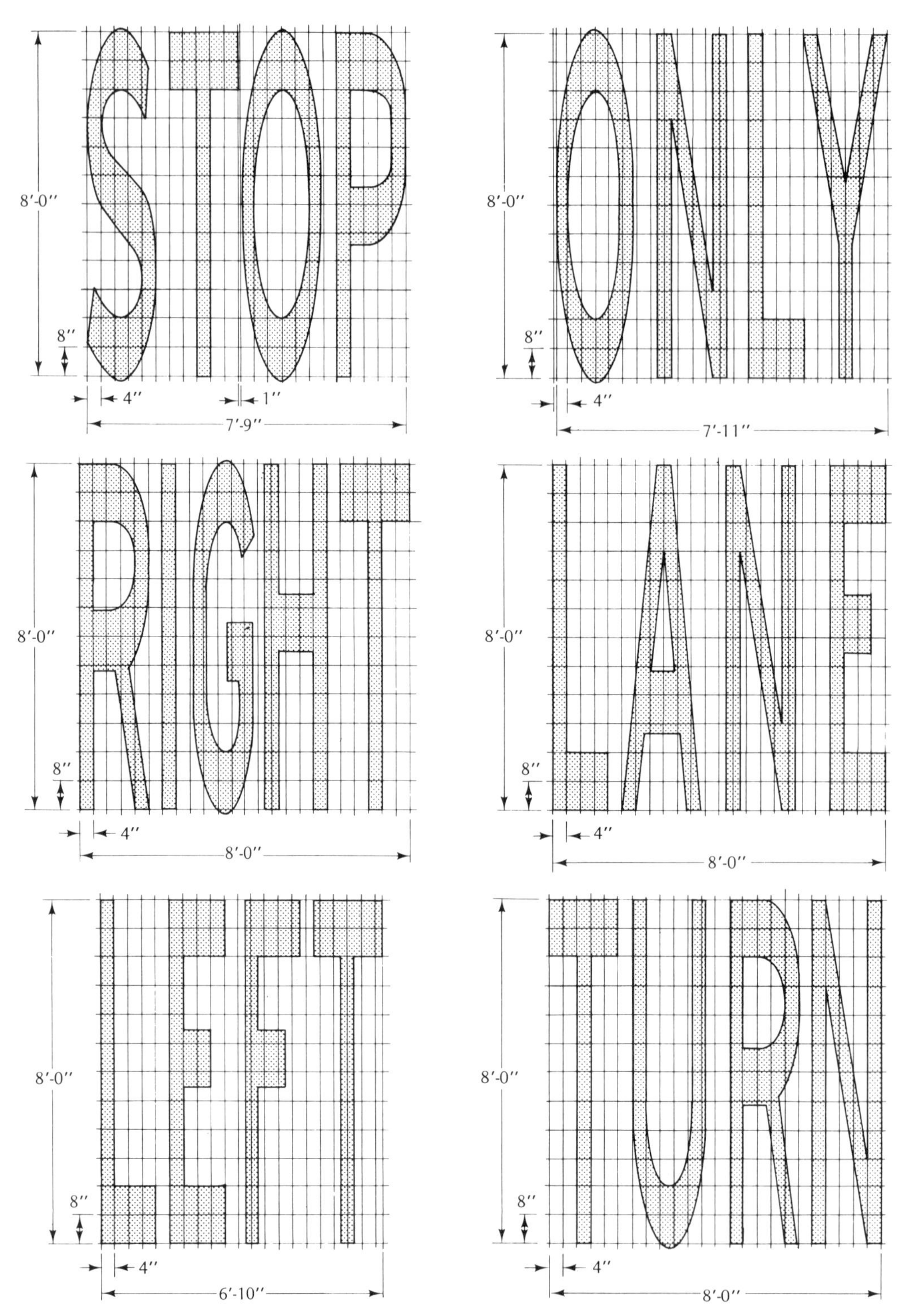

SOURCE: Federal Highway Administration, *Manual on Uniform Traffic Control Devices*, Washington, D.C.: U.S. Government Printing Office, 1971.

FIGURE 10-8
LETTERS FOR PAVEMENT MARKINGS

(a) Center of two-lane road with continuous center line

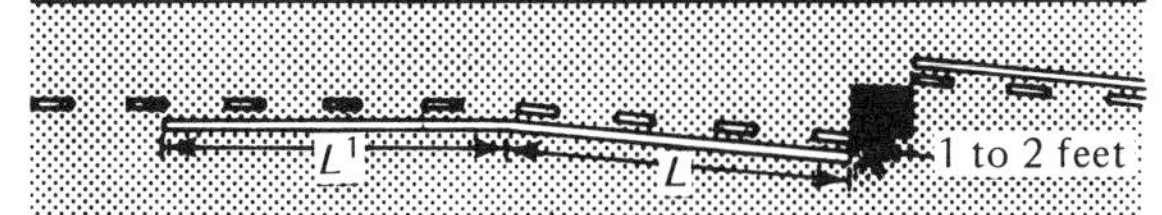

(b) Center of two-lane road without continuous center line

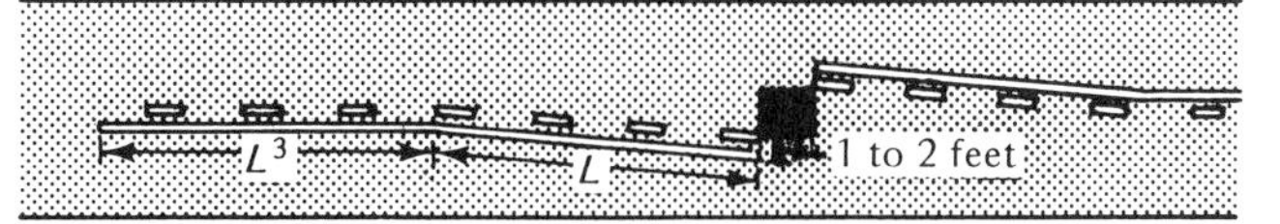

(c) Center of four-lane road

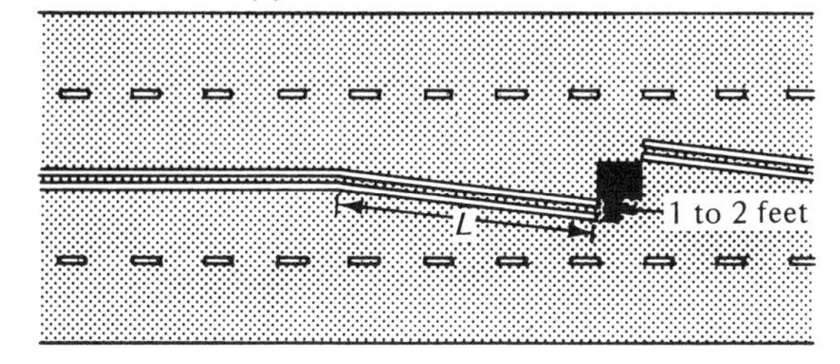

(d) Traffic passing both sides of obstruction

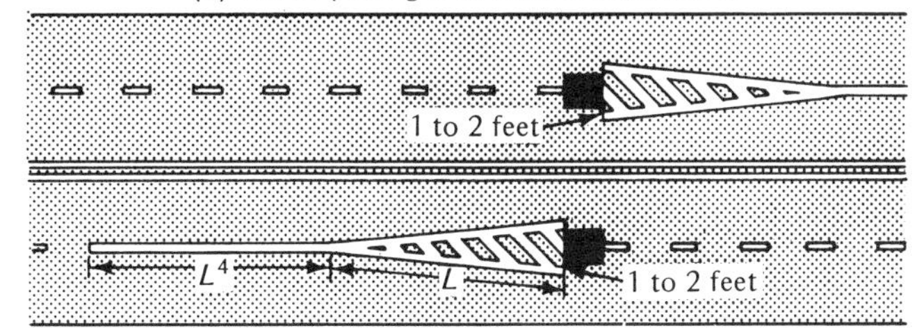

$L = S \times W$
L = Length in feet
S = 85th percentile speed in miles per hour
W = Width of the obstruction in feet (offset distance)
Minimum length of L = 100 feet in urban areas, 200 feet in rural areas

SOURCE: Federal Highway Administration, *Manual On Uniform Traffic Control Devices*, Washington, D.C.: U.S. Government Printing Office, 1971, p. 197.

FIGURE 10–9
APPROACH MARKINGS FOR ROADWAY OBSTRUCTIONS

TRAFFIC SIGNALS

One of the most important traffic control devices is the traffic signal. These control tools are electrically timed devices that assign right-of-way to one or more traffic streams at a time to safely and economically move traffic. Traffic signals are basically of two types, pretimed and traffic actuated, as discussed in Chapter 11 and below.

History of Traffic Control Signals

The earliest known use of traffic control signals dates back to 1868 in London, England, where the police force introduced semaphore arms that could be raised or lowered to control intersecting traffic. Three arms were used—one for "GO," one for "STOP," and one for "CAUTION."

By 1930, all the essentials of modern traffic control equipment as known today were available. The three-color signal sequence had become more or less standard. The first successful traffic signal, consisting of a small motor driving a rotor switch, had been placed in operation, and the traffic-actuated signal had been introduced. Since then, numerous refinements have been developed. Almost revolutionary advances in electronics have led to tremendous improvements in traffic signal components and hence in flexibility. Figure 10–10 is an illustration of typical traffic signal control hardware nomenclautre.

To understand the operation of a traffic signal, it is necessary to become familiar with the terms frequently used in traffic signal control. The *Transportation and Traffic Engineering Handbook* [Ref. (14)] gives a detailed description of these terms, which may be summarized as follows:

1. *Cycle.* The time period required for one complete sequence of signal indications.
2. *Phase.* A part of the signal cycle allocated to a traffic movement or a combination of traffic movements.
3. *Interval.* Any one of several divisions of the signal

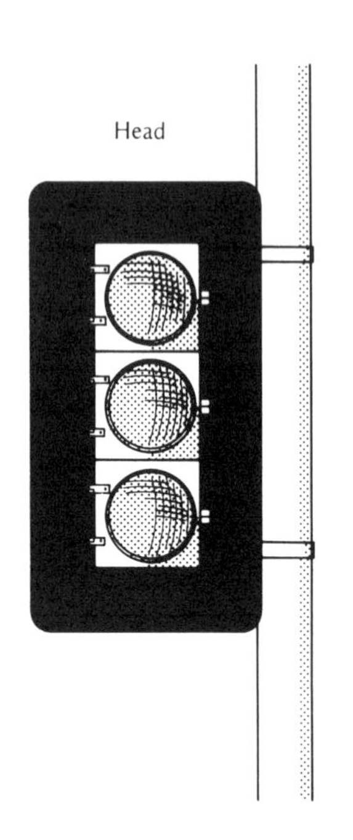

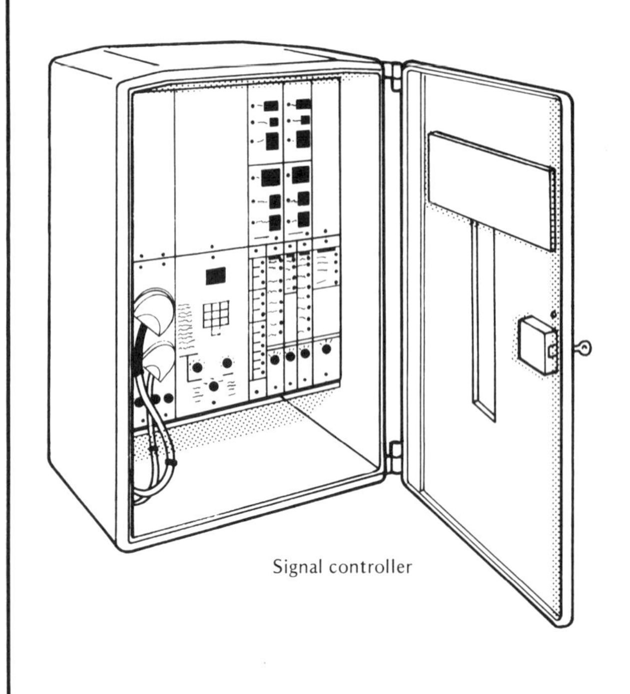

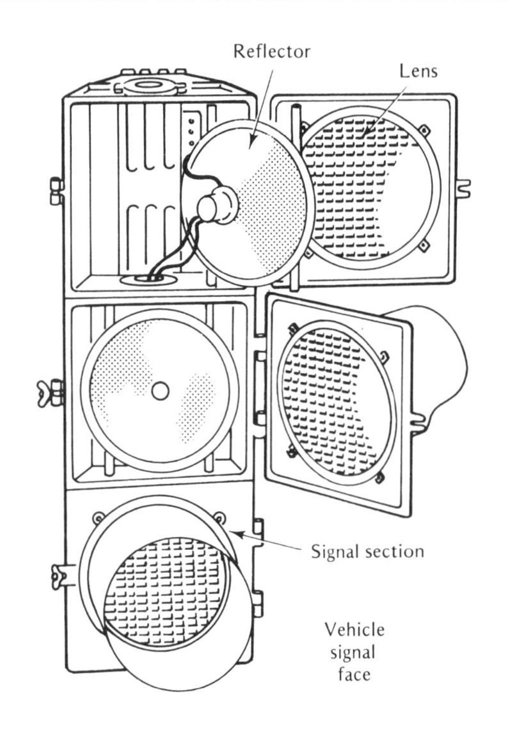

FIGURE 10-10
TRAFFIC SIGNAL EQUIPMENT

cycle during which signal indications do not change.

4. *Offset.* The time (in seconds or percentage of the signal cycle) when the green indication appears at a given signal following some reference time.
5. *Progression.* A time relationship between adjacent signals permitting the continuous operation of groups of vehicles at a planned rate of speed through the signal system.
6. *Through band.* The time in seconds between the passing of the first and the last possible vehicle in a group of vehicles moving in a progressive signal system.

Pretimed Signals

A pretimed controller is essentially a timing device that repetitively cycles the signals through red and green and yellow in a constant preset time sequence. Controllers of this type do not respond to the moment-by-moment demands of traffic at the individual intersection. It is, however, possible to provide a pretimed traffic controller that will have more than one timing routing (the so-called "multidial" signal controller; see Fig. 10-10). The various timing routines are implemented on the basis of a time clock or manual instruction rather than as a direct response to the instantaneous traffic needs.

A traffic signal controller that varies the apportionment of green in accordance with actual instantaneous traffic flow at the intersection is known as traffic-actuated or traffic-sensitive.

Traffic-Actuated Controller Operation

Traffic-actuated control units permit the cycle length and splits to vary in accordance with moment-by-moment traffic counts. These are recorded by vehicle detectors placed in or near the roadway on the approaches to the intersection. These detectors are of two types: (1) pressure-sensitive, magnetic, and induction loop devices imbedded in the pavement; or (2) radar, sonic, and infrared units mounted above or at the side of the road. The latter type emit beams or waves that detect vehicles. Actuated controllers also respond to pedestrian push buttons.

Generally, actuated control signals are not operated in a linked or interconnected manner with other signals along the route. They usually are isolated in the sense that they respond only to the traffic conditions existing at the approaches to the intersection being controlled.

Actuated traffic controllers fall within two basic categories: fully actuated and semiactuated.

The fully actuated control responds to traffic flow through vehicle detectors on each approach. The time assigned to each phase is proportional to the needs of traffic during the phase. The cycle time is the sum of the individual phase times determined by traffic demand.

Fully actuated controllers vary in their degree of sophistication. They normally work between a minimum and maximum permissible green time to prevent unduly short green intervals, which may trap pedestrians or slow vehicles, and to prevent unusually long intervals, which may create unnecessary delay for waiting motorists. More sophisticated controllers modify the green interval on the basis of the number and time of arrival of waiting vehicles facing the red indication and the flow of traffic on the approach with the green light.

Semiactuated controllers respond only to traffic flow on the minor street. This type of control ensures the major street a fixed minimum green interval each cycle. The major street green interval is terminated only after the minimum green time has expired and if there is traffic on the minor street waiting for a green signal. The minor street traffic will retain the green for an interval that will vary in accordance with the demands of its traffic. However, it is usual for a maximum limit to be imposed to prevent excessive delay to major street traffic. Semiactuated controllers can be used in a pretimed progressive system.

Two other types of actuated control are discussed in a recent publication [Ref. (16)] :

1. Queue control. The duration of the green for each phase depends on the queue lengths at the beginning of the phase and provision of sufficient time to discharge the queue.
2. Detector-occupancy control. The presence of vehicles in a given area (near the stop bar) is detected. The green phase is continued when vehicles are present, and terminated when the detection area is cleared, subject to predetermined phase maximums.

Traffic-actuated control minimizes unnecessary stop time and usually reduces delay. However, under heavy traffic flow conditions the maximum limits on each green phase occur frequently, and operations approach those of a pretimed controller. This type of control does not adapt too satisfactorily to an interconnected or linked system of route control. Some success has been achieved in linking by operating a series of actuated controllers with a common background cycle and permitting each green phase to vary within this framework.

Vehicle Detectors for Traffic Signal Operation

Several types of detectors are available to detect the passage (motion) and/or the presence of a vehicle. Each type has particular advantages and disadvantages. Most detectors can be arranged or designed for directional detection. A discussion of the various types is found in a 1974 publication by the National Advisory Committee on Uniform Traffic Control Devices [Ref. (17)]. About twenty different methods are listed in the *Transportation and Traffic Engineering Handbook* [Ref. (14)]. Only a listing of the major types of detectors is shown here. The interested reader will find detailed discussion in References (14), (16), and (17).

1. Pressure-sensitive detector
2. Magnetic detector
3. Loop detector (inductance)
4. Radar detector
5. Sonic detector
6. Magnetometer
7. Miscellaneous types
   a. Sound
   b. Infrared
   c. Pneumatic
   d. Electric eye (photoelectric)

For effective operation, proper location (distance from the intersection) and installation are very important and are discussed elsewhere [Refs. (14), (16), and (17)].

Special Signals

Other traffic control devices that can be classified as special signals include:

1. Automatic railroad crossing protection devices (flashing red lights, etc.).
2. Flashing signals (used as warning and stop sign control supplement).
3. Drawbridge signals (which may be regular traffic control signals).

Additional discussion of these devices can be found in the *Uniform Manual* [Ref. (1)] or the *Transportation and Traffic Engineering Handbook* [Ref. (14)].

REFERENCES

1. Federal Highway Administration, *Manual On Uniform Traffic Control Devices.* Washington, D.C.: U.S. Government Printing Office, 1971 (See also the new revised edition 1978.)
2. National Committee on Uniform Traffic Laws and Ordinances, *Uniform Vehicle Code and Model Traffic Ordinance.* Washington, D.C.: 1968 (with 1972 and 1975 supplements).
3. Borg, T.M., "Evaluation of the 55 m.p.h. Speed Limit," final report, Joint Highway Research Program. Purdue University, West Lafayette, Indiana: March, 1975.
4. Labrum, W.D., "The 55 m.p.h. Speed Limit and Fatality Reduction in Utah," *Traffic Engineering* (Sept., 1976), pp. 13–16.
5. Council, F.M., et al., "An Examination of the Effects of the 55 m.p.h. Speed Limit on North Carolina Accidents." North Carolina Highway Safety Research Center, Chapel Hill, N.C.: April 1975, 99 pp.
6. Rankin, W.W., *Highway User Quarterly.* Highway Users Federation for Safety and Mobility, Washington, D.C. (Sept. 1974), pp. 11–17.
7. Weckesser, P.M., et al., "Implications of the Mandatory 55 m.p.h. National Speed Limit," Report of the ITE Metropolitan Section of N.Y. & N.J., Subcommittee on 55 m.p.h. Speed Limit, *Traffic Engineering* (Feb. 1977), pp. 21–28.
8. Usborne, T.G., "International Standardization of Road Traffic Signs," *Traffic Engineering,* Vol. 37, No. 10 (July 1967), pp. 20–23.
9. American Automobile Association, *Sportsmanlike Driving.* Washington, D.C., 1968.
10. Zuniga, Jose M., "International Effort Toward Uniformity on Signs, Signals, and Markings," *Traffic Engineering* (May 1969), pp. 32–39.
11. Shoaf, R.T., "A Report on the United Nations Conference on Road Traffic," *Traffic Engineering* (May, 1969), pp. 16–31.
12. Federal Highway Administration, "Standard Alphabets for Highway Signs," Washington, D.C.: U.S. Government Printing Office, 1971.
13. Highway Research Board, "The Changeable-Message

Concept of Traffic Control," *Special Report 129*, Washington, D.C.: Institute of Transportation Engineers, 1972.

14. Baerwald, J.E., editor, *Transportation and Traffic Engineering Handbook*, Institute of Transportation Engineers, Englewood Cliffs, N.J.: Prentice-Hall, Inc., 1976.

15. Robertson, R.N. and J.D. Shelor, "The Applicability of Using Encapsulated Lens Reflective Sheeting on Overhead Highway Signs," paper presented at the Transportation Research Board Meeting, Washington, D.C., Jan. 1977.

16. Federal Highway Administration, *Traffic Control Systems Handbook*. Washington, D.C.: U.S. Government Printing Office, June 1976.

17. Federal Highway Administration, *Traffic Control Devices Handbook: An Operating Guide*, prepared by the National Advisory Committee on Uniform Traffic Control Devices, Washington, D.C.: Dec. 1974.

11

TRAFFIC CONTROL TECHNIQUES

The level of traffic control required depends on (1) safety, (2) economics, (3) level of service, or (4) a combination of these.

Intersection control seeks to resolve the right-of-way conflicts between crossing traffic streams. The grade separations described in Chapter 7 eliminate intersectional conflict, but for most streets and highways the physical separation of crossing streams is difficult to justify from a practical or economic standpoint. Therefore, it is necessary to rely on time separation of the different streams to avoid collisions.

This chapter also includes discussion of nonintersection controls such as curb use, lane use, one-way streets, and reversible flow.

INTERSECTION CONTROL

Ways of achieving time separation vary from no control to a grade separation. The *Uniform Vehicle Code* [Ref. (1)] and most state traffic codes contain a regulation establishing the right-of-way rule for uncontrolled intersections. This rule states that each vehicle in the traffic stream approaching an intersection can enter only at such time as there are no vehicles within or closely approaching the intersection to the right of the vehicle. If traffic volumes are too high to achieve an acceptable service level with this rule, regulatory signs may be used. Finally, a dynamic control device, the traffic signal, may be used to assign right-of-way to conflicting traffic streams.

Control by Signs and Regulations

As indicated, the simplest control is by the right-of-way rule. Sign control begins with the use of a Yield sign where one road is assigned right-of-way priority, but where minor road stops are often not necessary and safe approach speeds are at least 10 to 15 miles per hour. They are also used where free right turns are permitted but acceleration lanes are not provided.

Stop signs are installed at intersections of minor roads with major roads, and on minor streets intersecting arterial streets, at unsignalized intersections in a signalized area, or where accident experience or sight restrictions indicate the advisability of stop control. They provide control of intersection right-of-way through the assignment of right-of-way to all vehicles on the major route. These devices require that all vehicles approaching the intersection on the minor route come to a complete stop and enter only when no major route vehicles are within the intersection or approaching so closely as to constitute an immediate hazard.

A four-way (all-way) stop sign installation provides more complete right-of-way control at an intersection. With this control all legs have equal priority. The system has a relatively low capacity, because every vehicle must stop, and the drivers must decide when to proceed. Since all vehicles must stop, additional fuel is used and additional air pollution is generated compared to two-way stop controls. Four-way stop control is most effective when the flow on the approach streets is moderate, is approximately equal, and is in single lanes. For low to moderate volumes, a four-way stop controlled intersection can be as efficient as traffic signals. Such stop control is also used as an interim measure before signals are installed or where accident experience warrants right-of-way control but no signalization.

Turning Controls

A large portion of the delay and accident experience at intersections can be attributed to turning conflicts involving both vehicles and pedestrians. It is natural that engineers use traffic regulation to increase turning capabilities or to prohibit turning completely.

Among the permissive controls are those allowing a vehicle that does not have the right-of-way to turn and merge with a stream of traffic as long as it does not cross a conflicting stream. The rule that permits this type of movement for right turn on red at signalized intersections is an example of this. Another permissive control involves allowing more than one lane of traffic to make the same turn. This increase in capacity is obtained by requiring that the "inside" lane make the designated turn while the "second" turning lane has an option.

The most commonly used turn control is "NO LEFT TURN," which is used to increase safety and capacity at an intersection where accident experience appears to favor such regulation or control or where demands of conflicting traffic movements exceed the capacity of the intersection. Turn controls of this type can only be effectively applied after considering systemwide effects, since the prohibition of certain turns may create enough additional travel to compound the problems at other locations.

TRAFFIC SIGNAL CONTROL

As discussed briefly in Chapter 10, traffic control signals are those electrically or electronically controlled devices by which vehicular and pedestrian traffic is alternately directed to stop and proceed. They allow the engineer to effectively control all movement in an intersection. It is the responsibility of the engineer to ensure that traffic signals are safe, effective, and reasonable in their operation. The consequences of traffic signal failure can be severe, and governmental units may be held liable for property damages and personal injuries caused by such failure. It is imperative, therefore, that signals be well maintained.

Properly installed traffic control signals will:

1. Provide for orderly assignment of right-of-way and facilitate the movement of traffic and increase the traffic-handling capacity of most intersections.
2. Reduce certain types of accidents—most notably the right-angle collision.
3. Provide for a substantial flow of vehicular traffic at a reasonable speed along a roadway when coordinated with each other.
4. Provide for safe crossing of heavy traffic.

There is general agreement that traffic signals are not a panacea for all intersectional control problems. The decision to install a traffic control signal must be based on careful evaluation of available traffic data and on sound professional engineering judgment.

As more is learned about traffic flow, traffic signals are being given uses other than merely alternating the right-of-way between intersecting flows. Measures such as forming compact platoons of vehicles on a route or regulating the input of vehicles on controlled sections of roadways have been used to improve the quality of traffic flow since the early 1960's and are now rather widely accepted measures.

Criteria For Signal Control

The principal function of traffic control signals is to permit crossing streams of traffic to share the same intersection by means of time separation. Therefore, the major emphasis in criteria for signal control is the volume of traffic entering the intersection.

Generally, traffic control signals are not needed if traffic is sufficiently light and if adequate gaps appear at frequent intervals to permit all intersecting traffic to enter or cross the intersection with very little delay or inconvenience. Therefore, the criteria governing installation of traffic signals are determined by the probable frequency of adequate gaps and by the safety requirements of those using the intersection. The criteria related to signal installation are discussed in detail in the *Manual on Uniform Traffic Control Devices* and are referred to as "warrants" [Ref. (2)]. They are

1. Minimum vehicular volume
   a. For major street approach
   b. For minor street approach
2. Interruption of continuous flow
3. Minimum pedestrian volume
4. School crossing
5. Progressive movement
6. Accident experience
7. Systems warrant (e.g., the intersection is part of a system)
8. Combination of warrants

The need for a traffic control signal at a location must be carefully determined by an engineering study, involving not only the application of these warrants, but also an analysis of the physical conditions at the site and the total system needs.

Signalized intersections are among the most critical elements of a street system, since they exert a major influence on overall traffic operation. Signals must provide for safe, efficient movement of both motor vehicles and pedestrians. They can reduce capacity on roadways; however, this reduction can be minimized by the application of sound principles of signal design and operation.

The engineer's traffic signal design problem is to determine a plan described by the sequence and duration of individual phases that will serve all approaching traffic at a level of service consistent with individual motorists' desires as constrained by the needs of other traffic using the intersection.

Capacity

The capacity of a signalized intersection is affected by physical, operational, and control factors. Some of the physical factors include:

1. Width of roadway
2. Number of lanes
3. Geometric design of the intersection

Some of the operational and control factors are

1. Number of turning movements
2. Number and size of commercial vehicles and buses
3. Pedestrian traffic
4. Variation in demand during the peak hour
5. Metropolitan area population
6. Abutting land uses
7. Parking regulations
8. Traffic signal timing characteristics
9. Turn controls

An understanding of how these factors affect capacity is important in establishing the design of a signalized intersection. The effects of the above factors are thoroughly covered in the *Highway Capacity Manual* [Ref. (3)] and are briefly described later in this chapter.

An important factor that forms the basis for the design of signalization at an intersection is the manner in which vehicles enter and depart an intersection. When a signal first turns green, there is usually an initial time lag or starting delay affecting the first four or five vehicles in each lane, following which there are nearly equal time gaps between the remaining vehicles in the queue—approximately two seconds per vehicle. At an urban intersection this lost time may total from two to four seconds each time the signal turns green. In effect, there is a loss of capacity of one to two vehicles per approach lane at the beginning of every green period due to starting delay.

A further loss of capacity is experienced during the yellow interval when no traffic should be entering the intersection. However, in practice, part of the yellow interval is used by entering traffic. In any case, the greater the number of phases in a cycle, the greater the cumulative loss due to the starting delay and the greater the loss due to yellow interval. It is, therefore, evident that the greatest capacity can be achieved with the least number of phases in any cycle. It is also true that greater capacity may be achieved by using longer cycles; that is, by having fewer starting time delays and yellow interval delays during a given hour.

The capacity of an intersection is materially affected by vehicle turning movements. Left-turning vehicles will take more time to clear an intersection than straight-through movements. This is particularly true if their movement conflicts with other vehicles. They will also block vehicles in their own lane unless a separate left-turn lane is provided. The amount of this blocking depends upon the volume in opposing through lanes. When this conflict becomes excessive, it may be necessary to provide a separate left-turn phase. It has been found that the average left-turning vehicle requires 1.6 times as much time to clear the intersection as is required by a straight-through vehicle.

Another factor that affects the capacity of an intersection is the presence of buses, trucks, and tractor-trailer combinations. Because of their length, relative lack of maneuverability, and general lower acceleration ability, these vehicles require more time to clear an intersection than do passenger cars. Their effect has been observed to be equal to as much as five passenger cars. The effect of heavy trucks becomes more pronounced where the approach is on an upgrade and is least pronounced where the approach is on a downgrade or where the intersection forms part of a signal system that minimizes the probability that the truck must come to a stop before entering the intersection. It has been suggested that a medium-sized commercial vehicle on level grade is equivalent to 1.6 passenger cars approaching a signalized intersection.

Level of Service

The level of service as described in Chapter 4 is a measure of the quality of traffic flow. The following three factors are probably the most significant measures of the level of service provided at the intersection:

1. Vehicle delay
2. Queue length, or the number of vehicles backed up
3. The probability that a vehicle will enter the intersection during the first green phase after arrival.

Each of these measures is described in one of the following subsections.

Vehicle delay. When an intersection is converted to signalized control, it is evident that the traffic stream which previously had the preferred right-of-way will experience greater delay following the installation of signals. On the other hand, the traffic stream that formerly had restrictive control may now have less total delay. Therefore, in designing traffic signals for an intersection, the delay criterion to be used should consider the total delay for the intersection as well as the distribution of delays to individual approaching vehicles.

There has been much research on the delay to vehicles at signalized intersections. These studies have investigated the variables that affect vehicle delay and have established how they are interrelated. An important variable over which the traffic engineer has complete control is the cycle length that is used. It has been found that there are cycle lengths that minimize total or average delay. Figure 11-1 shows the form of this relationship, with C_o representing the optimum cycle length based on average vehicle delay [Ref. (4)]. It can be seen that cycle lengths that are too short create delays substantially greater than cycle lengths that are too long. Above the optimum cycle length the total delay at an intersection increases directly as the cycle length is increased.

Queue length. A second measure of level of service of a signalized intersection involves the number of waiting vehicles that accumulate during each signal cycle. This becomes most important in urban areas where intersections are closely spaced. Long lines of waiting vehicles may block cross traffic at adjacent intersections "upstream" as well as block access to exclusive left- or right-turn lanes that may be provided for traffic in the same stream. Thus, an important consideration in maintaining quality of flow at intersections involves keeping both the average and maximum queue length on each approach short. As with delay, it has been found that queue length increases almost directly with increased cycle length for cycle lengths greater than the optimum.

Probability of only one cycle delay. The third consideration of level of service of an intersection deals with the probability of entering the intersection during the first

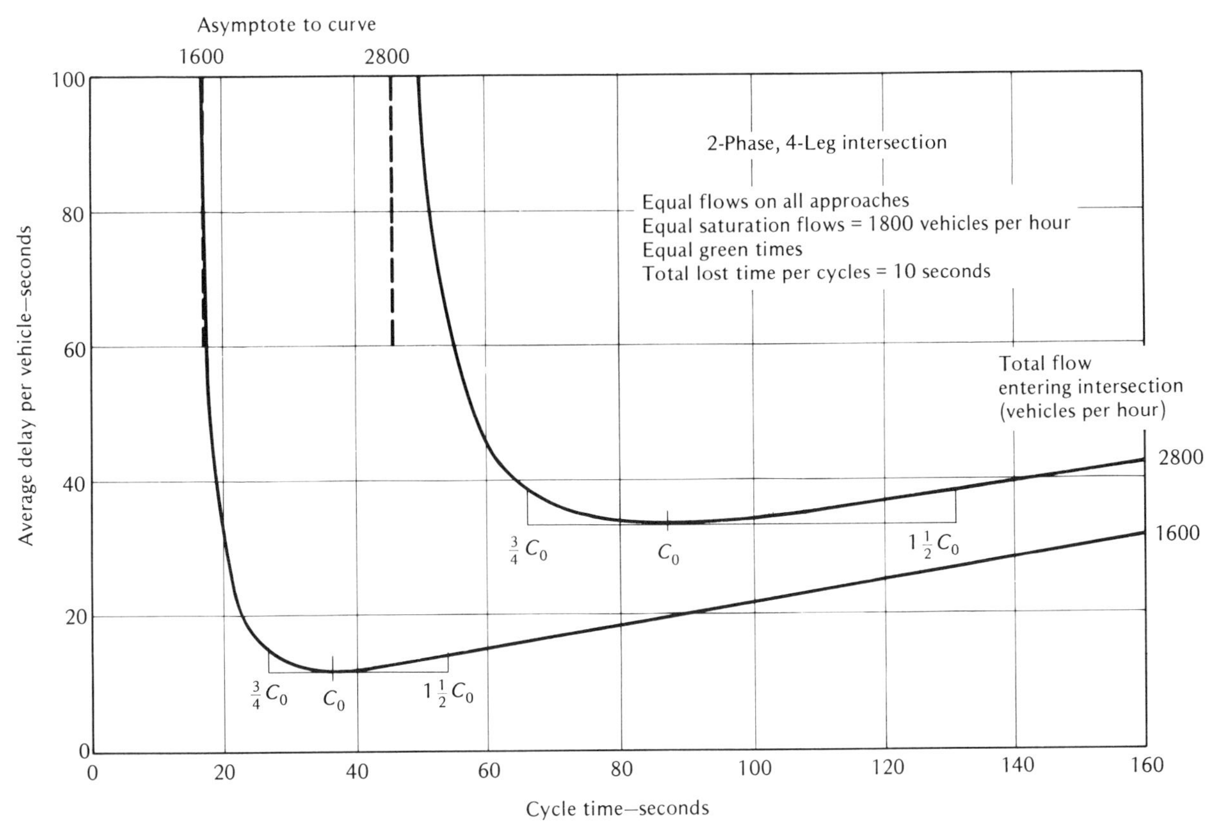

SOURCE: Adapted from F. V. Webster, *Traffic Signal Settings*, Road Research Technical Paper No. 39 Scientific and Industrial Research, Road Research Laboratory, 1958, p. 13.

FIGURE 11-1
DELAY AND CYCLE LENGTH

green phase to appear after the arrival of the vehicle. There is an obvious desire on the part of most motorists to have traffic continue to move. For many drivers, movement, no matter how slow, seems preferable to standing still. If they are delayed at a traffic signal for two or more cycles, they become increasingly annoyed at the delay. It should, therefore, be the aim in designing signal operation to make the green interval long enough so that all drivers will get through on the first green phase a large percentage of the time. A desirable, though not always attainable, goal would be a 95 percent probability that all approaching motor vehicles would clear on the first green. Cycle lengths resulting from adopting this policy are substantially longer than those that minimize total delay or queue length.

In summary, in the design of a signal-controlled intersection the objective should be to provide sufficient capacity for the volume of traffic approaching the intersection. Adequate consideration should be given to minimizing total delay, building short queues, and providing a high probability of passing through the intersection on the first green for most users.

Capacity Manual Method of Determining Signalized Intersection Capacity

The technique of calculating the capacity of signalized intersections described in the *Highway Capacity Manual* [Ref. (3)] is credited to O.K. Norman, who devoted many years to highway capacity and developed this intersection capacity methodology. This method treats each approach to the intersection separately and determines the final capacity iteratively, adjusting the level of service provided on the individual approaches as needed.

"Load factor" is the variable used to reflect the level of service provided to the intersection users. Since it is the percentage of phases during which the capacity of the approach is fully utilized, it is actually the inverse of level of service. Since arrivals at an approach vary widely, full utilization results in long delays for individual vehicles and long queues. Therefore, a low load factor is associated with a high level of service.

Having determined (in the field) the load factor, it then becomes necessary to explicitly describe the physical and operating conditions prevailing on the approach. These include:

1. One- or two-way operation
2. Width of approach
3. Parking regulation within 250 feet of the intersection on both sides of the street
4. Peak hour factor
5. Metropolitan area population
6. Adjacent land uses
7. Bus stop locations
8. Approach lane markings
9. Fraction of signal cycle allocated to approach
10. Traffic demand characteristics

The peak hour factor reflects the relative magnitude of the 15-minute traffic demand peak, which occurs within the peak hour and can be estimated from data obtained at the intersection or studies made elsewhere in the metropolitan area.

With the above information one enters tables similar to that shown on Fig. 11–2 and supplementary nomographs to correct for the other variables to determine the approach capacity.

Although this method of determining intersection capacity has been in use since 1965, there is extensive dissatisfaction and a need for research to develop an improved methodology. In fact, a major research effort in this area began in late 1977 [Ref. (5)].

Signal Control of a Single Intersection

How time sharing at an intersection is accomplished depends on the type of signal control used. A pretimed controller is normally used, and the green time is apportioned to each direction in accordance with the relative traffic demands determined from field observations. When a traffic-actuated controller is used, the green time is assigned to each direction on the basis of the traffic requirements determined from approach detectors.

When it has been determined that a traffic control signal should be installed, the traffic engineer is required to determine the type of control unit that he will use. He must first determine what function this signal will perform. The following are some of the questions that should be answered: Is the basic purpose of this signal to reduce vehicular delays due to relatively high volumes on two intersecting streets? If such is the case, a pretimed controller may be indicated. Or will it provide for an interruption of an extremely heavily traveled route by a much more lightly traveled route that would otherwise be delayed excessively? A traffic-actuated type of control would then be indicated. Or could the primary purpose of this signal be related to the needs of pedestrians?

Furthermore, one must consider the consistency of traffic flow. Are the traffic volumes predictable within reasonable limits, or are the variations in traffic volume erratic and highly unpredictable? If the traffic volumes are reasonably predictable and consistent over a substantial period of the day, a pretimed controller could be selected for which cycle and time splits could be calculated to give a fair distribution of time to the conflicting traffic stream. If not, a traffic-actuated controller may be required.

Pretimed Controller Operation

Many engineers prefer to use pretimed controllers because of their simplicity of operation and maintenance. They also are easily interconnected or linked to adjacent signal controllers to provide a relationship for progressive through traffic flow. On the other hand, this means of assigning right-of-way is disadvantageous when there are substantial short-term volume variations. It may introduce undue delay to main street traffic during low-volume periods. The economy and simplicity of the pretimed controller must be balanced against its inability to respond to short-term changes in traffic flow.

To illustrate, consider a simple case. An intersection at which the traffic arrives randomly was recorded as having an average of 600 vehicles per hour arriving on the major approach. An apparently logical conclusion would lead to the establishment of a green phase sufficient to pass 10

Adjustment for peak-hour factor and metropolitan area size							
Metropolitan area pop. (1,000's)	Peak-hour factor						
	0.70	0.75	0.80	0.85	0.90	0.95	1.00
Over 1,000	0.99	1.04	1.09	1.14	1.19	1.24	1.29
1,000	0.96	1.01	1.06	1.11	1.17	1.22	1.27
750	0.93	0.99	1.04	1.09	1.14	1.19	1.24
500	0.91	0.96	1.01	1.06	1.11	1.16	1.21
375	0.88	0.93	0.98	1.03	1.08	1.13	1.18
250	0.85	0.90	0.95	1.00	1.05	1.10	1.15
175	0.82	0.87	0.92	0.97	1.02	1.07	1.12
100	0.79	0.84	0.89	0.94	0.99	1.04	1.09
75	0.76	0.81	0.86	0.91	0.97	1.02	1.07

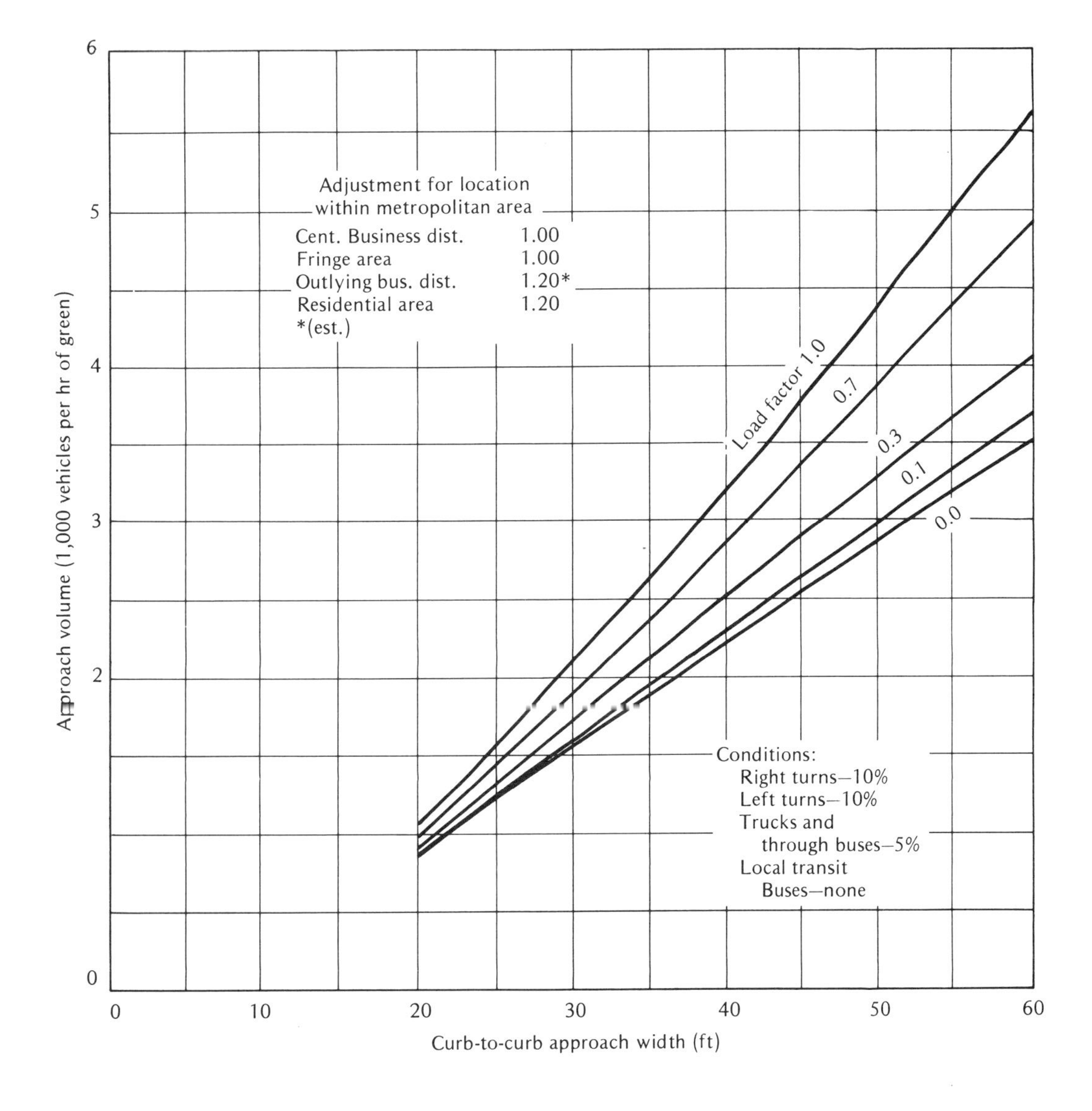

SOURCE: Highway Research Board, *Highway Capacity Manual—1965, Special Report 87,* Copyright 1966, p. 134.

FIGURE 11-2
APPROACH CAPACITY—ONE-WAY STREET WITH PARKING ON ONE SIDE

vehicles per minute. However, if a signal were timed for this rate of flow, the timing would be inadequate during those cycles when the arrivals exceed 10 per minute. For Poisson or random flow, 40 percent of the cycles would have demands exceeding 10 vehicles. The resulting delays would be substantial, queues would become longer and longer, and the probability of entering the intersection during the first green phase would be low. Timing the signal to provide capacity for more than 10 vehicles per minute results in unused green time on the approach.

It can, therefore, be seen that, where maximum efficiency is desired to accommodate all vehicles that may arrive at a signalized intersection without excessive delay, particularly when there are sharp, short-term changes in traffic flow, pretimed signals may not be as effective as one would desire.

Where there is a series of traffic signals at reasonably close intervals along a route, it is most advantageous to have these signals interconnected or linked. With this interconnection it is possible to provide a fixed offset at each succeeding signal so that a platoon of cars traveling along the route can be expected to arrive near the beginning of the green at each succeeding intersection. To accomplish this fixed offset, a common cycle length must be used at each of the signals involved in the system. In such an application, pretimed signals have a definite advantage; it will be illustrated later in this chapter that techniques are available whereby some flexibility and variation in phase length can be accommodated within a fixed cycle.

Traffic-Actuated Controller Operation

A description of traffic-actuated controller types as well as types of detectors is contained in Chapter 10, and a thorough description of the operation of actuated signals is continued in the *Traffic Control Devices Handbook* [Ref. (6)]. In general, each phase begins with a pretimed minimum green interval (initial green period plus an allowable time interval between actuations). At the end of the minimum time, the timing stops with the green indication remaining "on," if there has not been a detector call from another phase. If there has been a call, or when one is received, the green indication will continue as long as the allowable time interval or gap between vehicle actuations on the green phase is not exceeded, *or* until the end of a set maximum period. Thus, the demand of traffic is sensed by the controller. Traffic-actuated signal control is especially suited to locations where the traffic flows vary during the day.

SIGNAL SYSTEM OPERATION

With increasingly valid traffic flow theories it becomes possible to develop ways to make dense traffic move more freely by introducing gaps in the traffic stream, thus forming groups of cars into platoons of limited dispersion. These platoons can be funneled along a road so that they arrive at intersections and pass through during a green light interval while the gaps between platoons occur at the intersection during a red interval.

The interrelation of signals at adjacent locations designed to influence traffic flow and develop a control pattern is referred to as "system operation." In this section, some of the possible combinations that make up signal operation are considered.

A signal system is classified according to the street pattern and the traffic flow it controls. When the system is concerned with a single route and only with the interrelation of signals upon that route, it is referred to as a "linear system." If, on the other hand, it involves a geographic area, including more than one street with the possibility of such streets crossing one another, the system is referred to as a "network."

Linear Systems

Ideally, a linear system results in tight, compact groups of vehicles proceeding smoothly. They arrive during the green interval and clear the intersection before the red. However, it would require strict discipline to restrain platoon dispersion as the distance between signals becomes greater and if vehicles join the platoon from cross streets.

To assist in maintaining platoon discipline, the spacing of traffic signals must be sufficiently close so that the effect of dispersion does not become exaggerated. However, even close spacing of signals does not eliminate the disruptive effect on platoons of turning vehicles at intersections and other stream interference, including friction at non-signalized intersections along the route.

The maximum distance for maintaining tight platoons will vary with the character of the route, and the amount of turbulence and friction that the platoon will meet. However, it is estimated that spacings of signals should not exceed approximately one-half mile if compact platoons are to be retained.

Progressive Systems

On a one-way street a platoon should theoretically be able to travel at a constant rate, the speed of the advancing green light. This situation may be illustrated graphically by plotting the time required to travel between signals against the distance or spacing between signals. The slope of this chart represents the speed of traffic. This diagram is referred to as a "time-space" diagram (see Fig. 11-3). By indicating the green interval at each signal location as a time span, two lines can be drawn, one at the beginning of the green, representing the first car in the platoon, and one at the end of the green, representing the last car that can be accommodated. The time between the two lines is referred to as the "band width." However, when a progressive system is applied to two-way streets, the interaction of control schemes with the geometry of the street becomes most apparent. Most cities are not neatly laid out as rectangular groups of streets; intersections are not only irregular in spacing, but often occur at intervals that make a rhythmic progression in both directions at the same time almost impossible. Therefore, a time-space diagram representing traffic flow in two directions will indicate either a preference to traffic flow in one of the directions, or the best compromise to equally benefit both directions (see Fig. 11-4).

Simultaneous System

The progressive system that has been mentioned in the preceding paragraph establishes the time relationship of succeeding greens in a manner that would closely approximate the speed of the platoon along the route. Some systems, however, have all the green lights on a route commence at the same moment. This system releases all platoons simultaneously at all intersections, and on the red signal stops all movement simultaneously. Under certain intersection spacings it is possible for a simultaneous system to closely approximate a progressive system.

The major advantage of this system is that, because all traffic along the route is stopped simultaneously, cross traffic at unsignalized intersections within the system can enter or cross the main route more easily.

Alternate System

Another common control technique is referred to as the "alternate system." With this system, some signals turn green simultaneously, and the remaining ones turn red at the same time. The advantage of this system is that it provides the same speed of progression in either direction and, therefore, equalizes the advantage of any progression for both directions of traffic at times when there is little justification to arbitrarily assign an advantageous progression to one direction exclusively. Figure 11-4 illustrates an alternate system.

Network Systems

The individual streets or routes that make up a network system are treated in much the same manner as the single street or route of a linear system. However, the network system has the added complexity of the need to coordinate the timing at the signalized intersections that the routes share in common. Here, where progression often must be coordinated for four directions rather than two, it is much more difficult to design a system providing for uninterrupted movement.

The design of a system for a grid of intersections covering a relatively large area is virtually impractical with such manual techniques as the time-space diagram. Some success has been achieved in the solution of this sort of optimization problem through the use of computers. Improved computer techniques will make it possible to develop more effective progression patterns responding to the needs of traffic.

CONTROL SYSTEMS

There are three basic control systems used with signal systems. In "pretimed" signal system control the cycle splits, lengths, and offsets between intersections are controlled by a time clock, which establishes control based on previously measured conditions deemed typical. In the traffic-adjusted or traffic responsive control one of a number of predetermined system control alternatives is also applied. However, in a traffic-adjusted system, current traffic flow characteristics in the network, as determined by vehicle detectors at a number of locations, are the basis for the system control assignment. Computer-controlled systems are characterized by almost complete measurement by detectors of traffic flow throughout the system and by the use of a computer programmed to provide the optimal control for any set of measured conditions for the system.

Pretimed Control

The simplest form of signal system coordination involves the use of a "master" controller supervised by an electrically operated program timer, which determines the time of day and the day of week that given controls will be in effect at the individual intersections. Interconnection is

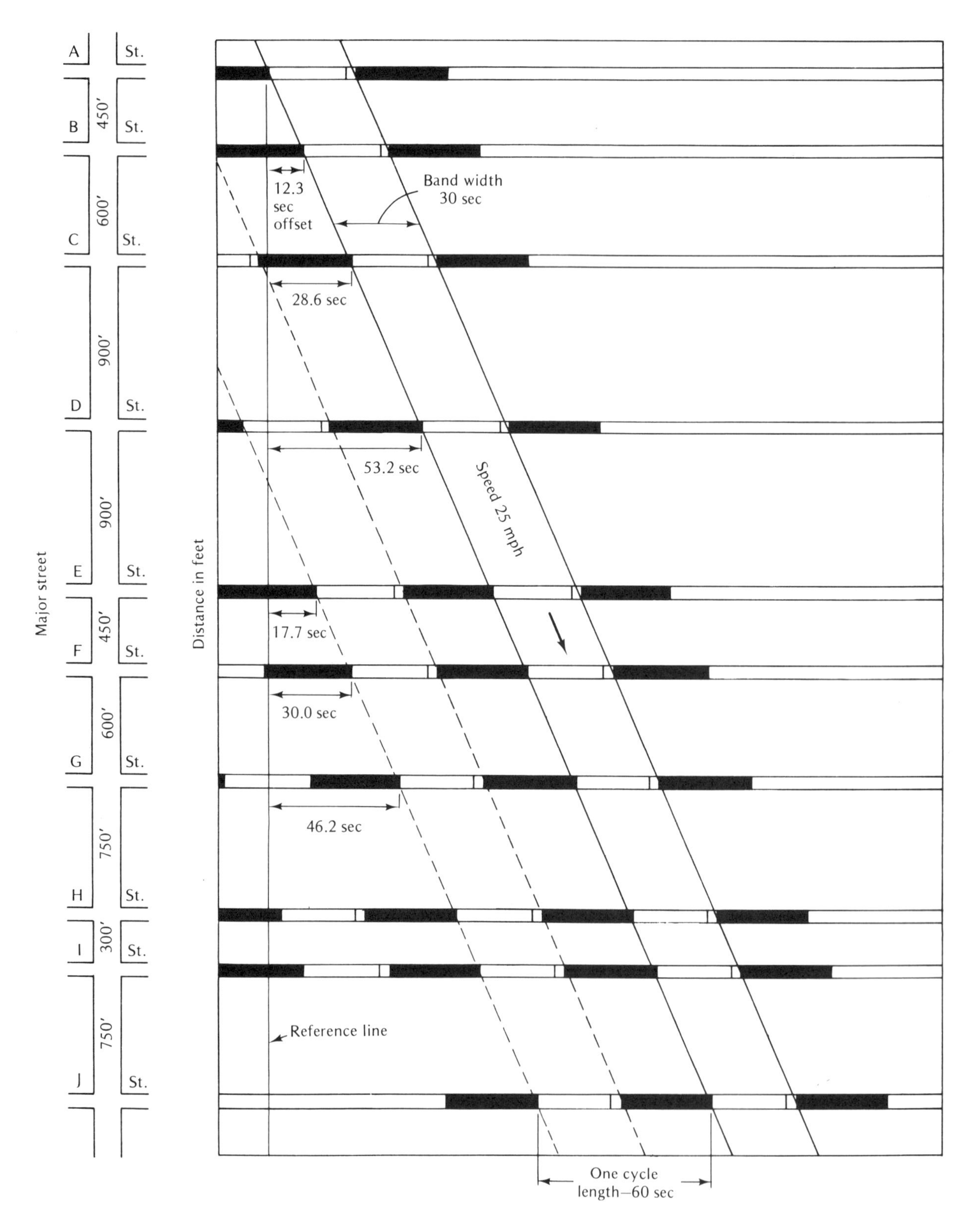

FIGURE 11-3
TIME—SPACE DIAGRAM FOR MAJOR ONE-WAY STREET

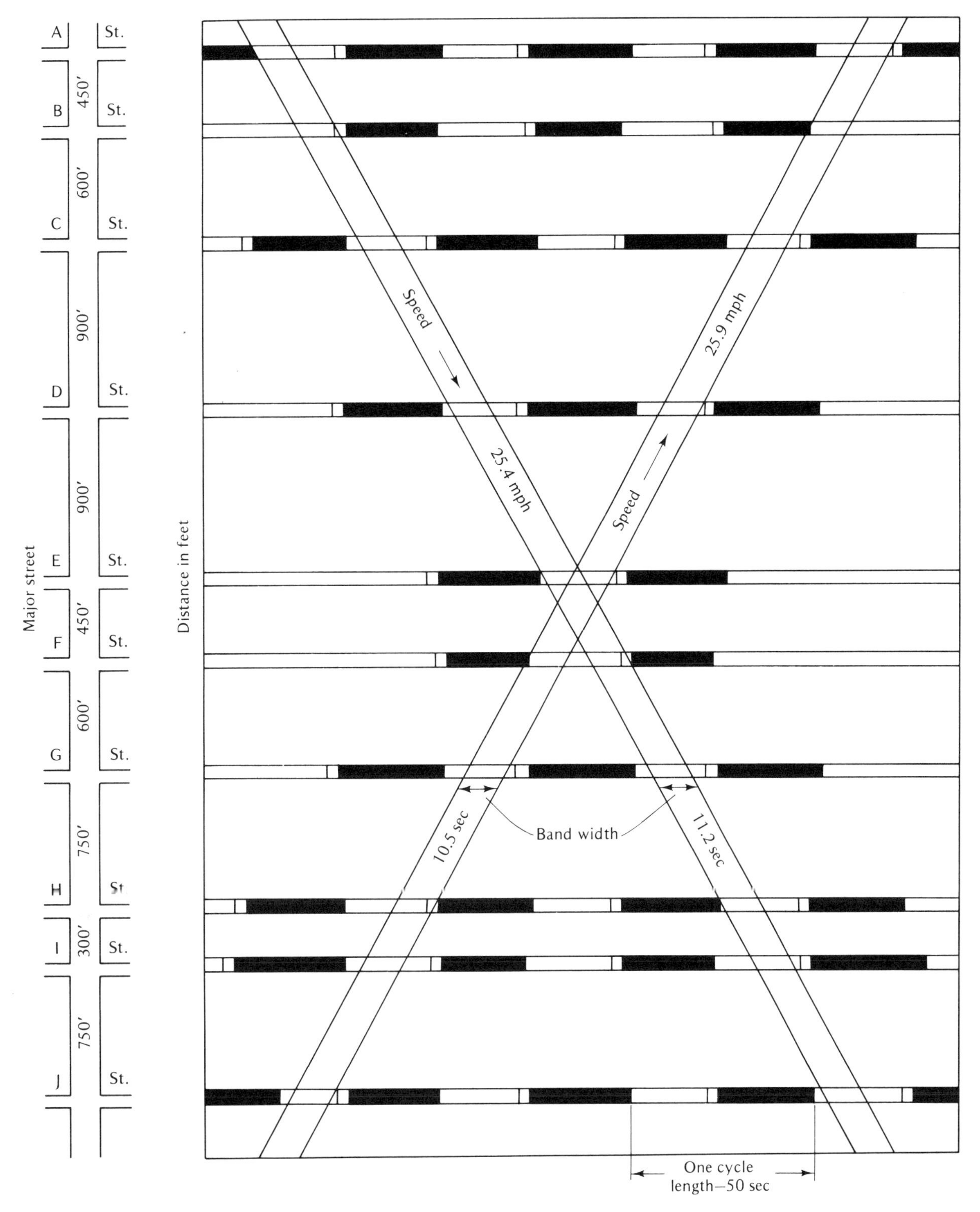

FIGURE 11-4
TIME—SPACE DIAGRAM FOR MAJOR TWO-WAY STREET

provided by cable, leased telephone circuits, or radio control.

Traffic-Adjusted Control

The majority of signal systems in use are linear. Of these, a great many control two-way traffic. It was illustrated earlier that usually spacing between signals is irregular and a two-way progression, providing optimum conditions for both directions at the same time, is uncommon. Furthermore, as traffic volumes and traffic densities change, the speed of progression and size of platoons will change. A system has, therefore, been devised whereby a few carefully located vehicle detectors sample the volume and/or density of traffic flow on the system route. The detectors also provide the data necessary to determine the predominant direction of traffic flow. Then, by means of a preprogrammed schedule of cycle lengths, splits, and offsets, a selection is made of the appropriate set of parameters for conditions during the sampling period. In this manner, the cycle length and splits can be selected to closely match the platoon requirements from time to time. In a like manner, a decision is made regarding the direction of flow to be favored and the speed of progression. This system is referred to as "traffic adjusted."

The limitations introduced by a system of this sort are the limitations imposed by the adequacy or inadequacy of the detectors and the capabilities of the control apparatus. These can vary from a simple arrangement involving a relatively few sample detectors, and the capabilities of selecting a few preplanned programs, to a system that contains a large number of vehicle detectors and a control apparatus permitting an extensive set of programs.

Computer Control

To obtain greater flexibility and control in complex system operation, traffic engineers have turned to the computer. The vast number of variables that must be accounted for in designing a "real-time" control program for a system of traffic signals requires that the computer have the capability to absorb vast quantities of data and determine, by calculations and predetermined standards, the optimum scheme that will suit traffic conditions that exist at that moment. The greater the flexibility of a computer to control traffic in a system or at an individual intersection, the greater the benefit that can be derived.

The most desirable computer control system should offer the possibility to divide the green time available at each intersection in accordance with the traffic demands on each approach. It should relate the green interval at successive intersections in a manner that will minimize the number of stops and delays to through traffic. It should be capable of detecting congestion at any location and applying remedial procedures. Such procedures could involve the deliberate reduction of traffic inflow to the congested area and the simultaneous change of cycle and time splits at the congested location to permit the maximum outflow of the congested approach.

Digital computer traffic control systems have been installed in many cities. One of the most comprehensive systems is the UTCS system installed in Washington, D.C. The *Traffic Control Systems Handbook* [Ref. (7)] also contains a thorough discussion of the various computer-controlled system concepts; two of the better-known algorithms for computer control of signals are:

1. SIGOP. Takes a description of the signalized street network and determines the optimal plan of cycle lengths, phase splits, and offsets.
2. TRANSYT. Consists of: (a) a traffic model that calculates the performance index for a given set of signal timings and (b) an optimization procedure that modifies signal timings (phase splits and offsets) to improve the performance index.

Future signal control will, no doubt, involve even more use of the more and more readily available digital computer.

Traffic Metering

Traffic metering or entrance ramp control is the most widely used form of freeway traffic control. As urban freeways became more and more congested in the late 1950's and in the 1960's, traffic engineers sought ways of improving flow. The desire, as discussed in Chapter 4, is to maintain level of service operation at approximately D or better, rather than to achieve level E and then collapse to level F. By maintaining or at least reducing noncongested flow on the freeway, the efficiency of the operation of the freeway corridor is increased. A good discussion of this topic is in the *Traffic Control Systems Handbook* [Ref. (7)]. Figure 11-5 is a schematic layout for a pretimed entrance ramp metering system. The types of ramp metering are:

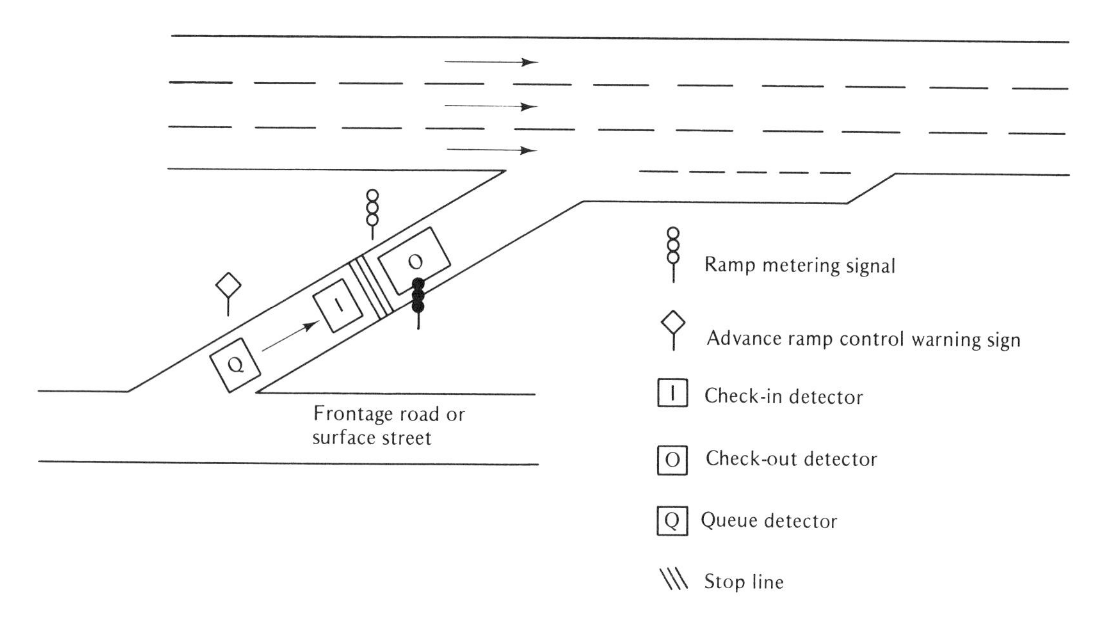

SOURCE: Federal Highway Administration, *Traffic Control Systems Handbook.* Washington, D.C., 1976, p. 131.

FIGURE 11-5
LAYOUT OF PRETIMED, ENTRANCE RAMP METERING SYSTEM

1. Pretime Control. The metering rate is fixed according to clock time (time of day).
   a. Single entry. Traffic signal allows only one vehicle per green phase.
   b. Platoon metering. When metering rates are greater than 900 vehicles per hour, platoons of two or more vehicles are released per green interval.
2. Traffic-responsive Control
   a. Demand-capacity control. Metering rates are selected on the basis of a real-time comparison of upstream volume and downstream capacity.
   b. Occupancy control. Predetermined metering rates are selected on the basis of real-time occupancy measurements taken upstream or downstream.
3. Merge Control. Attempts to achieve optimum use of freeway gaps by enabling the maximum number of entrance ramp vehicles to enter without causing significant disruptions to freeway traffic.
4. Integrated Ramp Control. Application of ramp control to a series of entrance ramps, accounting for the interdependency of ramps, thus leading to overall prevention of congestion.

Metering has also been applied to freeway exit ramp control and at the entrance to tunnels and bridges.

Special Controls

This category of signal controls includes all applications other than regular or standard controls such as:

1. Directional Controls. For off-center or reversible lane operations.
2. Priority Operations. High-occupancy vehicles or buses only.
3. Bus Preemption Systems. Buses are allowed to extend the green indication upon approach to an intersection.

A brief discussion of these measures is included in the next two sections.

NONINTERSECTION CONTROLS

There are several control techniques used by the transportation engineer to achieve safe, efficient, and convenient flow on existing facilities. Application of these techniques requires that the engineer carefully evaluate all aspects of the solutions under consideration, since these control schemes significantly affect activities other than traffic. These controls include curb usage control and one-way street systems. The engineer is also involved with the establishment of highway speed limits and lane controls.

Curb-use Control

Vehicles stopped on the street or roadway at the curb have several effects on traffic flow. First, the space occupied by the vehicle prevents effective use of the traffic lane immediately adjacent to the curb. When parking is intermittent, vehicles using the curb lane as well as the lane nearest the curb parking lane create a disturbance as they move to the left as they approach the parked vehicle. Vehicles moving into and out of parking spaces interfere with moving traffic. Drivers of the parked vehicles frequently exit from the vehicle on the left side, thus creating more conflicts with passing vehicles. Studies have shown that significant improvements in capacity result when curb parking is prohibited. It also is apparent that the smoothness of flow is maintained by prohibiting curb usage.

On the other hand, streets must also provide land access to abutting property and nearby facilities. Goods movements require truck loading zones. Bus and taxi patrons must be loaded and unloaded at locations reasonably near their destinations. Private-vehicle street parking, although not a legal right, is generally considered acceptable when other street use needs do not predominate. The removal of this privilege without the provision of off-street parking creates serious problems for activities dependent on automobile-borne patrons.

Therefore, the engineer must carefully study the supply of parking facilities, the demand for parking space and other curb uses, and the needs of moving traffic, in developing a curb-use control plan. One must carefully weigh these competing demands for the limited curb space available as curb-use control is developed. When curb parking is permitted, one must also allocate the space by establishing time limits to encourage appropriate usage.

The most common technique for control of curb parking is signing, but, in order to be effective, it must be coupled with an enforcement program (e.g., towing of vehicles parked during hours of prohibition). Meters also provide a degree of control, encouraging turnover, but additional small signs may be required.

Lane Controls

The specific control of use of a lane on a facility may have several advantages. One form of control is to reserve a lane for use by transit vehicles only, as shown in Fig. 11-6. With this control, bus movements are expedited through exclusive bus use of the curb lane during periods of peak traffic flow [Ref. (8)].

Another control technique used by the engineer to improve operations on the existing street system is to take advantage of unbalanced demands and allocate additional lanes to the heavier movements at the expense of opposing traffic lanes. These techniques have proved highly effective on major streets, particularly during peak periods.

The effective control of specific lanes is accomplished through curb-mounted signs, overhead signs, changeable message signs, and through lane-use signals. The most frequent technique for lane control is the application of lane-control signals that are centered over the particular lanes they control, normally between intersections, at toll booth areas, or over specific lanes on freeways. Green arrows, red X's, and (optional) yellow X's designate the proper direction of travel. Whenever a steady green arrow is displayed, a steady red X is simultaneously displayed to traffic approaching from the opposite direction. A flashing yellow X may be used to permit both directions of flow to use a center lane for left turns. Travel in the lane for any distance is not intended.

Reversible lane operation is the most common application for lane-use control signals. It should be employed only after the need for such control can be demonstrated by a competent engineering study, and only if the planned operation can reasonably be expected to operate safely and efficiently. It may be appropriate to convert to reversible lane-control operation when vehicular traffic flow in one direction on a two-way street, highway bridge, or tunnel shows the following characteristics:

1. The level of service during peak periods decreases to Level D, as described in the 1965 edition of the *Highway Capacity Manual* (the facility is experiencing congestion).
2. A predominant volume of traffic flowing in one direction is present for only a specific period during a day. This flow may be complemented by a similar predominant flow in the opposite direction during a different time period. Reversible lane operation can provide additional capacity for the predominant flow direction

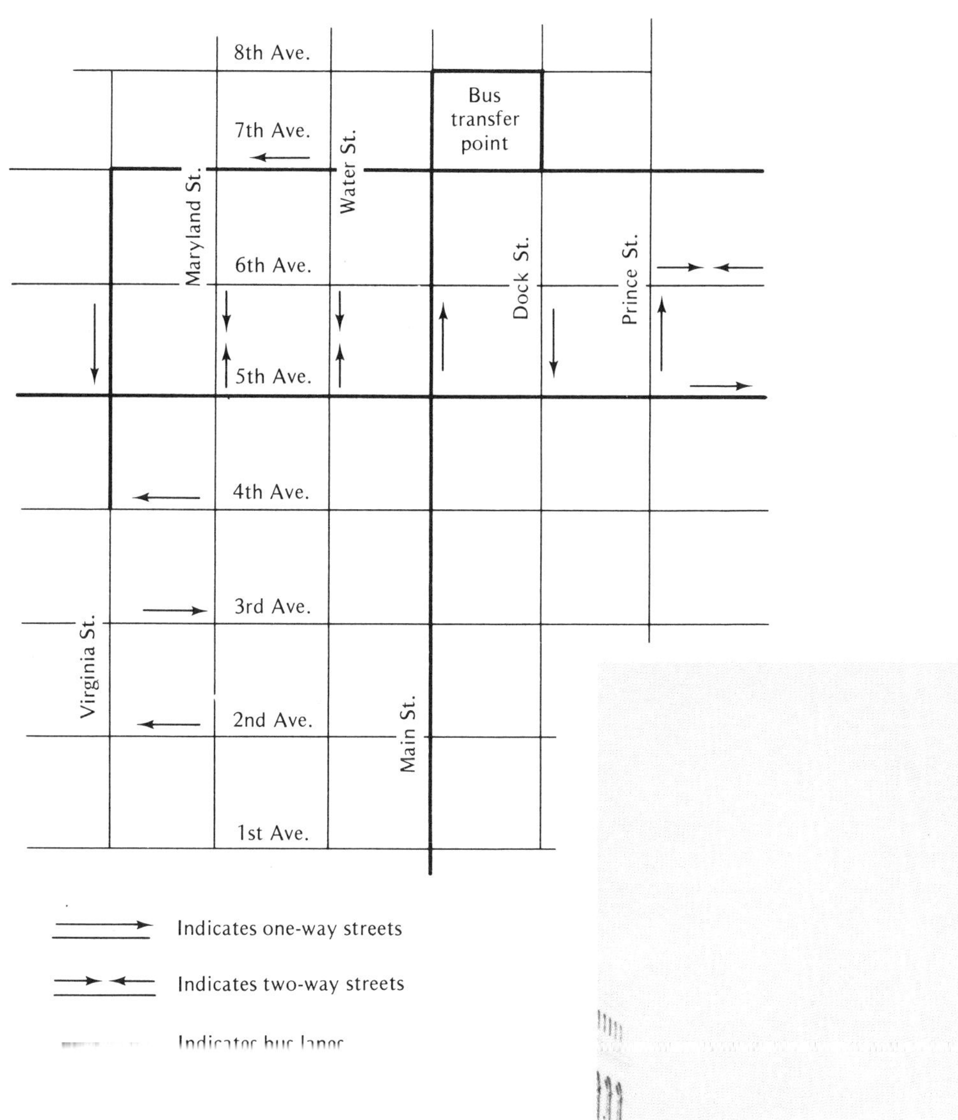

FIGURE 11-6
EXCLUSIVE BUS LANE PLAN

by reversing a lane normally used for the lower volume flow to use for the higher volume flow. Lane-control signals are often used to accomplish the reversing.

Striping for an extra lane in one direction can be used effectively for permanent unbalanced flow when:

1. An extra lane can be gained because of cross-section dimensions.
2. Signals are timed for one-way flow, but the opposite direction can not be prohibited.
3. The area near the beginning or end of one-way street operation requires additional capacity.

One-Way Streets

One of the oldest forms of control is a one-way street in which all opposing traffic flow is prohibited. The disadvantages of such streets are additional travel distances, some difficulties with transit routing, and the loss of two-directional access to properties. However, the advantages are substantial.

The removal of the constraint of using only half the road often makes it possible to introduce cross-sectional geometrics that result in an additional lane of traffic flow or more curb parking than previously existed. One-way systems usually provide two or more parallel moving lanes, which free traffic in congested areas from the delays involved with very slow-moving vehicles and double-parked vehicles. Intersectional capacity is improved by simplifying and reducing conflicting movements. Accidents at both mid-block and intersections are generally reduced. Travel speeds are increased to desirable levels in congested areas as the flexibility from the increase in the number of lanes and the easier signal timing that can be achieved with one-way operation are realized.

The general criteria for one way streets are:

1. A traffic problem can be relieved by one-way operation.
2. One-way operation is more desirable than other alternatives.
3. Parallel streets are available for pairs of one-way streets.
4. Suitable transition areas exit at the ends of the one-way system.
5. Any existing two-way public transit route can be handled.
6. The one-way street fits into the overall comprehensive street plan.
7. A significant increase in capacity will occur.
8. Safety will be improved through reduction in vehicular-pedestrian conflicts and vehicular-vehicular conflicts.
9. Improved operations through reduced travel time, etc. will result.
10. Improved economy in providing additional needed capacity at small cost will result.

The development of a one-way plan requires particular attention to detail on the signing and control for satisfactory operation as well as to public information and public relations, especially concerning the residents and businesses along the streets affected. The *Transportation and Traffic Engineering Handbook* [Ref. (9)] and the *Uniform Manual* [Ref. (2)] contain more extensive discussions of one-way streets.

Reversible Streets

A special type of one-way operation is reversible flow, which involves reversing the entire street for part of the day. The types of reverse include:

1. Street is inbound during part of the day and outbound for the remainder of the day.
2. Street is inbound during the morning peak, outbound during the evening peak, and two-way at other times.

The advantages and disadvantages of reversible streets are similar to those for one-way streets, except that a parallel street for operation as a pair is not required. The manual on uniform traffic control devices contains a thorough discussion of signal controls as well as signs required for reverse flow operation [Ref. (1)].

Truck Routes

Another control option is the establishment of specified routes for through trucks. Thus trucks can be restricted from some congested routes at specified times. Basically, proper signing and trucker cooperation are required for implementation.

TRANSPORTATION SYSTEM MANAGEMENT (TSM)

Beginning with a joint UMTA/FHWA announcement in September 1975, a new federal program was launched in an attempt to merge traditional traffic engineering (or short-range transportation solutions) with the longer-range transportation planning process [Ref. (10)]. Most of the procedures applicable to this new program, called Transportation System Management (TSM), are traditional traffic engineering techniques. However, the major emphasis is on coordination with the long-range plan, on improved treatment of public transportation, energy efficiency, and better overall utilization of the total existing surface transportation system. The objectives of TSM are [Ref. (10)] :

1. To make more efficient use of existing highways and transit systems.
2. To make best use of limited financial resources.
3. To bring about a more rational organization and better balance between the various transportation subsystems.
4. To emphasize service improvements that can benefit urban areas in the immediate future.

The joint regulations require that the urban transportation planning process contain a TSM element. This element should include both low-capital transportation improvements (LCTI) and some medium-capital intensive projects. Some of the TSM items may include priority for high-occupancy vehicles [Ref. (8)] (transit and carpool lanes), preemption of signal phases by transit vehicles, parking controls, changes in work schedules, and traffic engineering improvements to increase capacity and reduce congestion, and hence to benefit the entire traffic stream.

REFERENCES

1. National Committee on Uniform Traffic Laws and Ordinances, *Uniform Vehicle Code and Model Traffic Ordinance.* Washington, D.C.: 1968 (with 1972 and 1975 supplements).
2. Federal Highway Administration, *Manual on Uniform Traffic Control Devices.* Washington, D.C.: U.S. Government Printing Office, 1971. (See also the new edition 1978.)
3. Highway Research Board, *Highway Capacity Manual,* Special Report 87. Washington, D.C.: 1965.
4. Webster, F.V., *Traffic Signal Settings.* Road Research Technical Paper No. 39, Scientific and Industrial Research, Road Research Laboratory, 1958.
5. Transportation Research Board, NCHRP Project 3-28, "Development of an Improved Highway Capacity Manual," Three-phase study that began in late 1977.
6. Federal Highway Administration, *Traffic Control Devices Handbook: An Operating Guide.* Prepared by the National Advisory Committee on Uniform Traffic Control Devices, Dec., 1974.
7. Federal Highway Administration, *Traffic Control System Handbook.* Washington, D.C.: U.S. Government Printing Office, June, 1976.
8. U.S. Department of Transportation, "Priority Techniques for High Occupancy Vehicles: State-of-the-Art Overview," Nov. 1975, Washington, D.C.
9. Baerwald, J.E., editor, *Transportation and Traffic Engineering Handbook,* Institute of Transportation Engineers. Englewood Cliffs, N.J.: Prentice-Hall, Inc., 1976.
10. UMTA and FHWA Joint Regulations, 23 CFR, Part 450, *Federal Register,* Sept. 17, 1975.

T 1 2 LANHAM
metrobus
M
metro

12

MASS TRANSIT SYSTEMS

Mass transit is the basic common carrier system in urban areas. Common carriers—which also include airlines, railroads, and intercity bus systems in the passenger market—provide services on authorized routes and fixed schedules at published fares for all travelers who wish to use the service.

The most important distinction between a highway department and a transit agency is that the former furnishes a system on which users travel or ship freight in their own vehicles and with their own drivers, whereas the latter provides the vehicles and drivers but, unless exclusive right of way is involved, has no responsibility for the roadway. The primary concern of transit agencies, therefore, is to operate a transportation system rather than to construct it. The technical skills needed are in the area of systems analysis, operations, and maintenance. Some civil and structural engineering may be required for planning and designing yards and shops on bus systems, and there is a major engineering challenge in creating or extending rapid transit networks.

Highway and traffic engineers working in urban areas are concerned with optimizing the total transportation system of which transit is a part. They must, therefore, understand how transit operates, how passengers decide on using it, and how it may be constrained by physical, operational, or financial factors.

TRANSIT NETWORKS

The shape of transit networks reflects the demand patterns. A brief review of these patterns is, therefore, necessary. For further discussion, see the section on "modal split" in Chapter 13.

Transit users can be classified in two groups—"captive" and "choice" riders. Captives are those who have no alternative transportation mode available; they have no access to an automobile either as drivers or passengers, nor is walking or bicycling a viable alternative for those trips for which transit is used. The majority of this group depend on transit so frequently that they must choose their home, work location, shopping, and other facilities in relation to well-established transit lines. Their travel patterns, therefore, are concentrated in the older, more central parts of an urban area, within which trips radial to the central business district (CBD) predominate, but journeys not involving the CBD are also common.

Choice riders are attracted to transit if they feel that the travel time and cash costs involved are competitive with those of the automobile alternative. This is most likely to occur for peak period trips with one end in the CBD because of the likelihood of congestion at the approaches to the downtown area and the high parking costs within it. Almost all the desire lines of choice riders, therefore, radiate from the CBD and, when superimposed on those of the captive riders, produce a total pattern that is largely radial.

As a result, transit networks show a strong tendency to radiate from the CBD. Many of the routes have antecedents dating back to the nineteenth and early part of the twentieth century, reflecting the relative stability of living and travel patterns of nonautomobile households. Additions to the network are primarily more radial lines serving newly developed, generally low-density outlying areas and connecting them either to the older network or directly to the central area of the city.

Routes

Figure 12-1 shows the basic elements of such a transit network. Routes with the prefix A are the basic *arterial* lines, radiating from the CBD, serving both captive and choice riders. These routes produce the most revenue per vehicle-mile. There is appreciable demand in off-peak and weekend periods. The maximum load point usually occurs at the edge of the CBD. In some cases, the demand falls rapidly with increasing distance from downtown, in which case branching of the routes can be considered (Routes A-1 and A-2 in Fig. 12-1).

Where the demand between outlying areas and the central district has grown sufficiently large, *express* routes (Route E-1 in Fig. 12-1) may augment the arterial routes. The principal motivation for their establishment is to attract choice riders by offering competitive travel times. However, the transit system can also benefit from the more productive utilization of its equipment at higher speeds. Express routes often operate only during weekday peak periods because of the lack of demand at other hours.

There are often outlying areas that cannot be served directly from the CBD, either because through-routing and branching are not possible, or occasionally because topographic or street conditions require the use of smaller vehicles than are useful on arterial routes. In that case *feeder* lines are added to the network (Routes F-1 and F-2 in Fig. 12-1). These lines often operate during the same hours as the arterials, and are most commonly found in support of rapid transit lines. Taken by themselves, they appear economically least viable, but they make a contribution to the patronage of the arterial routes which must not be overlooked. The maximum load point on a feeder route always occurs at its intersection with the arterial lines.

In smaller cities the demand for transit linkages that do not involve the CBD—mostly from captive riders—is insufficient to warrant special consideration. Such trips must be made via the downtown area even when both origin and destination lie on the same side of it. In larger cities, *circumferential* or *crosstown* routes can be justified to reduce the length and inconvenience of such trips (Route C-1 in Fig. 12-1). Planning and scheduling such routes presents some difficulties, since there may be several maximum load points at intersections of different arterial routes, and since an effort must be made to have vehicles of the circumferential route meet those of the radial network at transfer points.

Occasionally a transit network will include *shuttle* routes to connect two major traffic generators in fairly close proximity to each other. Figure 12-1 shows one such route (S-1) serving a fringe parking lot intended to relieve parking shortages in the CBD.

Right-of-Way

The right-of-way on which transit service operates can be either shared with general traffic, semi-exclusively reserved for transit with minor interference from other vehicles, or exclusively assigned to transit (Table 12-1).

Shared rights-of-way are available to transit service at no cost other than trivial expenditures for bus stop signs and perhaps benches and shelters. However, the transit system is subject to congestion and delays and to the possibility of disruptions and accidents that are beyond its

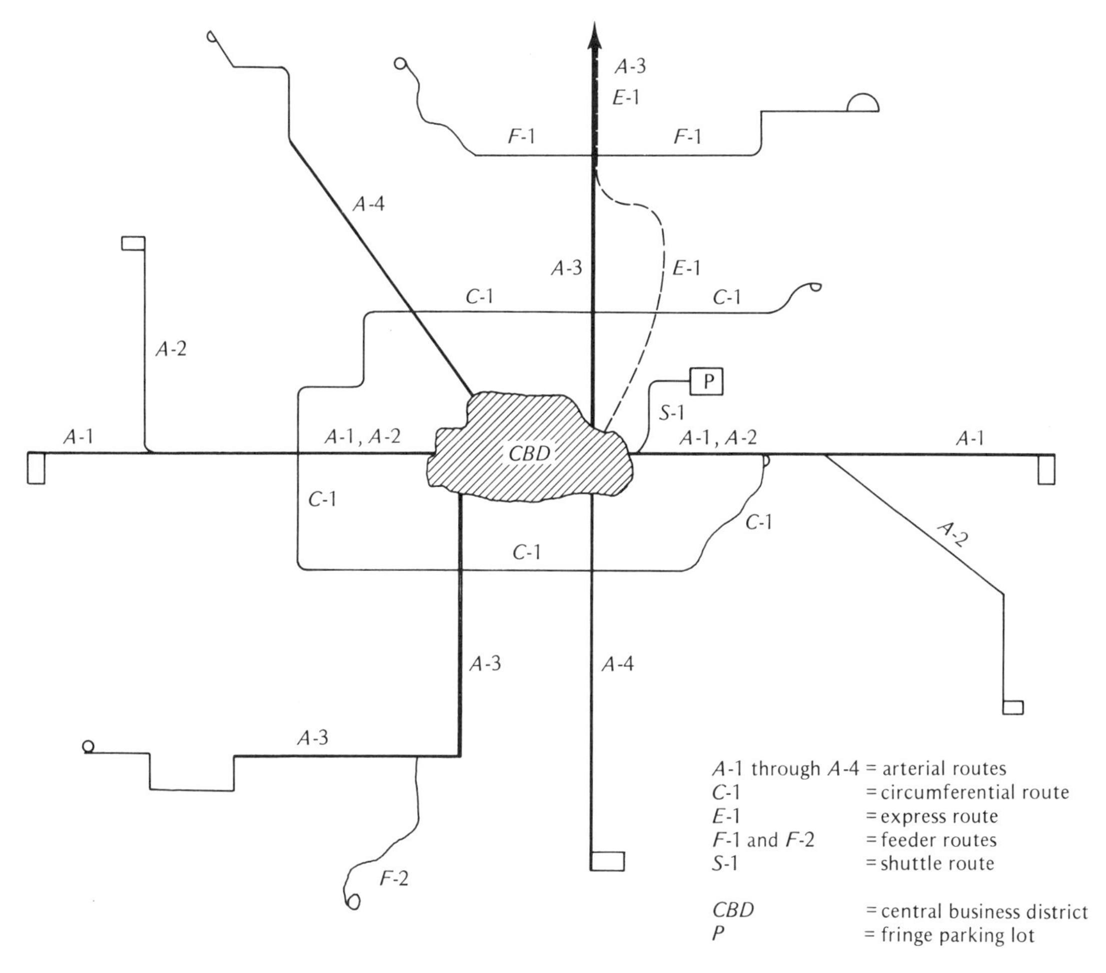

FIGURE 12-1
SCHEMATIC TRANSIT NETWORK

control. Operating reliability increases considerably on a *semi-exclusive* right-of-way with only minor capital expenses; however, in the case of bus lanes taken from other traffic there may be a considerable increase in congestion for motor vehicles in the remaining lanes. *Exclusive* rights-of-way involve a quantum jump in capital expenditure and are, therefore, not often justified. However, they are the only type of transit facility over which operators have complete control, allowing them to select operating parameters and giving them freedom from external sources of delay or disruption.

Rapid transit systems are constrained to exclusive right-of-way. While this offers the opportunity for maximizing reliability, it limits route flexibility. Buses are the only current class of vehicle that can freely move from shared to semi-exclusive or exclusive right of way and whose infrastructure can be improved in small increments.

Access Points

Every transit trip involves an access segment between origin and the system and another between the system and the destination, as well as the transit segment itself. In striving to minimize the total trip time, trade-offs must be made between optimizing the access segments and the transit segment. This is done by the manner in which access points—stops and stations—are located.

On a single route, frequent stops will minimize access times but increase the travel time of the vehicles, and vice versa. Toward the outer terminals of a route, accessibility may be enhanced by splitting the line into branches; however, the service frequency on each branch will necessarily be less than on the basic route, and what is gained in reducing access time will be partially or wholly offset by less convenient schedules. Route extensions and new

TABLE 12-1
CLASSIFICATION OF TRANSIT ROUTES BY TYPE OF RIGHT-OF-WAY

RIGHT-OF-WAY	GUIDEWAY	VEHICLE	COMMENTS
Shared	Street	Bus, trolley coach, streetcar	Basic local transit service, subject to traffic interference.
	Freeway	Bus	Radial express routes.
Semi-exclusive	Bus lane on street	Bus	In downtown and on its approaches. Shared with turning vehicles. Contraflow lanes on one-way streets.
	Bus lane on freeway	Bus	On radial freeways for express routes. Shared with car pools.
	Medians in Street	Streetcar	On radial boulevards outside downtown for "semimetro" service.
Exclusive	Bus lane	Bus	On radial freeways; also occasionally exclusive roads or ramps. Express routes.
	Tracks or other fixed guideway	Trains of vehicles	Rapid transit service for longer trips on radial routes.
	Tracks	Streetcar	Downtown distribution system of "semimetro."

routes into previously unserved territory will improve accessibility unqualifiedly.

Area coverage or accessibility is, therefore, a function of the spacing of both routes and stops. Standards, such as those shown in Table 12-2, can be used to design adequate transit networks. These standards are usually modified to provide better accessibility (closer spacing of routes) where terrain hampers walking, and they do not apply to special services for older or handicapped persons.

LOCAL TRANSIT SYSTEMS

Six technologies appear in the history of local transit, and five of these have largely or completely disappeared. The first—the horse omnibus—and the last—the bus—took full advantage of streets as they existed without requiring their modification and, therefore, without investment in the infrastructure of the network. The horse omnibus, however, languished because of the poor condition of nineteenth-

TABLE 12-2
ACCESSIBILITY STANDARDS FOR URBAN TRANSIT

POPULATION DENSITY (THOUSANDS/MI^2)	AVERAGE SPACING (MILES)		ROUTE MILES PER MI^2
	Radial Routes	Crosstown Routes	
Over 12	0.40	0.60	4.00
10-12	0.50	0.75	3.33
8-10	0.60	0.90	2.67
6-8	0.80	1.20	2.00
4-6	1.00	1.50	1.67
2-4	1.00	. . .	1.00
Under 2	2.00	. . .	0.50

SOURCE: Adapted from: Massachusetts Bay Transportation Authority, *Service Policy for Surface Public Transportation.* Boston, Mass.: 1975.

century paving, and therefore was replaced by the horse car, which solved the problem by using tracks embedded in the paving stones or rising above the dust and mud. The cable car was invented for routes too steep for the horse, and electricity made the streetcar vastly more efficient than horsepower. However, these three rail-bound technologies lacked flexibility. So, as soon as internal combustion engines were developed for heavier vehicles, buses made their appearance. An electric version, the trolley coach, was competitive against gasoline buses, even though it, too, was confined to those streets above which power distribution lines had been strung. Finally, the development of diesel buses offered sufficient advantages of efficiency, flexibility, and minimal capital cost that these have become predominant in local transit.

It is important to understand this technological situation at the outset of a discussion of local transit, because it defines the character and potential of such systems. Bus systems can utilize the ubiquitous network of streets, excepting only very narrow, steep or dead-end streets, although environmental constraints may reduce the options. Experimental routes are easily installed and, if unsuccessful, as easily abandoned. Vehicles of different dimensions and performance specifications can be mixed in the fleet of one system.

However, the disadvantages of transit operations on shared right-of-way, mentioned earlier, must be weighed against these advantages. Also, as will be shown later, the capacity of buses, trolley coaches, and streetcars in a corridor is somewhat below that of rapid transit trains.

Routes

Local transit networks include most or all of the types of routes illustrated in Fig. 12-1. Most commonly, these routes operate in mixed traffic—shared right-of-way—on suitable streets and freeways. For environmental reasons, transit routes are generally confined to major arterials and collector streets. However, to serve outlying neighborhoods adequately, some residential streets may have to be used.

Expeditious transit service is not only attractive to potential riders but also reduces operating costs for the transit agency. Where alternate routes are under consideration, the one with smoothest traffic flow may be chosen. Special traffic engineering measures can enhance transit operations: signals can be timed to favor continuous transit flow or, in special circumstances, can be modified with a preemption capability that allows the transit vehicle to demand a green indication on approaching. Buses are often excepted from turn prohibitions applying to other traffic.

The next level of route quality involves semi-exclusive rights of way in the form of bus lanes or streetcar reservations. Bus lanes along the curb are most beneficial when this lane has been previously used for parking or loading. However, where a curb lane is already subject to "No Stopping" regulation, and where buses are scheduled in sufficient quantities to warrant a bus lane, other traffic tends to avoid the lane in any case. If a lane is designated for transit only, other vehicles wishing to make right turns are permitted to enter the lane just ahead of the intersection.

Transit lanes in the center of the street are generally used for streetcars, although buses are sometimes also routed on them. Other traffic is kept out by paint markings or raised curbs, and often is prohibited from making left turns anywhere along the street. At transit stops special pedestrian islands must be fitted into the street cross section.

In one-way street systems it is sometimes difficult to route transit so that important traffic generators or transfer points are served in both directions. One solution is a contraflow transit lane in streets of adequate width. Parking must be eliminated on what was the left side of the one-way street, and a lane of about 14 ft (4.25 m) delineated by means of standard markings and supplemental signs. Taxis and emergency vehicles are often also permitted to travel in these lanes.

On freeways, buses can be given semi-exclusive rights of way in median lanes, provided that no left-side ramps occur. Such lanes can operate either with or against the general traffic flow. In either case, buses must weave across the outer freeway lanes when entering and leaving their special lane, and within it are not separated from other traffic by physical barriers. Car pools are generally allowed to share bus lanes that move with the traffic flow to reduce the adverse impact caused by the capacity reduction for general traffic and to encourage multioccupancy use of automobiles. Contraflow lanes are only rarely feasible. Traffic demand in the nonpeak direction must be low enough so that one lane—preferably two to include a buffer lane—can be assigned to the contraflow operation. Cuts must be made in the median where buses enter and leave the contraflow section. At the beginning and end of each peak, traffic cones or other devices must be installed and removed.

Bus roadways are the most elaborate form of right-of-way for these vehicles. They may be assigned exclusively to buses or shared with car pools. The El Monte Busway in Los Angeles is such a facility, complete with two bus stations en route and special access ramps (Fig. 12-2). The center roadway of the Shirley Highway outside Washington, D.C. was originally intended as a reversible freeway and is used in this manner by buses and car pools—inbound in the mornings and outbound in the afternoons and evenings (Fig. 12-3).

Stops and Terminals

Transit stops are commonly provided at the sidewalk curb so that waiting, boarding, and alighting passengers are separated from moving traffic. Stops must be designated by informative signs to guide occasional riders. If headways between vehicles exceed about 10 minutes, some comfort features such as benches or shelters are important. Shelter walls offer additional space for transmitting information from the transit management to the passenger, such as maps and schedules. The curb face at bus stops is painted with the appropriate markings to indicate parking and stopping prohibition for motor vehicles.

Stops may be located either in the approach ("near side") or the exit ("far side") of an intersection or in midblock locations. The choice is made on the basis of turning movements of buses at the intersection, major turning movements of other traffic, the status of the cross street, transfer flows, if any, and the location of very large traffic generators.

Near-side stops are preferred where more traffic joins the street than turns off it, at intersections with one-way streets moving from right to left, and at locations where the buses will make a right turn. Far-side stops are suggested where there are heavy turning movements off the street, at intersections with one-way streets moving from left to right, and where buses make left turns. However, the final choice also depends on signal timing, the relation of stops of different routes meeting at transfer points, and the adequacy of the sidewalks at the alternate locations under consideration.

Midblock stops are more rarely used, and never at major transfer points. They may be justified in long blocks at the center of which major traffic demand is generated, or where insufficient locations at adjacent intersections exist. The latter can happen if the number of buses per hour exceeds the capacity of a single stop, and duplicate stops for some of the routes are needed.

Minimum lengths of curb bus stops are given in Table 12-3. The need for stops holding more than one bus at a time can be calculated from the capacity formula given later. However, the capacity of the second position at a stop is only about 75 percent and of the third position, 50 percent of the capacity of the front position, since the exit from these additional loading locations may be blocked by vehicles stopped in front.

Occasionally special bus stops are needed in freeway rights-of-way. Many express routes use freeways only as fast connections between downtown and outlying areas and are not designed to provide service along the freeway. However, some routes to outer suburbs, where low density of demand makes branch routes unfeasible, may have need for freeway bus stops. Additional locations for such facilities occur where freeway routes intersect important circumferential routes, usually near the downtown perimeter.

Where diamond interchanges exist, bus operators prefer to have buses leave via the off-ramp, stop at the intersecting street, and then reenter the freeway. This minimizes walking distances for passengers. At more complex interchanges, where buses would lose considerable time in reaching the street, special stops with approach and exit lanes and pedestrian path connections to the intersecting street are fitted into the interchange layout [Ref. (1)].

Off-street bus stops range from small turning circles and loading areas at the ends of routes—often remnants from streetcar operations—to large transportation terminals. Probably the largest in the world is the Port Authority Bus Terminal at the Lincoln Tunnel in New York City (Fig. 12-4), which handles about 7000 buses with 200,000 passengers on a typical weekday, and has an hourly capacity in one direction of about 750 buses and 33,000 passengers. Design of bus terminals is discussed in Chapter 9.

Vehicles

As already mentioned, the predominant local transit vehicle is the diesel bus. Since it moves on public streets, it must conform to the maximum dimensions and weight limits imposed on all motor vehicles by state laws. In most states these maxima are:

- Length 40 ft (12.2 m) for single vehicles
60 ft (18.3 m) for articulated buses
- Width 8.5 ft (2.6 m)
- Height 13.5 ft (4.1 m)—only of concern for double-deck buses

FIGURE 12-2
EL MONTE BUSWAY, LOS ANGELES (COURTESY OF THE CALIFORNIA DEPARTMENT OF TRANSPORTATION, DISTRICT 7) *PHOTO BY HAL SPRING*

FIGURE 12-3
BUS ROADWAY IN SHIRLEY HIGHWAY

TABLE 12-3

MINIMUM DESIRABLE LENGTHS FOR BUS CURB LOADING ZONES

TYPE OF STOP	LENGTH OF LOADING ZONE[a]			
	One-bus Stop		Two-bus Stop	
	ft	m	ft	m
Near Side[b]	$L + 65$	$L + 20$	$2L + 70$	$2L + 21$
Far Side[c]	$L + 40$	$L + 12$	$2L + 40$	$2L + 12$
Midblock[c]	$L + 100$	$L + 30$	$2L + 100$	$2L + 30$

Note: L = length of the longest buses using the stop.
[a] Based on side of bus positioned 1 ft (30 cm) from curb. If bus is to be positioned 6 in. (15 cm) from curb, 20 ft (6 m) should be added to near-side stops, 15 ft (4.5 m) to far-side stops, 35 ft (11 m) to midblock stops.
[b] Add 15 ft (4.5 m) where buses are required to make a right turn, or 30 ft (9 m) if there is also heavy right-turn volume of other traffic.
[c] Based on roadways 40 ft (12 m) wide; add 15 ft (4.5 m) to length if roadway is only 32 ft (10 m) wide. These dimensions allow bus to leave loading zone without passing over street centerline.

SOURCE: Based on Highway Research Board, *Highway Capacity Manual*. Special Report 87. Washington, D.C.: Transportation Research Board, 1965, Table 11.8.

- Weight 20 short tons (18,100 kg)—2-axle bus
  30 short tons (27,200 kg)—3-axle bus

The interior furnishings of buses is determined by finding the optimum trade-off between capacity and comfort. Special consideration is given to the trip lengths to be served; where most trips are over 30 minutes in length, most or all passengers should be seated; on short shuttle routes, very few seats need be provided. Comfort is maximized—and capacity reduced—if seats of ample dimensions are installed, standee space is reduced to little more than the aisles required to reach the seats, and only one door is provided, since space next to doors cannot be used for seats. (One or more emergency doors may also be required, but seats may be placed next to these.) Conversely, if capacity is crucial, fewer and smaller seats will be specified, additional doors will be provided in the rear of the vehicle, and wider doors will be called for to speed up loading and alighting at stops.

FIGURE 12-4

MANHATTAN BUS TERMINAL OF THE PORT AUTHORITY OF NEW YORK AND NEW JERSEY (ARCHITECT'S RENDERING COURTESY OF THE PORT AUTHORITY OF NEW YORK AND NEW JERSEY)

In smaller bus systems it is generally advantageous to acquire a single bus type suitable for all routes. Larger systems can afford specialized high-capacity buses for short routes and high-comfort vehicles for service to more outlying areas. Specific examples of bus capacities will be discussed later (Table 12-4).

Except for motive power, trolley coaches have similar characteristics to buses. Streetcars, however, have usually been exempted from motor vehicle size and weight limits. The standard PCC car built in the United States and, by license, in other countries, is 46.5 ft (14.2 m) long and 9 ft (2.75 m) wide. Dimensions of articulated streetcars are found in Table 12-4.

Auxiliary Facilities

Local transit systems require a number of supporting facilities. Vehicles must be stored when not in use and must be maintained and serviced. Employees must be dispatched, supervised, and trained. A central staff, which administers, plans, and schedules the operation, must be accommodated.

The most important auxiliary facility is the vehicle yard and shop area. In small systems, only one such installation is needed. However, as networks get larger, servicing all routes from a single location requires excessive amounts of "dead-head" running between the yard and some of the routes. Experience also indicates that yards holding more than 400-500 vehicles become unwieldy. The storage function requires considerable land; hence yards are located in areas of low land values and, for environmental reasons, in industrial areas if possible. Since many buses are idle in the middle of weekdays after bringing commuters downtown, any space that can be purchased or leased near the CBD, such as under elevated freeways or rapid transit lines, is valuable.

Besides bus parking, each yard includes a transportation building at which operators report, are assigned vehicles, are dispatched, and are checked in at the end of shift. Space is also provided for extra operators who are standing by to replace sick personnel or to run special trips. Routine servicing facilities at each yard include fuel tanks and pumps, bus washers, lubrication lanes with inspection pits, and a strong vault for storing fare boxes. One yard in larger systems is designated for heavy maintenance and will include tools for rebuilding engines, transmissions, and other vehicle components. In small bus operations, heavy maintenance may be contracted to an outside firm.

The operating headquarters may be located at one of the yards or in a separate office building. Central administration includes supervision, planning, scheduling, financial control, accident claims, marketing, labor relations, and similar functions. Many modern local transit fleets are equipped with two-way radios, requiring a central control room as another supporting facility.

SEMIMETROS

In the period from 1920 to 1960 many cities abandoned streetcar systems. In the largest metropolitan areas of the world they were replaced by rapid transit, and in small towns by buses. However, a group of cities, mostly in the population range of ½-2 million, found no ready alternative, causing some of them to upgrade their networks to *semimetros* (also inaccurately called "light rail transit") by providing exclusive rights-of-way underground in the city center. In Amsterdam, for example, sections remaining on the surface in outer areas are sometimes placed in reconstructed medians of wide streets or relocated on exclusive surface alignments in connection with the development of new suburbs.

Although semimetros do not offer the maximum capacity and reliability of rapid transit, they permit inclusion of portions of existing streetcar trackage at little or no cost. Extensions on the surface are relatively inexpensive, since stops are simple, route junctions are at grade, and grade crossings with streets are allowed. It is, therefore, feasible to retain an elaborate and accessible network at a cost far below that of a new rapid transit project.

RAPID TRANSIT SYSTEMS

The first urban railroad on exclusive right-of-way was the Metropolitan Railway of London, opened in 1863. By 1920 most of the major cities of the world had built subways and/or elevateds. After a 40-year lull, during which only a few cities, including Toronto and Cleveland, inaugurated rapid transit systems, new projects opened in Montreal, the San Francisco Bay Area, Washington, D.C., Mexico City, and São Paulo, all in the Americas, in less than 10 years. Since about 1960 there have also been major new systems or extensions to existing networks in many other metropolitan areas around the world.

Routes

Almost by definition, rapid transit systems require exclusive rights-of-way. The horizontal alignment of routes is generally radial, and is more precisely determined by the optimum station locations, which in turn are decided by accessibility to major traffic generators in the route corridor and to the tributary areas via the street system and other transit lines. In areas of high land values, routes are located as much as possible on land that is already owned publicly, namely below or above streets.

The decision on vertical alignment is made primarily so that the sum of land, construction, and environmental costs is minimized, although irregular topography also plays a role. Construction costs are least at grade, but this requires permanent occupancy of a strip of land varying from 40 to 100 ft (12 to 30 m) in width, and building of grade separations for all important cross streets. It is, therefore, generally feasible only in outlying areas, where land is inexpensive, or in freeway medians, where the cost of land and grade separations can be shared with the highway project. Elevated construction costs are perhaps twice as much as facilities at grade, but the land below can be used for a street, for parking or industrial uses, or even for linear parks. In densely built-up areas, including CBDs, the adverse impacts of noise and reduction of daylight militate against elevated construction. Another doubling, or even tripling, of costs occurs if underground alignments are considered. However, land costs are minimized by the use of street rights-of-way, easements under private property instead of outright purchase, and development of air rights over those parcels that have to be acquired for one reason or another. Environmental deterioration is minimized, although it must be recognized that underground travel is not as attractive to passengers as moving in daylight.

Most rapid transit systems in the world use steel-wheeled cars running on steel rails. For over two centuries (starting with horse-drawn mine railroads of eighteenth-century England) this has proved reliable and mechanically efficient. The use of rubber tires on wood or concrete tracks, supplemented with steel wheels and steel rails for switches and emergencies, was developed by the Paris Metro and is also used in Montreal and Mexico City. The rapid transit system of Sapporo, Japan involves rubber tires on concrete without the additional steel components.

Two commercially operating lines are monorails. The Wuppertal Schwebebahn, built in 1900, hangs from a single rail, the optimum solution for a location in a narrow valley, partly straddling a river, in days when steel technology was not yet far advanced. The Tokyo Airport line is, strictly speaking, a monobeam, since the cars straddle a longitudinal concrete beam. All other urban transit technologies put forward in recent years—other forms of monorail and air cushion cars—have not been employed in full-scale commercial versions.

Stations

Rapid transit stations are major civil and transportation engineering projects. The main components—platforms for boarding and alighting passengers, stairs, escalators, elevators, and passageways for circulation, fare collection installations, and supporting space and equipment—are designed to provide adequate capacity for movement and accumulation of peak traffic flows for the design year. Connections to the street system, parking lots, and, perhaps, the basements of adjacent buildings are important factors affecting accessibility. Design inputs include some of the human characteristics described in Chapter 3. Figure 12-5 shows a rail transit station with a large parking lot to allow access by automobile.

The first generation of transit systems, generally confined to the denser parts of cities, were designed to relieve surface streets of heavy horse omnibus and streetcar traffic. In order to attract fairly short trips, stations were spaced on an average one-half mile (800 m) apart. No special provisions were made for any but pedestrian access. In many cases it was possible to place all facilities under the street, and little or no private land had to be utilized.

By contrast, systems designed since 1950 extend into suburbs with the intention of serving longer trips. Station spacing is longer—an average of 1 mile (1.5 km) in many networks and even more in some. Suburban stations provide space for off-street bus stops for feeder lines, parking areas for up to 2000 cars, and space for dropping off and picking up auto passengers. Major land acquisition is involved. Such stations are bound to have an impact on their environment because of their size and the traffic generated by them.

Vehicles and Trains

Since rapid transit vehicles operate on facilities especially designed for them, their dimensions and performance characteristics are integrated with the design of the fixed facilities to produce the desired capacity and average speed. The important vehicle parameters are length, width, and height of single cars, the maximum length of a train, top speed capability, and maximum rates of acceleration and deceleration.

The choice of width and height bear directly on the required cross section of tunnels and underpasses, as well as

FIGURE 12-5
RAIL TRANSIT STATION WITH PARKING AREA FOR PATRONS—ORINDA STATION, BART SYSTEM, SAN FRANCISCO, CALIF. (COURTESY OF TUDOR ENGINEERING CO., SAN FRANCISCO, CALIF.)

on the lateral and vertical clearances above ground. The length of a single car, together with the placement of the trucks on which it rides, defines the amount of overhang at curves and the concomitant need for extra clearance between tracks and between cars and tunnel walls or other lateral obstructions. These constraints are traded off against the weight advantages of longer cars.

The total train length, which together with the vehicle width determines capacity (discussed later), fixes the length of stations. For a given capacity, a wider, shorter train would increase tunneling costs but decrease station costs, and might be chosen for a network with little underground mileage. Conversely, a narrower, longer train might be selected if costs of underground construction are major items, while somewhat higher station costs can be tolerated.

Maximum rates of acceleration and deceleration are limited to 3 mph/sec (5 kmh/sec) to safeguard standees. Top speed capability is a function of this constraint and of station spacing. For example, between stations one-half mile apart a train cannot exceed 62 mph (100 kmh) at acceptable rates of change of speed. When all station spacings and the longer sections of climbing grades are analyzed, it is generally found that rapid transit trains need not have a maximum speed capability in excess of 70–80 mph (110–130 kmh).

The interior arrangement of trains is subject to the same policy decisions concerning comfort and the same capacity analysis that has been described for buses. The main difference is that, since fares are not collected on trains, more and wider doors can be provided. In conjunction with high platforms, which eliminate steps for boarding and alighting, such doors minimize the time during which trains have to stop at stations and contribute to maximizing the capacity of the route.

Auxiliary Facilities

Besides normal administrative facilities, rapid transit systems are supported by yards for storing trains, shops for maintaining them and the fixed facilities, and a control center.

Every transit line needs access to a yard with which is usually combined a routine maintenance shop. Because of the limited maneuverability of rail vehicles (minimum curve radius, maximum gradient), yards are best located on a parcel of land that is long and relatively narrow. If a site cannot be located adjacent to the line, a connecting link must be built at extra cost that is not productive in terms of passenger service or revenues. Since employees begin and end their workday at yards, a transportation building is provided. This houses the dispatcher's office, crew rooms, and lockers. Routine maintenance facilities are located in a separate building with inspection pits under the tracks and access to the necessary tools and machines. A train-washing machine is also needed.

Heavy maintenance, including major motor overhauls and wheel grinding, is usually concentrated at just one point in the system. If there are several separate routes in the network, track connections must be made so that all vehicles can be serviced at the heavy maintenance shop when needed. Should this prove unfeasible, the cars may have to be moved over the road on flatbed trucks, or duplication of heavy maintenance capability may be necessary.

Roadbed and track maintenance is also concentrated at a single point, if possible, with track connections between otherwise separate lines of a network also serving maintenance trains. This activity is often, but does not need to be, combined with a yard and shop area. Essential components include spur tracks for parking maintenance equipment, materials storage areas, and access to the site for delivery of these materials, preferably by railroad spur.

A rail yard with heavy maintenance shop is shown in Fig. 12-6.

The degree of train control has increased greatly during the history of rail transit. Manually operated signals were succeeded by centralized traffic control, and, more recently, by semiautomatic or fully automatic train movement control, often involving computers. However, an extended period of debugging automated systems, during which time service interruptions or minor degradation of service quality can occur, must be anticipated. A control center is usually located in or near the administrative headquarters. Here supervisors receive information on the current status of the system from display boards, computer printouts, and telephone contact with train operators.

URBAN RAILROADS

In some large metropolitan areas, such as New York, London, Paris, and Tokyo, a large burden of urban transportation is borne by intercity railroads. Many of the lines involved were located for intercity travel and, in generating urban land uses along their corridors, found themselves serving increasing numbers of local trips. Up to about 1920 some railroads added routes in the suburbs, where they found sufficient demand. However, since that time, many routes have been abandoned because of the high costs of railroad operation and the reluctance by managements to allow long-distance passenger and freight shippers to subsidize urban transportation.

A major problem of urban railroad services is the location of the central terminals. There is almost always only one downtown access point per corridor, and this is often located at the edge of or outside the CBD. Typically, it is a stub terminal with inherent limitations to the number of train movements that can be handled per hour.

In some cities with extensive electrified suburban lines, such as Brussels, Munich, and Frankfurt, these drawbacks have been overcome by connecting terminals located on opposite sides of the CBD with underground tracks and one or more intermediate stations for direct access to the city center. Capacity was increased because trains no longer need be reversed at crowded terminals. In Paris, suburban rail lines were diverted before reaching their old terminals into a new network of lines across the inner city. The resulting operations in these cities are a mixture of rapid transit in the city center and urban railroad outside it.

PARATRANSIT

Paratransit may be defined as those forms of mass transit that do not fully meet the definition of common carriers given at the beginning of this chapter. Some do not follow fixed schedules, such as the jitney services common in many countries of Latin America and Asia. Others follow neither schedules nor fixed routes, including taxis and demand-responsive buses. The published tariff does not apply to "club bus" and van pool arrangements, which also are available only to a restricted clientele instead of to all interested travelers. Charter travel meets none of the common-carrier definitions.

FIGURE 12-6
RAIL TRANSIT YARD AND MAINTENANCE SHOPS—SOUTH HAYWARD YARD, BART SYSTEM, SAN FRANCISCO, CALIF. (COURTESY OF TUDOR ENGINEERING CO., SAN FRANCISCO, CALIF.)

The contribution made by jitneys in large cities of developing nations is very large. In Caracas, for example, one-third of all urban transit trips are made by "por puestos," and in Hong Kong the proportion is even higher. In the United States, jitneys were forced out of business (except in Atlantic City and San Francisco) by transit companies in the years before the Second World War.

Taxis provide a chauffeured service equivalent in comfort and speed to the private car, but at considerable expense. However, it is shown in Washington, D.C. that fares can be reduced considerably if shared taxi riding is an available option. The utilization of taxis in this manner deserves study in other cities.

Demand-actuated bus programs—also called "Dial-A-Bus"—have had mixed success. They provide door-to-door service in minibuses within sectors of urban areas in response to calls received by a central dispatcher or to standing orders placed for days and weeks in advance. They have been quite successful in Canada, especially in cities with cold winters when walking to and waiting for a regular bus is not an attractive prospect. In Stratford, Ontario, a demand-actuated service takes over from regular route service after 6 P.M. weekdays and all day on weekends. This has resulted in a reduction in the financial losses that had occurred during these hours. In the United States, Dial-A-Bus has usually been superimposed on normal service and has run in competition with it. Also, fares have often been at the same rate as those of the standard service, even though a premium fare might be justified. As a result, revenues have covered as little as 10 percent of operating costs, and several projects have had to be abandoned.

Club buses are organized by employers or employees working in areas not well served by normal transit routes, provided one or more bus loads of riders commute along the same corridor. Fares are paid weekly or monthly, and the bus is scheduled along a fixed route convenient to the riders. If the number of commuters desiring such service is too small to make a club bus feasible, the employer may organize van pools by purchasing vans and leasing them to employees, who assume responsibility for taking groups of fellow workers to and from work. Each rider pays a weekly or monthly fee either to the driver or to the employer, depending on the financial arrangements.

Charter service permits a group of persons making a nonrepetitive journey to avail themselves of one or more vehicles of a common carrier. In urban transit, chartering is confined to buses, except for a rare occasion when a group of rail fans may wish to charter streetcars or a rapid transit train.

TRANSIT SYSTEMS IN MAJOR ACTIVITY CENTERS

Major activity centers, such as airports, regional shopping centers, university campuses, and large amusement parks, experience a unique form of demand for transit service, characterized by large quantities of trips in the range of one-half mile to little more than a mile. A variety of specialized short-haul transit systems has been developed to meet this need. Examples are found primarily in airports and amusement parks, because financing has been more readily available at such locations. Central business districts are a form of major activity center, and the U.S. government is considering the feasibility of short-haul systems that would improve circulation in downtown areas.

Most of the systems in this category—known by a variety of trade names—operate on exclusive right-of-way. A simple version is a "horizontal elevator"—a single vehicle shuttling on a track between two or more stops at an airport. Slightly more elaborate is a loop system with several vehicles operating in a single direction. Complex systems, such as the one installed at the Dallas-Fort Worth Airport, consist of complete networks with switching capability, and with cargo as well as passenger vehicles. All systems are electric, but some have motors on the vehicles, whereas others are propelled from the track bed. Because of their scale, they operate at only moderate speed and use vehicles of only medium capacity.

OPERATING PARAMETERS

The performance of a transit system is determined by a number of parameters, fixed by the design of system components, by current service criteria, by the system environment, or by a combination of these. Capacity and travel speed may be considered physical characteristics not subject to day-to-day adjustment by transit management. Level of service (the quantity of space actually offered), comfort, convenience, frequency of service, reliability, and safety from accidents and crime can be partially or totally controlled by operating criteria and their execution.

Capacity

The capacity of a transit route is the product of the passenger capacity per vehicle or train and the maximum number of vehicles or trains that can travel on the route. The last term almost always is the capacity of the busiest stop or station on the line.

The trade-off between vehicle capacity and comfort has already been discussed. Table 12-4 gives examples of such capacities, using comfort levels specified either by the vehicle manufacturer or the buyer.

An approximate formula for the capacity of a bus stop is

TABLE 12-4
EXAMPLES OF TRANSIT VEHICLE AND TRAIN CAPACITIES

TYPE OF VEHICLE OR TRAIN	LENGTH		WIDTH		TYPICAL CAPACITY[a]			REMARKS
	ft	m	ft	m	Seats	Standees	Total	
Minibus—short haul	19.5	5.95	7.7	2.35	18	12	30	
Transit bus	30.0	9.15	8.0	2.45	36	19	55	
	35.0	10.65	8.0	2.45	45	25	80	
	40.0	12.20	8.5	2.60	53	32	85	
Articulated transit bus	54.1	16.50	8.2	2.50	48	124	172	European model
	60.0	18.30	8.5	2.60	69	41	110	U.S. specifications
Streetcar train	140.0	42.70	9.0	2.75	177	198	375	3-car P.C.C. train,[b]
Light rail car train	220.0	67.00	8.9	2.70	204	366	570	3-car train, 6-axle car, U.S.
	170.0	51.80	7.7	2.35	128	372	500	2-car train, 8-axle car, Europe
Rapid transit train	605.0	184.40	10.0	3.05	500	1,700	2,200	10-car train, New York IND.
	600.0	182.90	10.3	3.15	616	2,000	2,616	8-car train, Toronto
	700.0	213.35	10.5	3.20	720	1,280	2,000	10-car train, BART, San Francisco

[a]In any transit vehicle the total passenger capacity can be increased (and passenger comfort decreased) by removing seats and making more standing room available, and vice versa.
[b]Presidents' Conference Cars.

SOURCE: Wolfgang S. Homburger, *Notes on Transit System Characteristics.* Berkeley, California: University of California, Institute of Transportation Studies, Information Circular 40, 1975.

$$N_{max} = \frac{1800}{D} \quad \text{or} \quad H_{min} = 2D \qquad (12.1)$$

where

N_{max} is the maximum number of buses per hour.
H_{min} is the minimum headway between buses in seconds.
D is derived from one of the following formulae:

$$D = aA + bB + C \quad \text{for two-way flow (busiest door)} \qquad (12.2)$$

$$D = \max\begin{Bmatrix} aB \\ \hline bB \end{Bmatrix} + C \quad \text{for one-way flow through doors} \qquad (12.3)$$

where

a = average alighting time per passenger.
A = total number of passengers alighting.
b and B are similar terms for boarding passengers.
C = clearance time—the time lost in opening and closing doors, or to traffic delays when ready to leave.

Values of a and b vary with the number of stairs to be negotiated, the type of fare transaction, if any, and whether doors are designed for single file or two-abreast movement. Typical values of a for a single file are 1.5–2.0 seconds. In standard situations with fares collected by the driver, values of b range from 2.5 to 3.0 seconds; however, this value is reduced if a substantial proportion of the passengers is using prepaid passes. If fares are collected at a point before the bus is boarded, a value of 2.0 seconds can be assumed for b. The value of C must be measured at the bus stop being analyzed, since the main component is the delay when waiting for green lights at near-side stops or for a break in traffic at far-side stops. At far-side stops in bus lanes C approaches zero.

In major bus terminals, all passengers alight or board. Since buses must also await their scheduled departure time, an additional component is added to the clearance time. Hence, the average dwell time (D) at such terminals is at least 5 minutes.

On rapid transit routes a similar analysis can be made, using formula (12.2) to calculate the dwell time. Because of high-level platforms, values of a and b are about 1 second per file. The clearance time includes the opening and closing of doors, including delays caused by passengers holding doors open when the train has been cleared for departure by the control system. The capacity of the line through this station is computed from:

$$N_{max} = \frac{3600}{D + S} \quad \text{or} \quad H_{min} = D + S \qquad (12.4)$$

where S is the minimum separation between succeeding trains in seconds, as determined by the signal or train control system, and where N and H have similar meanings, as in Eq. (12.1).

Some counts of actual traffic on transit routes operating at capacity are listed in Table 12-5.

Speed

Average speed along a route is a function of stop spacing and either the top speed capability of the vehicle or train or the speed limit imposed by law or physical conditions. It was stated in our discussion of networks generally that choice riders seek the shortest total travel time door to door, and that maximizing the speed of transit routes at the cost of accessibility is not necessarily advantageous. However, if accessibility remains about the same, increases in speed are beneficial to users. Transit management also benefits from higher speeds, since it is able to utilize vehicles and staff more productively.

Take, for example, a route 15 miles (24 km) long where the current average speed is 10 mph (16 kmh). Assume also that 15 minutes must be allowed at each terminal for recover and layover time, 8 minutes of this time for driver rest and 7 minutes to guard against late arrival

TABLE 12-5
MAXIMUM OBSERVED TRANSIT PASSENGER VOLUMES

TYPES OF TRANSIT OPERATION	LOCATION	PEAK HOUR FLOW			REMARKS
		Vehicles or Trains	Persons	Persons per Veh./Train	
Buses on streets	Hillside Ave., Queens, New York City	170	8,500	50	Reserved curb lane plus use of adjacent lane
	Market Street, Philadelphia	143	8,300	58	Reserved curb lane plus use of adjacent lane
	Washington Blvd., Chicago	108	3,800	35	Reserved lane in center of one-way street
	K St., N.W., Washington, D.C.	130	6,500	50	No reserved lane
Buses and streetcars on streets	Market St., San Francisco	155	9,900	—	Streetcars in center of street, buses in curb lanes
Buses on freeways and toll facilities	Lincoln Tunnel, New York	735	32,560	64	Preferential r.o.w. at entry
	Bay Bridge, San Francisco[a]	360	14,920	41	Preferential passage through toll plaza
	N. Lake Shore Dr., Chicago	80	4,000	50	No preferential r.o.w.
Buses on exclusive freeway lanes	I-495, New Jersey	490	21,600	44	Counterflow lane
	Shirley Hwy., Washington, D.C.	110	5,550	50	Separate roadway in median
Streetcars	Hannover Fair, Germany	40	18,000	450	Train length: 168 ft (51 m)
Rail rapid transit	IND Queens Line, NYC	32	61,400	1,920	Train length: 660 ft (201 m)
	IND Eighth Ave. Express, NYC	30	62,030	2,070	Train length: 600 ft (183 m)
	Yonge Street, Toronto	28	35,166	1,260	Train length: 480 ft (146 m)

[a]Some intercity buses also used this facility, but are not included in totals.

SOURCE: Institute of Traffic Engineers, *Capacity and Limitations of Urban Transportation Modes.* Washington, D.C.: 1965. Herbert S. Levinson, *Bus Use of Highways: State of the Art.* Washington, D.C.: NCHRP Report 143, 1973. Vukan R. Vuchic, *Light Rail Transit Systems.* Washington, D.C.: U.S. DOT Report DOT-TSC-310-1, 1972. University of California, Institute of Transportation and Traffic Engineering. *Traffic Survey Series A.* Semiannual.

because of traffic congestion en route. One vehicle will then complete a cycle in 210 minutes, and 42 vehicles will be required if demand calls for headways of 5 minutes. If by means of traffic engineering measures, such as a bus lane in the most congested part of the route, the average speed can be increased to 12 mph (19.6 kmh) and the reliability component at the terminals can be reduced to 4 minutes, the cycle time becomes 174 minutes, and only 35 buses will be needed to provide the same level of service. Perhaps the faster service will attract new patrons, so that the decision will be to reduce the assigned fleet to 39 buses and dispatch these at headways of 4½ minutes.

The example uses realistic average speed figures, as Table 12-6 shows. In very large cities peak hour speeds can drop to 5 mph (8 kph) and are seldom more than 10 mph (16 kmh) unless special provisions are made to expedite transit vehicles.

Frequency of Service

Frequency of service is determined by demand at the maximum load point along a route. However, certain minimum frequencies (maximum headways) are usually set as a matter of policy for lightly patronized lines. In larger metropolitan areas all routes operate at least twice an hour except late at night, and in smaller cities a frequency of one bus per hour is considered minimum.

To make coordination between different routes at transfer points feasible, base period (off-peak) headways are usually set as a multiple of either 7½ or 10 minutes on routes with relatively infrequent service. This also assists the public in memorizing schedules. Where demand requires 10 or more buses per hour, transfer coordination becomes less crucial, and exact headways are chosen to maximize vehicle utilization.

Fares

Transit operations are affected by the level and types of fares charged the passengers. Although the level of fares is ranked below travel time by choice riders in making modal decisions, it does have an effect. Captive riders respond in some degree to fare levels by adjusting their total travel. When taken together, price elasticity of both groups combined has for many years been measured at 0.25–0.33; i.e., for every one-percent increase in fares, a patronage reduction of ¼–⅓ percent can be anticipated. The process is not necessarily reversible, and fare reductions do not always bring corresponding increases in passengers. The relative importance of travel time to fares is sometimes illustrated when a fare increase coincides with an improvement in service. The result usually is an increase in traffic.

Two types of fares usually appear in a tariff. One type distinguishes between different groups of users—children, students, senior citizens, and the handicapped. A good

TABLE 12-6
TYPICAL AVERAGE TRANSIT SPEEDS

TYPE OF SERVICE	AVERAGE SPEED Peak mph	kph	Off-Peak mph	kph
Local bus on collector street—small city	10	16	12	20
—large city	5	8	7	11
Local bus on arterial street	10-11	16-18	13-15	21-24
Local bus on arterial reserved lane	15	24	17[a]	27[a]
Express bus on freeway	30	50	45	70
RAPID TRANSIT FACILITY	**AVERAGE SPEED mph**	**kph**	**STATION SPACING miles**	**km**
Yonge Street, Toronto, Canada	17.6	28.3	0.5	0.8
IRT Lexington Ave. Express, New York	19.6	31.5	1.0	1.6
IND 6th/8th Ave. Express, New York	24.5	39.5	1.3	2.1
Cleveland Rapid Transit, Ohio	28.0	45.1	1.2	1.9
BART Concord-San Francisco Route	43.0	69.0	2.9	4.7
BART Fremont-San Francisco Route	46.0	74.0	3.1	5.0

[a]Estimated

SOURCE: Reference (2), and data from BART, Oakland, Calif.

argument can be made against the requirement for the transit agency to bear the burden of discounts offered special groups for social reasons. In many countries it is accepted that education and welfare budgets should reimburse transit operators for these discounts. The other tariff stratification is by frequency of travel. Discounts for daily use of transit are common, even though such use occurs mostly in peak periods when the marginal cost of transporting the additional traffic attracted by such discounts is highest. An increasingly popular tariff component is the systemwide pass valid for a week or a month. This draws the attention of transit users away from the cost of the trip they are about to make, just as car drivers seldom consider the cost of gasoline consumed in a single trip.

In most countries in the Americas, a single "universal" fare is charged for trips of any length within a system. However, if the area covered is very large, it is sometimes divided into zones, and an extra charge is made for crossing each zone boundary. Other systems charge a nominal sum for the privilege of transferring between routes, on the assumption that riders using two or more routes travel greater distances than those who do not. However, this often rewards the transit agency for inefficient routing of its lines and penalizes those passengers who are offered an inferior service.

It has been mentioned that the manner of fare collection plays a role on dwell time at stops and, hence, in total travel time and stop capacity. The less fare collection takes place on the vehicle, the greater is the efficiency of the system. This can be achieved by several methods. Promotion of passes speeds up boarding, since the driver need only glance at the pass. If buses with wide doors are purchased, boarding can proceed in two files. At major stops and terminals, if an enclosed waiting area can be laid out, fares can be collected at the entrance to this area by turnstiles. A third method, used in some European countries, is to sell all types of tickets and passes at stores, banks, and vending machines installed at the more important stops of the network. Tickets are time-stamped just before use, and any or all doors of the vehicle can be designated for entrance. Drivers are relieved of all fare collection functions, although they and a team of roving inspectors make random checks for invalid or missing tickets.

Fare collection procedures on rapid transit systems must be efficient to minimize the number of collection machines needed in relation to peak flows passing through stations. Turnstiles operated by single coins or, if the fare cannot be paid by a single coin, by tokens, are most common. For systems that charge fares related to the trip length, such as London, Paris, and San Francisco, magnetically coded tickets have been introduced. These must be inserted in gates at both ends of the trip. On entry, the station of origin and the time are coded. The exit gate checks the entry information and, if the ticket contains sufficient funds, subtracts the cost of the trip. Passengers pass through gates at a rate of about 30 per minute. Near the exit gates an "Add Fare" machine is installed at which passengers can make up any deficit in their ticket balance.

Some work has been done to devise analogous ticket systems for buses, so that larger local transit systems could also charge distance-based fares or could collect zone fares automatically. However, the cost of furnishing the required equipment for each bus is not likely to be recovered by the extra revenue that a new fare system might produce and has been a deterrent to development of such a technology.

Other Operating Parameters

Convenience describes the ease with which the system can be used. It is enhanced by adequate channels of information between transit management and potential passengers and by keeping the various steps of using the system simple. Information can be imparted at bus stops, by publishing easily read maps and schedules, and by maintaining an efficient telephone information service. Simplicity is desirable in the tariff structure as well as in all aspects of network design and operation.

Reliability, ranked very high by passengers, can be adversely affected by service irregularities or vehicle breakdowns. The system operation is supervised to detect major schedule deviations and to order such adjustments as will return the system to normal conditions as quickly as possible. This can be done by field inspectors, by a central dispatch/control operation via two-way radio, or by automatic vehicle-monitoring devices. A well-planned and -executed vehicle maintenance program minimizes breakdown of vehicles while in revenue service or the cancellation of trips because of the lack of enough vehicles in operating condition to meet schedule requirements. Both breakdowns and missed trips reflect negatively on the user's perception of reliability.

Safety from accidents can never be completely assured. However, good driver training programs can reduce accident rates considerably, thus again contributing to reliability as well as to the positive aspects of the financial balance sheet.

Safety from crime is an environmental problem. Where a crime problem exists, a major cooperative effort between the transit management and the law enforcement authorities is required. Rapid transit systems generally must establish an internal police force to patrol trains and stations. On local transit, good communication between law enforcement and the central dispatcher—who receives emergency messages from drivers by radio or "silent alarm"—is essential.

MANAGEMENT ASPECTS AFFECTING SYSTEMS PERFORMANCE

A discussion of all aspects of managing transit systems is beyond the scope of this book. However, reference to certain major activities and problems is essential in understanding the capabilities and performance of urban transit. This includes coordination within an urban area among several systems, labor costs, and total financial resources.

Coordination

Only in rare instances, as in Washington, D.C., Pittsburgh, and St. Louis, have all urban transit services been merged into a single administration. More commonly, there are several major networks, mostly in public ownership, and perhaps some private lines. Arrangements among such systems for joint fares, reduced transfer costs, or even coordination of schedules are rare. Even when systems have been merged, vestiges of their previous independence still remain. Thus, passengers may be puzzled in New York as to why they can transfer between some routes of the major transit system freely, while having to pay a second fare when other routes are involved. Separate networks were merged at the top, but not at the level where passengers are concerned.

Mergers are often not possible. There may be physical reasons, such as the impossibility of combining the urban portion of large railroads with local transit management. More often, institutional barriers intervene. Directors of public transit agencies, responsive to a specific constituency, may wish to preserve the role played by their electorate and themselves. Labor union locals may not wish to merge. Individuals may fear for their jobs.

Coordination short of merger can be accomplished in several ways. As a start, the various transit agencies in an area can cooperate in preparation and distribution of a single regional transit map, and in coordination of their telephone information services. Further steps include an arrangement for joint fares or transfers between different systems, coordinated schedules where systems interchange passengers, joint planning of new routes, and agreements to eliminate route duplications. A final step, taken in several European metropolitan areas, is a single tariff with tickets valid on the routes of every carrier in the region. Long before these final steps are reached, it is necessary to create a formal organization to supervise this cooperative effort. Regional transportation planning agencies are showing increasing interest and concern in moving toward such solutions.

Labor Costs and the Peaking Pattern

Salaries and wages represent perhaps 75 percent of all transit operating expenses, and most of these accrue to operating personnel. Maintenance and office staff can be scheduled for straight shifts, although some shifts may have to be at night. However, because of the patterns of demand on weekdays, it is impossible to do the same for vehicle operations. Figure 12-7 shows the number of transit vehicles in service by time of day for weekdays. It can be seen that perhaps 40 percent of these vehicles can be staffed in straight shifts, but that the operators of the remaining vehicles are needed for a few hours in the morning and evening peaks only.

Labor unions, in striving for better working condition for their members, have obtained contracts that generally limit the proportion of all operators who may be assigned to split shifts. Also specified are the maximum number of unpaid hours between the two pieces of work of a split shift, the number of hours after reporting to work after which overtime is due, and the maximum "spread" or time between start and end of the shift. It is, therefore, quite common that operators on split shifts, whose productive time in terms of serving passenger demand is perhaps 3 to 4 hours, will receive basic wages for eight hours and overtime for two more, or the equivalent of eleven hours of wages.

Therefore, it becomes extremely important that vehicle schedules be refined to reduce the need for split shifts and overtime payments, and to meet all the conditions of the labor contract. Some of this work has now been computerized, but details in labor contracts vary widely and change often, requiring constant modification of procedures used to develop optimum schedules.

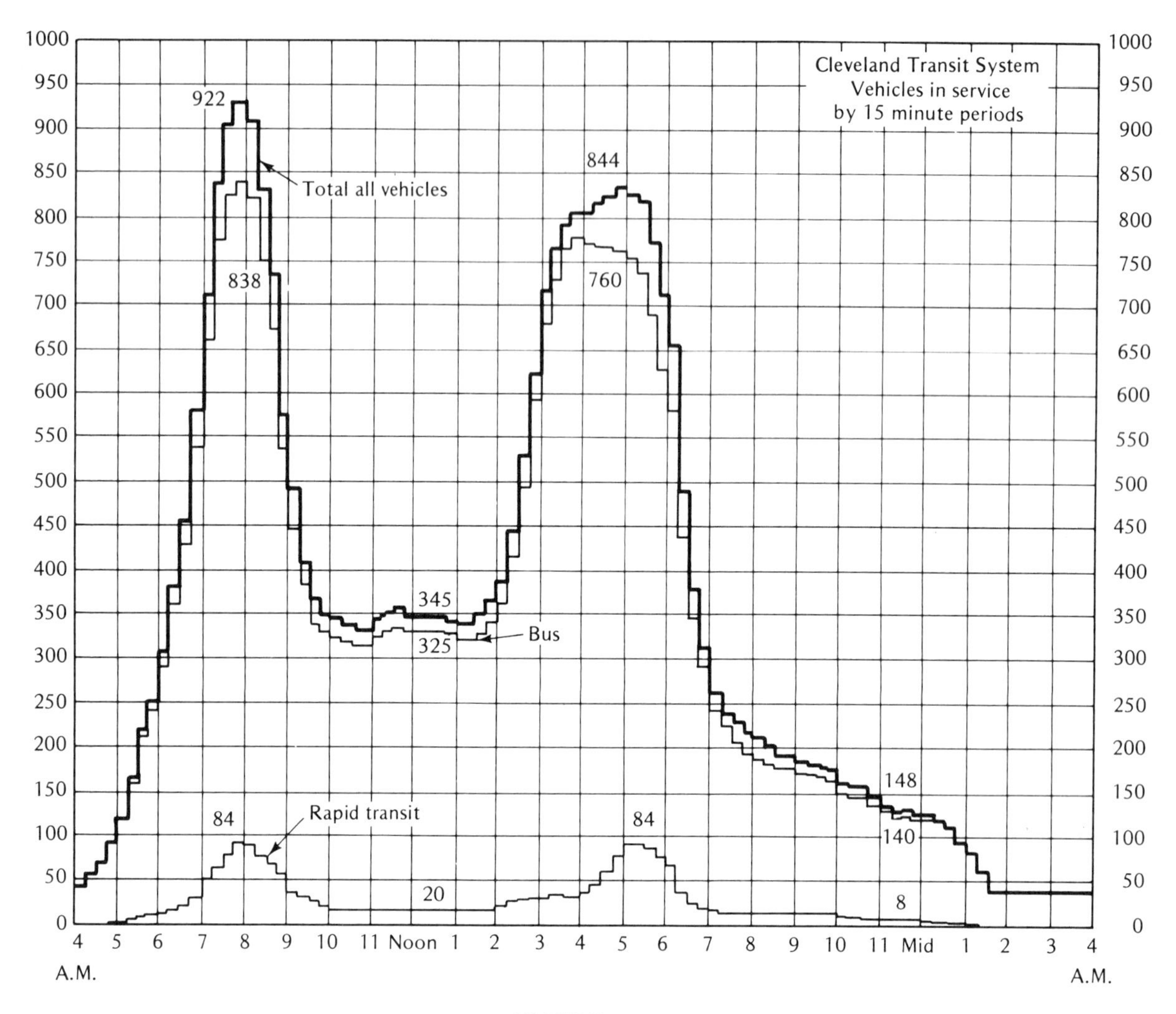

FIGURE 12-7
TRANSIT EQUIPMENT REQUIREMENTS (CLEVELAND)

It can be seen that, from the point of view of transit management, there is nothing attractive about increasing peak period ridership, when labor costs are highest. (Fleet size also increases with peak demand, and fleet productivity decreases because the additional vehicles are not useful off-peak or on weekends. The capital costs involved are another, though lesser, burden on the balance sheet.) However, an increase in off-peak patronage would utilize idle operators, who are being paid in any case, and idle vehicles. For these reasons, transit managers would support policies of staggering work hours. In Washington, D.C. and Louisville, Kentucky, to cite two examples, premium fares are charged in peak periods to promote a patronage shift toward off-peak periods.

Financial Resources

The financial history of urban transit can be divided into three phases. The first exists where there is little competition from other modes and a large captive market. Profits can be made in sufficient quantities to attract private enterprise and even assure some competition. Some government regulation is a usual part of this phase to ensure an orderly supply of services and prevent exploitation by transit monopolies. This phase ended in Europe well before the Second World War, and in the U.S. and Canada soon afterwards, but still continues in many developing countries. At the start of the second phase, public ownership of transit becomes necessary to prevent abandonment of services. At the beginning, the relief from many taxes and from the need to pay dividends is sufficient to balance the books. Later, as red ink threatens, it is agreed that capital investments represent benefits to the entire community, not only to the present riders, and these are, therefore, subsidized by taxes from local sources or from a higher level of government. Fares are set so that the revenues produced cover at least all operating expenses, with perhaps a little contribution to capital costs. This phase ended for many U.S. systems around 1970 as a result of rapidly rising labor costs and of the reluctance to raise fares correspondingly because of environmental and other public concerns. In the third phase, subsidies are also made available for operating expenses.

Setting the level of fares was relatively simple during the first two phases just described. In the first, fares were set to produce a profit and were, in any case, closely scrutinized and sanctioned by regulatory commissions. In the second, they had to match operating expenses. A policy on fare levels is yet to be found for systems receiving some operating subsidy, since there is no obvious formula or rationale.

A simplified monthly balance sheet for a bus transit system is shown in Table 12-7. The income includes revenues from the operation of the system, mostly from fares, subsidies allocated by local, state, and federal sources, and minor miscellaneous items. Expenses are largely under the heading of operating costs, i.e., costs that vary roughly

TABLE 12-7
MONTHLY REVENUES AND EXPENSES FOR A TYPICAL BUS SYSTEM

	THOUSANDS OF DOLLARS		PERCENTAGE OF TOTAL EXPENSES	
Operating Revenue				
Fares	1306		31.7	
Charter services	67		1.6	
Contract services	57		1.4	
Interest on investments, deposits	30		0.7	
Other, including advertising	37		0.9	
Total operating revenue		1497		36.4
Other Receipts				
Local property taxes	1516		36.9	
Local sales taxes	671		16.3	
Federal operating assistance	496		12.1	
Total other receipts		2683		65.3
Total of all revenues and receipts		4180		101.7
Operating Expenses				
Transportation	2297		55.9	
Maintenance	457		11.1	
Welfare and pensions	437		10.6	
Administration and general	283		6.9	
Operating taxes and licenses	151		3.7	
Traffic solicitation, advertising	69		1.7	
Operating rents	18		0.4	
Terminal expenses	13		0.3	
Total operating expenses		3887		94.6
Capital expenses				
Depreciation and amortization	127		3.1	
Interest on bonded debt	13		0.3	
Principal on bonded debt	83		2.0	
Total capital expenses		223		5.4
Total of all expenses		4110		100.0
Excess of receipts over expenses		70		1.7

with the quantity of service furnished. Capital costs are understated, because capital grants for purchase of vehicles neither show up as income nor are charged off as depreciation.

Perhaps the most important message hidden in such balance sheets is that urban transportation planners cannot look to transit to assume additional tasks until additional subsidies have been found or unless the added costs can be covered by fare increases.

REFERENCES

1. American Association of State Highway Officials, *A Policy on Design of Urban Highways and Arterial Streets.* Washington, D.C.: 1973, pp. 486–494.
2. American Public Transit Association, *Transit Fact Book.* Washington, D.C.; Annual, approx. 32 pp.
3. De Leuw, Cather & Co., *Characteristics of Urban Transportation Systems.* Washington, D.C.: U.S. Dept. of Transportation, Report URD. DCCO. 74.1.4, May 1974, 111 pp.
4. Institute for Defense Analysis, *Economic Characteristics of the Urban Public Transportation Industry.* Arlington, Va.: 1972, v.p. (Available from U.S. Govt. Printing Office, Stock No. 5000–0052.)
5. Lea (N.D.) Transportation Research Corp., *Lea Transit Compendium.* Huntsville, Ala.: Annual, 9 vols.
6. Levinson, Herbert S. et al., *Bus Use of Highways.* Washington, D.C.: Transportation Research Board, National Cooperative Highway Research Program.
   a. *Report 143: State of the Art,* 1973, 410 pp.
   b. *Report 155: Planning and Design Guide,* 1975, 161 pp.
7. Quinby, Henry D., "Mass Transportation Systems," *Transportation and Traffic Engineering Handbook.* Arlington, Va.: Institute of Transportation Engineers, 1975, Chapter 6.
8. Shortreed, John H., editor, *Urban Bus Transit: A Planning Guide.* Waterloo, Ont.: University of Waterloo, Dept. of Civil Engineering, The Transport Group, 1974, 360 pp.

M
metro

13

URBAN TRANSPORTATION PLANNING

Planning is a process of looking ahead, a means of preparing for the future. How far ahead the planner looks determines whether the planning is to be considered short- or long-range. The engineer's program for optimizing the use of an existing transportation system by installing various traffic control devices such as signs and signals is an example of short-range planning. The urban transportation planning process, on the other hand, involves planning for 20 to 25 years ahead and is an example of long-range planning.

There is a continuing need for new transportation capacity in urban areas. All over the world, cities are growing through migration from rural areas and natural population increase. With rising incomes and higher standards of living and more leisure hours, the urban resident demands more and better urban services, including improved levels of transportation service. A rising trend in family incomes and increased automobile ownership have created a corresponding trend toward low-density development in the suburbs. This trend, along with decentralization of industrial, commercial, and retail activity, has resulted in an increase in travel generally, and especially by private automobile.

There is, however, also a need to plan the response to major changes in economic and environmental conditions. Especially in the United States, cities developed in an era of plentiful land and cheap energy. Urban spatial development and the concomitant transportation systems reflect this history. However, both land and energy are rapidly becoming scarce resources. While broad transportation goals may remain relatively unchanged—the provision of mobility adequate for a high standard of urban living—the means to achieve such goals may shift away from an emphasis on providing new facilities toward one of maximizing the use and efficiency of existing transportation networks.

The objective of urban transportation planning, therefore, is to develop orderly programs under which an integrated transportation system can be fully developed and its operation and management optimized. This includes the highway and mass transit networks and their terminal facilities. Such planning must consider present and projected land uses and the resultant travel requirements for the movement of persons and goods during the next 20 to 25 years at levels of service acceptable to and within the financial resources of the community. The plan must conform to the goals of the region and to policies of the state and nation.

The need for continuing comprehensive planning was first recognized by the U.S. Government in the Federal Highway Act of 1962 and the Mass Transportation Act of 1964, and is now a requirement for federal aid applications in urban areas of more than 50,000 population.

ORGANIZATION FOR URBAN TRANSPORTATION PLANNING

A number of different urban transportation planning organizational patterns have been used in the past. In smaller areas the state transportation department may do the planning itself, or a temporary project group may be formed, supported in large measure by the state transportation department with participation by local elected officials, planners, and others. A second alternative is the assignment of the urban transportation planning task to a metropolitan or regional comprehensive planning agency. An example of this was the creation of the Southeastern Wisconsin Regional Land Use-Transportation Study within the framework of the Southeastern Wisconsin Regional Planning Commission. A third choice is the formation of a special regional transportation planning organization, such as the Tri-State Transportation Commission of New York City or the Metropolitan Transportation Commission of the San Francisco Bay Area. In any case, links to the regional council of governments must be maintained.

A typical study group's professional staff includes transportation planners and engineers, city planners, economists, geographers, demographers, sociologists, systems analysts, and computer programmers and is headed by a study director who reports to a policy and technical committee. This committee, composed of representatives from the four levels of government—cities, counties, state, and federal—provides policy and technical guidance. The policy committee also provides a channel of communication from the study group to the political structure of the community. The policy committee can provide judgments and appraisals of alternative transportation plans during the planning process.

Another group that is commonly found in the urban transportation study organization is the citizens' advisory committee. This group can be of great assistance to the study group and, along with the policy committee, should participate in the development and evaluation of alternative plans.

No plan is politically neutral, especially with metropolitan areas typically fragmented into many governmental units. The question of political acceptability cannot be safely left until the end of the planning process. The study staff must stay in effective communication with the responsible officials who have decision-making powers concerning the implementation of transportation plans.

Transportation planners fulfill a staff function. They are not decision makers. Their responsibility is to define the program, get facts, analyze them, design systems, and explain to the community the consequences of the possible solutions to the problem. They then present to the decision makers the results of the analyses and recommend what they consider the optimum solution.

Figure 13-1 is a chart of the urban transportation planning process described in the remainder of this chapter.

GOAL AND POLICY FORMULATION

One of the most difficult but also most important tasks of the transportation planning group is to determine the goals and aspirations of the community. These aspirations can be expressed in terms of land use patterns, the urban environment, and the level of standards of service to be provided by the transportation system. These goals must be set realistically within financial, political, and social limitations.

The task of goal determination begins with the establishment of the transportation planning process and the policy and citizen advisory groups. The professional transportation planning staff must work closely with the local elected officials representing the many governmental jurisdictions included in the study area and also with influen-

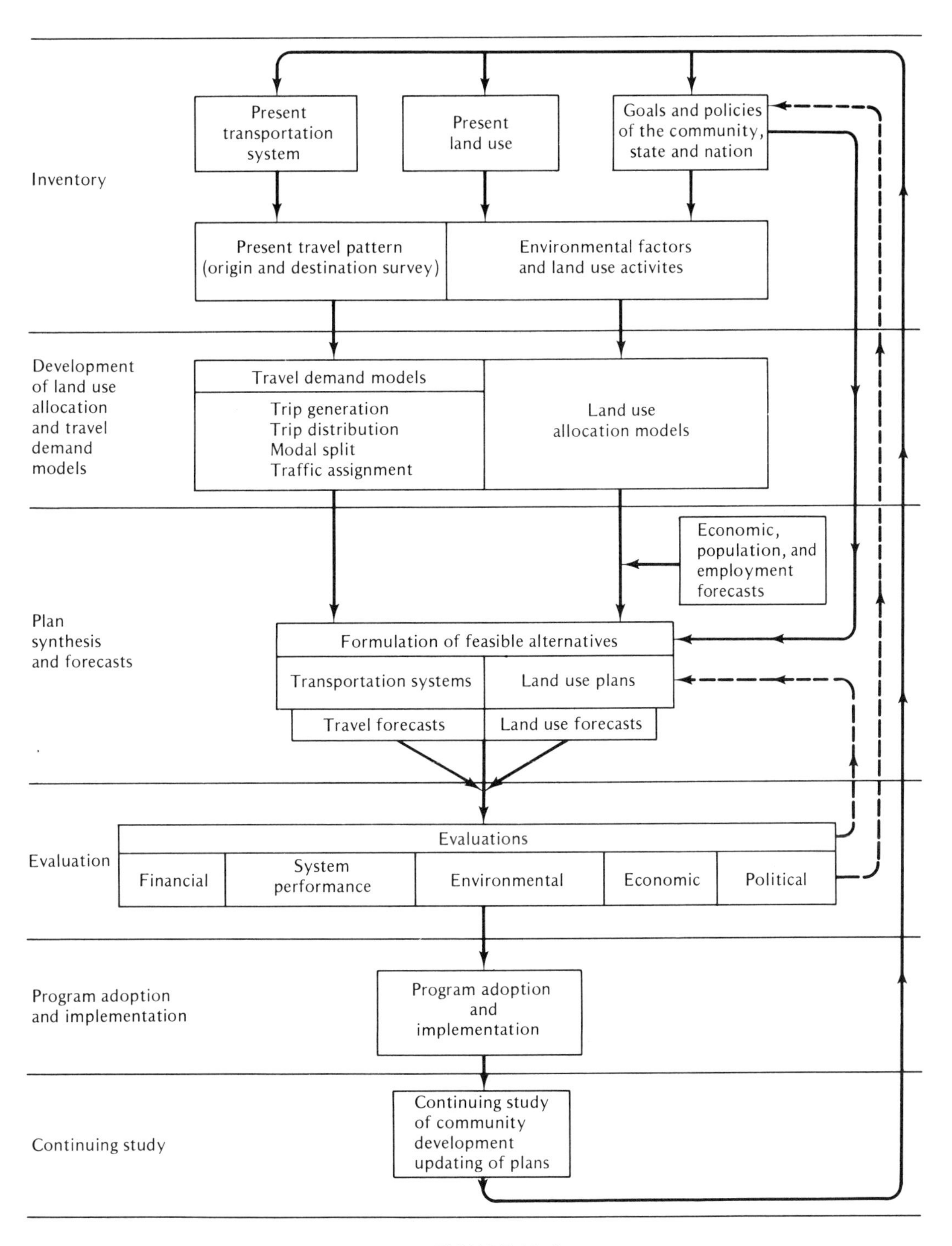

FIGURE 13-1
THE URBAN TRANSPORTATION—LAND USE PLANNING PROCESS

tial private groups and individuals who may have a strong voice in deciding what community projections and programs are to be carried out. If the goals and aspirations of the community can be determined, the transportation alternatives with their standards or levels of service can then be formulated by the professional staff for testing and evaluation. Following this, recommendations can be made to the community.

Policies and guidelines of the federal and state governments must be included in this stage of goal identification. Three major concerns of these higher levels of government are minimization of adverse environmental impacts, conservation of resources, and special consideration of disadvantaged groups within the population.

There will doubtlessly be conflicts between some policies and aspirations articulated by different parts of the community and different governments. Not all of these will be worked out at the start of the planning process. A major challenge for transportation planners is to anticipate how these conflicts are likely to be resolved and, even more difficult, how goals may change during the period of the initial study (short range) or by the horizon year (long range). Planners do have the consolation that, should they guess wrong about long-range developments, some corrections can be made during the continuing phase of the comprehensive planning process.

INVENTORY

The first work done by the transportation planning group is to conduct an inventory of existing land use and existing transportation facilities. This data collection and processing in large metropolitan areas may require two years to complete and up to 60 percent of the cost of the entire transportation study.

Use is made of records of various public and private agencies and aerial photographs supplemented by an extensive field survey to determine the present arrangement, parcel by parcel, of land use activities such as residential, commercial, industrial, public buildings, transportation, and recreational and public open space. The intensity of the activity is also recorded. Data are gathered concerning population distribution and economic activity.

The inventory of transportation facilities encompasses the physical and operating characteristics of all modes of transportation for moving persons and goods—highway, transit, air, water, and rail. For the highway system, a classification is made in which the highways are subdivided into categories such as freeways, arterials, and local streets. As described in Chapter 5, information gathered for each highway segment includes dimensions, type and condition of surface, capacity and level of service provided, operational control devices and regulations, and volume of traffic using the facility. Travel time studies for peak and off-peak periods, accident studies, and volume counts are made for the different segments in the highway system.

Highway terminal facilities also are studied. An inventory is made of both on-street and off-street parking facilities for passenger vehicles and on- and off-street loading and unloading facilities for trucks. This information is collected for the central business district (CBD) and other areas such as large outlying commercial districts and industrial centers where parking problems may exist. Information gathered includes location and number of spaces, type of operation, whether private or public, parking regulations, usage of all parking spaces (time occupied and time vacant), and charges made for their use.

Information gathered for the transit system includes routes of the various forms of transit (bus, rail rapid, commuter railroad); headways; operating speeds; overall travel time between terminals; schedules; types, amount, capacity, and condition of equipment, including rolling stock, traveled way, terminals, and maintenance facilities; vehicle-miles traveled; passenger volumes carried; and revenue by route.

The major source of information on present travel patterns is the home interview survey, in which household and trip data are collected. Trips generated outside the region are included by means of roadside interviews at the study area boundary. Origin-destination information is also obtained from transit riders and from a sample of taxis and trucks operating in the area. These studies are described in Chapter 5.

Special surveys may be made to obtain information on family reasons or preference for residential location, environmental factors of the community that are of importance to them, and reason or preferences as to choice of mode of travel for various trip purposes. Another type of special study of increasing interest among transportation planners is a goods movement survey on a more comprehensive scale than the standard truck travel survey.

Detailed interview-type surveys are sometimes made at the destination end of trips in areas of high land use activity. Results are helpful in developing attraction equations in trip generation analysis (see below). They also provide a measure of the intensity and characteristics of parking demand and an indication as to the desirable locations and size for parking facilities. In these surveys the trip purpose and destination of parkers in the study area are determined; the length of time parked is recorded, and the usage of all parking space is tabulated.

All data except network information are organized by geographic analysis zones, into which the area is divided. If these zones bear a close relationship to census tracts, social and housing data from the most recent census can be incorporated with survey results. All data obtained from sampling processes must be expanded to represent the entire population.

TRAVEL DEMAND MODEL FORMULATION

The data gathered in the inventory describing the socioeconomic characteristics, the travel pattern, the existing land use pattern, and the transportation system are analyzed to determine the relationships among these measurements. To give the data geographic structure, centroids are located in each analysis zone to represent the location of all origins and destinations in that zone. Transportation networks are described by nodes and links to simplify path-finding procedures and to calculate separations between zone centroids; the same descriptions are used in the traffic assignment stage. Extensive use is made of computers to handle the large quantities of data which must be processed during the analysis.

Travel demand models are then developed from this analysis. These mathematical models enable the planner to simulate the response of the community to various land use patterns and transportation systems so that travel demand for the future may be predicted. The mathematical models do not contain the relationships between all the variables that exist, since this would be practically impossible. The planner instead abstracts what appears from the analysis of the existing travel pattern to be the important variables so that reality can be closely simulated by use of the models.

The travel demand models developed include trip generation, trip distribution, modal split, and traffic assignment. A trip is defined as a one-way movement from an origin to a destination. Each trip has two trip ends: the origin and the destination.

Trip Generation

Trip generation models describe the basic tendency for making trips in relation to characteristics of the population, use of land, and transportation accessibilities. Trips are classified by purposes—for example, trips to and from work, shopping, business, or social-recreational activities—and are also grouped as either home-based or nonhome-based. In most urban areas, 70–80 percent of all trips have one end at the home of the trip maker. Because of this, home-based trips, whether the home is the origin or the destination, are analyzed together. The trips are defined as *produced* at the home and *attracted* at the nonhome trip end. Nonhome-based trips are defined as being produced at their origin and attracted at their destination.

Trip generation models most commonly are sets of equations derived by multiple regression. The dependent variable is the number of trip productions or attractions of a specific type of trip purpose. Independent variables for home trip ends may include such characteristics as family size, family income, automobile availability, occupations of family members, residential density, and locational factors, such as distance from the CBD or general accessibility to the rest of the region. For nonhome trip end equations, numbers of employees by industry type, commercial and industrial floor space, sales volumes, or gross land areas may be suitable independent variables.

However, such equations fail to explain trip generation precisely, because analysis zones are never quite homogeneous. In an attempt to overcome this problem, category analysis was developed for the London Travel Survey in 1967 and is being refined further. For home-based trips, households are placed into categories, according to such basic attributes as car availability and family size. For each cell in such a matrix an average trip generation rate is computed, and it is assumed that this rate will apply for future conditions. For nonhome-based trips, traffic generators can be categorized by type and intensity of activity. It would seem, however, that trip generation varies in part with transportation accessibility to the household or the land, and such a factor has not yet been incorporated in category analysis.

The inputs to the trip generation model, then, are the socioeconomic and locational characteristics of the trip makers and the land use activity. The output of the model is the number of trips generated per zone or, if category analysis is used, per household and per nonresidential unit of land use.

Trip Distribution

The trip distribution model will have as input the trip ends predicted by the trip generation formulation. However, when it is first developed, the trip ends calculated from the inventory are used. The function of the distribution models is to arrange these trip ends into groups of trips, each having a specific origin zone and destination zone. The output is a trip table in which the number of trips by purpose between any pair of zones is given.

The model distributes a proportion of the trips generated in any zone to all other zones on the basis of their relative attractiveness and in an inverse relationship to the separations between the zones. Travel time is the most descriptive measure of separation, but cash outlays for tolls, parking and transit fares, and the inconvenience of walking and waiting can be included.

A number of different trip distribution models have been used in transportation studies. A typical, widely used one is the gravity model. Trips are again stratified by purpose, and separate formulae are developed for each. The model hypothesizes that the number of trips between two zones i and j varies directly with their level of trip generation and inversely with their separation. The formulation is

$$T_{ij} = P_i \frac{A_j F_{ij} K_{ij}}{\sum_{k=1}^{N} A_k F_{ik} K_{ik}}$$

where

T_{ij} = the number of trips between i and j for a specific purpose.
P_i = total trips generated in i for that purpose.
A_j = total trips attracted to j for that purpose.
F_{ij} = inverse function of separation between i and j.
K_{ij} = a zone-to-zone adjustment factor used to take into consideration varying socioeconomic characteristics of i and j that affect travel between i and j.
N = total number of zones in the study area.

The F and K factors for a specific study area must be determined through a process called *calibration*. The factors are adjusted until it is possible to synthesize existing travel patterns by means of the model. The assumption is made that the F factors change with time only in response to changes in the transportation network (new separation values). The K factors may or may not remain constant with time depending on predicted changes in the socioeconomic characteristics of the zones in the future.

Modal Split

The purpose of modal split analysis is to predict the proportional use of two or more modes of travel. In urban areas, the basic split analyzed is between the use of mass transit and the private automobiles. However, submodal split models are sometimes developed to describe the relative use of different forms of mass transit.

In the early years of traffic estimation, modal split was considered an issue of choice among competing alternatives, and models formulated the relationships of the choice to characteristics of the trip maker, of the trip, and of the performance of each of the modes. Modal split was either used immediately after trip generation, by splitting trip ends, or after distribution by splitting zone-to-zone trip interchanges. The former method (the trip-end modal split model) was found to be computationally simpler and statistically on a sounder data base, while the latter (trip-interchange modal split model) was more sensitive to system performance, and therefore, was more useful in testing the consequences of proposed changes in the networks of each mode.

However, it is now agreed that modal split procedures must recognize the existence of two distinct groups of transit riders: *captives*, who cannot avail themselves of transportation by automobile, and *choice riders*, who can. The classic models are based on choice, and are therefore inadequate to describe the behavior of captive riders.

Most captive riders come from households where there is no car available. Their travel pattern can be described by generating transit trips through multiple regression or category analysis, and then distributing them among zones. Transit accessibility is a variable in the generation stage, reflecting the need for captive riders to live in zones with some transit service. In the distribution stage, separation is calculated for the transit network costs and times, and trips are distributed only between pairs of zones connected by transit service.

Choice riders are not separated from automobile drivers and passengers until after trip distribution. Once the trip interchange matrices for various trip purposes have been produced, a choice model is used to estimate the proportion of trips between any pair of zones likely to use transit. The trip-interchange modal split model is such a choice formulation. Some of the variables used are relative travel time by each mode, relative travel cost, relative inconvenience in terms of walking and waiting involved, and economic status of the trip maker. The dependent variable in the equations, or in sets of curves, is the proportion of the total trips that will use transit.

Other choice models developed in recent years use generalized travel cost differences between the trip by highway and the trip by transit, and then compute the probability that a user with specific characteristics and trip purpose would use transit. These are generally formulated not by using data aggregated to zone totals, but by summarizing trips for specific types of travelers.

The final step in this two-stage modal split analysis is the summing of the origin-destination trip matrices for the captive and the choice riders to produce the total transit riding pattern. The results must be used with some caution in future predictions. If the level of service on the transit system in the inventory period was much below what is contemplated for the future, the model may not be able to predict either captive rider trip generation or choice rider transit patronage with any accuracy. The experience of other urban areas with transit systems of the quality being planned may be worth studying in such situations.

Traffic Assignment

The last in the series of travel demand formulations, the traffic assignment model is used to allocate the zone-to-zone trips by a given mode to the various routes in the network of transportation facilities. The output of the traffic assignment model is the volume of traffic (either vehicles or transit passengers) on each section or link in the network. Separate assignments are made for the transit trips and the highway trips.

The description of each network (highway and transit) in terms of such link characteristics as length, speed of travel, cost, and capacity is at the core of this procedure. Mathematical techniques and computer programs have been developed which enable minimum time, distance, or cost routes between each pair of zones in the study area to be rapidly determined. The trips between each pair of zones can then be assigned to the shortest or least costly route between the zones.

Traffic assignment models for highway networks, known as *capacity restraint models*, consider the relationship of travel time on a link to the volume assigned to the link. The model reflects the fact that travel time on a link increases as the assigned load approaches capacity.

Another variation in traffic assignment models concerns the number of routes to which the trips between any pair of zones are assigned. The all-or-nothing technique assigns all the trips to one route—the shortest route. Another procedure allows the trips between a pair of zones to be assigned proportionally to several routes.

Demand and Usage

The four-stage process described in this section is usually called travel *demand* modeling, but the inputs are travel *usage* within the constraints of the transportation system as it functioned during the inventory period. Where transportation links are overcrowded or nonexistent, demand may be changed spatially or shifted in time, or trips may be suppressed entirely from the pattern that would be found in an adequately served region. The difference between demand and usage, the *latent demand*, is often ignored in transportation planning, but it appears as traffic wherever transportation availability improves. Since there has been no systematic development of methods for estimating latent demand, judgment, based on studies of areas with excellent transportation service and therefore presumably little latent demand, must be used. Special research into travel motivation can also lead to estimates of the amount of demand suppressed at the time of the inventory studies.

LAND USE ALLOCATION MODELS

The purpose of the land use allocation model is to distribute spatially predicted growth in urban activities such as residential, industrial, and commercial development, recreational facilities, and public open space. In the United States, most of the land development is done by private developers rather than the government. The pattern is controlled only by zoning regulations enacted by the government, which are always subject to change, and governmental development of such facilities as sewer and water systems, transportation facilities, and urban renewal.

A number of different land use allocation models have been developed. They range from very sophisticated mathematical formulations rivaling in complexity the makeup of the areas they attempt to describe to less realistic, but simpler procedures. Those in the first groups may require quantities and qualities of data which are unobtainable, whereas those in the latter can be operated but may not predict at the level of detail that planners might desire. In either type of model, some judgment may be part of the input. The classic Lowry model, for example, requires the planner to locate future basic employment in the region manually, after which the model will allocate the remaining land uses.

One input to all models is accessibility, which is a function of the transportation system. A feedback process, therefore, occurs, and the land use model must be operated in close conjunction with the testing of alternative transportation plans. Other information incorporated into this process includes the present land use pattern from the inventory stage, the holding capacity of each zone based on present development and zoning regulations, topography, availability of basic utilities, especially those furnished underground and therefore representing a major development commitment, and the cost of land.

All land use allocation models are relatively new and still undergoing major refinements. Since they have been calibrated for data spanning only 10-15 years, their ability to predict a future 25 years or more away with accuracy still remains to be proven.

ECONOMIC, EMPLOYMENT, AND POPULATION FORECASTS

The future of a region generally, and the future land use plan alternatives specifically, cannot be visualized without availability of forecasts of economic, employment, and population trends. Such studies are entrusted to economists and demographers, and the following brief summary falls far short of doing justice to the complexity of this work.

One of the most important and difficult forecasts used as input to the transportation planning process is the economic activity prediction for the study area. Economic activity is the primary reason for the development of an urban area. Land use activity and the resultant demand for transportation follow.

A number of methods have been used to predict the levels and types of economic activity that could be developed in an urban area. These range from a simple extension of past trends in the various components of economic activity to a sophisticated input-output analysis of the region. The latter requires detailed data on the interrelationships among the many independent sectors of economic activity.

Estimates of future levels of employment follow from the forecasts of economic activity, as do income and automobile ownership levels, all factors of importance to the transportation planner who must predict future travel demand.

Population predictions are based on past population trends and the most recent census data available. Trends in age distribution and family size are identified. Estimates are made of future birth rates, based on the number of women moving into and out of the child-bearing age range, death rates, and rates of migration into and out of the area. The migration patterns may be influenced by economic trends predicted by the economic forecast reports. Generally, several population forecasts are made, giving a range within which the actual trend is likely to be. Forecasts are checked against new census data and revised in the light of emerging trends.

FORMULATION OF FEASIBLE ALTERNATIVE LAND USE AND TRANSPORTATION PLANS

With all information and forecasts at hand, the planning process now reaches the stage where alternatives for the future are generated. This calls for creativity and imagination as well as for the ability, acquired through experience, of being able to estimate the likely feasibility of proposed ideas at an early stage, so that major effort is not wasted on completely unrealistic alternatives.

Land use and transportation plans must be developed side by side, even though the work may be assigned to different specialists in different agencies. Each possible land use alternative may suggest one or more transportation systems, which, in turn, will lead through the use of a land use allocation model to checking the feasibility of the proposed land use concept. The final product that moves forward to the evaluation stage consists of a land use plan and a transportation plan, the latter including components dealing with major highways, local streets, transit systems, terminal facilities, and traffic management.

The value of model development and of the extensive use of computers becomes recognized in that planners have the opportunity with these tools to prepare for evaluation a number of major and minor alternatives in a relatively short span of time. As they receive "feedback" from policy makers or the community, they can undertake the generation of still more variants with little additional effort.

Land Use Plan

A number of different land use patterns may be evaluated for an urban area. The feasible alternatives will depend, however, on the existing pattern of land uses because of the very considerable investment reflected in the existing arrangement. Another guide for the possible patterns stems from the goals and aspirations of the community. Some of the alternatives that might be considered are a single, concentrated central business district, a multicentered urban area with no one dominant central business district, a city in which low-density residential areas spread in all directions from the center of the area, or a corridor arrangement where high-residential-density developments are concentrated at nodes along the corridors that radiate from the central business district.

After feasible combinations of land use plans and transportation systems have been formulated, each is used as input to the land use allocation model. Other inputs to the model are the existing land use pattern, economic activity, employment and population forecasts, zoning regulations, and proposed sewer and water networks. The output of each land use allocation model is a future spatial arrangement of land use activities for the horizon year (20 to 25 years in the future). If these results do not conform to the land use planning goals on which the alternative is based, other transportation plans may be tested. It is, of course, possible that there is no feasible transportation solution that will promote the land allocation desired by the land use proposal.

Transportation Plan

Engineering design is involved in the formulation of alternative transportation facilities to meet the demand for transportation services required by future land use patterns. The design of future transportation systems is heavily influenced by the existing network of transportation facilities.

The transportation planner investigates various combinations of the different modes of transportation. The alternatives can vary from an automobile-highway-dominant system with a minimum mass transportation facility to the other extreme of a mass-transit-dominant system with minimum highway facilities.

The Major Highway Component

The objective of establishing a hierarchy of highways is to concentrate all but the land access segments of trips on a system of major highways designed and located for this service, and to remove them from local streets, where they would cause environmental nuisances and traffic operations problems. Figure 13-2 shows schematically the four major classes of urban roads. *Major highways* may be defined as those roads on which the function of expediting traffic is more important than that of providing access to adjacent land. Freeways have been designed with no access function at all; major arterial streets enhance through movement by right-of-way priority at intersections with minor streets, design standards for speeds somewhat above those possible on local streets, and parking restrictions (which reduce the access role of the arterial). In the residential subdivision sketched in the lower left section of Fig. 13-2, the street pattern has been arranged in such a fashion that no houses front on the major arterial streets.

The major highway system component includes all existing roads of this classification. Possible additions, whether new construction or upgrading of local streets, are tested by "loading" the major highway network with the traffic forecast produced by the land use plan. Instances of underutilization or inadequacies of capacity are identified and network adjustments are made until a reasonable traffic balance is achieved. It must be recognized that changes in the assumed major highway network will change accessibilities to and the separation between analysis zones, and that the traffic estimation models—and perhaps even the land use allocation model—may have to be repeated for each correction in the plan being tested.

The Local Street Component

No attempt is made when one is working at the scale of the urban region to plan in detail those streets whose primary or sole purpose is to provide access to land parcels—the *local streets*. To include them in the coded network used in traffic estimation and land use allocation would be unduly cumbersome, while not increasing the precision of the results. It is also felt that such local streets are not of direct regional significance and that, therefore, detailed

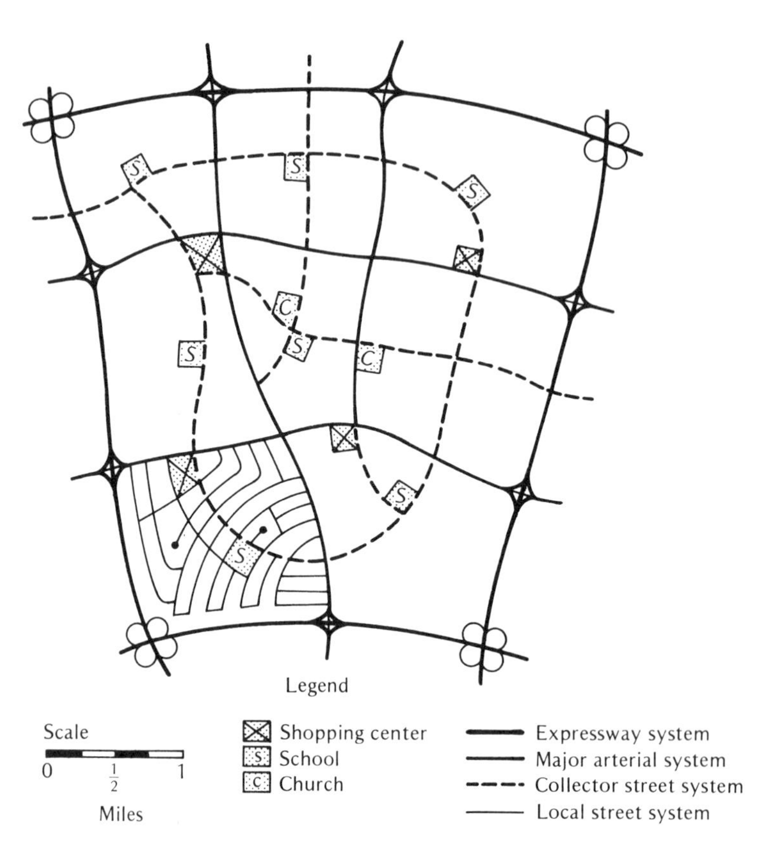

FIGURE 13-2
SCHEMATIC URBAN HIGHWAY NETWORK SHOWING HIERARCHY OF STREETS

decisions should be left to the municipalities within the metropolitan area.

The regional transportation plan, therefore, confines itself to the question of coordinating the local street system with the major highway and the transit plan and, perhaps, to setting design and operation standards. Some of the more important local streets, usually called *collector streets,* may be used for bus routes as well as for their primary function of collecting traffic within a zone for delivery to the major street system. Standards for these must be higher than for the purely local access streets, and a general statement may be desirable on the approximate spacing of collectors.

Local streets also include those giving access to downtown and industrial sites without performing arterial functions. Different sets of standards are appropriate for these.

The Transit System Component

Various transit networks will be tested for compatibility with planning goals and for effectiveness of the individual elements within the system. As adjustments are made to the major highway plan during the formulation and testing stage, transit routes doubtlessly will require adjustment also. Computer programs are available for going beyond the step of "loading" the transit network to estimating the capital and operating costs implied with each alternative as well as the revenues.

The types of transit networks put forward for analysis will reflect the planners' awareness of the community goals and environmental impacts (some of which are mentioned below). In regions where there is general satisfaction with transportation conditions and the state of the environment, only modest modifications to the present network may be proposed. Where a shift in emphasis toward more transit use is articulated, much work will be performed in creating new transit services and testing their effect on traffic and land use patterns.

The Terminal Facilities Element

A transportation network cannot be considered to operate adequately if insufficient provision is made for passenger and freight transfer, and for vehicle storage at the ends of each trip. In plan formulation the demand for terminal facilities is estimated per zone. For areas of high trip end densities, such as the CBD, the proportion of parking demand that will have to be met by off-street facilities must be calculated. Although a precise definition of each lot and garage is not needed, the analyst must show that the total requirement is realistic in terms of available resources, land, and community goals. Large bus and truck terminals may deserve detailed analysis at the regional level, but for smaller units sets of standards will suffice.

The Transportation Systems Management Element

Transportation systems can be improved considerably by optimizing traffic control and management. Such improvements may obviate the need for new facilities. Regulations of the Federal Highway Administration now require that a Transportation Systems Management (TSM) element be included in all regional transportation plans. The objectives of the TSM program are to reduce traffic congestion and facilitate the flow of traffic, and to preserve and revitalize mass transit systems. Specific actions to be considered in preparing the TSM include:

- Traffic operations improvements by traffic engineering methods.
- Preferential treatment for transit and other high-occupancy vehicles.
- Appropriate provisions for pedestrians and bicycles.
- Management and control of parking.
- Staggering work hours, and peak congestion pricing for transit riding and automobile tolls to encourage off-peak travel.
- Programs to reduce vehicle use in congested areas through establishment of auto-free streets or areas, encouragement of car pooling, and the like.
- Transit service improvements.
- Increase in transit management efficiency.

Special funding is available for implementing the TSM element. Obviously, not all actions listed above are appropriate in every regional plan. Some schemes may be of regional importance and, therefore, will be analyzed and described in detail. These may include freeway traffic surveillance and ramp metering, priority freeway lane designations, and traffic control near major activity centers. Other applications will be local in nature, and the plan will confine itself to setting standards for these. A further discussion of the TSM element is found in Chapter 15.

Supporting Analyses

Before any alternative can be advanced for evaluation, three additional feasibility checks need to be performed: scheduling, financing, and implementation. Too often the evaluation will be made on the basis of the picture of the region developed by each alternative for a future planning year. It is quite possible, however, that the optimum plan for 25 years hence is inferior for the short and intermediate term, or that it is likely to become inferior to some other choice in 30 years.

For any alternative it is, therefore, necessary to prepare schedules of improvement and to visualize the transportation system as it will perform at, say, five-year intervals between the present and the horizon year. In the process it may turn out that an alternative is unfeasible because under any sequence of progress that can be devised the intermediate-stage effects on land use are likely to lead away from the desired development goals. For other alternatives, it may turn out that some priority schedules are vastly superior to others. Again, the ability to estimate travel and land use patterns quickly by computer makes this type of analysis possible.

A parallel study involves financial resources. It is not so much a question of assembling a lump sum for the ultimate plan as determining the source and size of money flows during the entire planning period. The implementation analysis delves into the question of political and public acceptability of each alternative, again not only in the horizon year but at each stage through which the region must pass before then.

EVALUATION

Those alternatives which have survived the feasibility scrutiny of the transportation planning process now reach the evaluation stage. As is shown in Fig. 13-1, five broad areas may be identified: financial, system performance, environmental, economic, and political values. Some of the values are quantifiable, albeit in different systems of measurement, whereas others are not. However, a number of economic systems of evaluation have been developed that quantify as many values as possible in economic terms. Benefit-cost and cost-effectiveness analyses are the best-known of these methods. Rating frameworks are sometimes used for those aspects which defy translation into an economic scale.

The evaluation of the financial, economic, and system performance features of each alternative has already started during plan formulation, but generally in parallel fashion rather than in a single effort. Now they are brought together, and environmental impact studies are conducted for each alternative still under consideration. The political acceptability of each alternative is also considered with special reference to the many governmental units within the metropolitan area, many of which are likely to have differing and often opposing interests and goals.

Environmental Impact Studies

The U.S. Congress declared in the National Environmental Policy Act of 1969 as national policy "to use all practicable means and measures . . . to create and maintain conditions under which man and nature can exist in productive harmony, and fulfill the social, economic, and other requirements of present and future generations of Americans." The law requires that any proposal for major federal action (which includes the granting of federal aid) either be accompanied by an environmental impact statement (EIS) or by a declaration that the environment will not be significantly affected. The adequacy of an EIS can be challenged in court, and projects can be delayed as a result. However, an EIS is only an advisory document to the decision makers; it cannot make the decision for them. Many states within the United States and some other countries have similar legislation.

This requirement applies to transportation planning generally, and has become an integral part of urban transportation planning. It has served well not only as advice to the policy committee of the regional transportation planning agency but also to make the planners more aware of the relationship between their proposals and the environment. In preparing the EIS, they must consider not only the impacts, but what alternatives producing less adverse results might be available, what mitigating measures might minimize such undersirable effects, what irreversible or irretrievable commitments of resources are involved in their proposal, and how the local short-term uses of the environment relate to the maintenance and enhancement of long-term conditions.

In its Environmental Assessment Notebook Series [Ref. (10)], the U.S. Department of Transportation lists thirteen impact categories in three groups:

- *Social Impacts*
  1. Community cohesion
  2. Accessibility of facilities and services
  3. Displacement of people
- *Economic Impacts*
  4. Employment, income, and business activity
  5. Residential activity
  6. Effects on property taxes
  7. Regional and community plans and growth
  8. Resources
- *Physical Impacts*
  9. Environmental design, aesthetics, and historic values
  10. Terrestrial ecosystems
  11. Aquatic ecosystems
  12. Air quality
  13. Noise and vibration

When one is reporting on a single project, perhaps only a few items on this list of impact areas may need to be studied. For urban transportation plans in general, all deserve scrutiny. Some, especially in the economic area, have been considered for many years, whereas others are of such recent concern that study methods and evaluation criteria have not yet been fully developed. Noise and air pollution have been of such great concern that federal standards have been established for maximum acceptable levels of each of these types of nuisance. Although these standards are difficult—or even impossible—to meet, new projects should lead toward, rather than away from, their attainment. Guidelines for preparation of an EIS for highway projects, for example, are set forth by the U.S. Federal Highway Administration [Ref. (11)].

Stimulated by the recognition of future energy shortages, but in consonance with several of the other environmental objectives, urban transportation plans now often show increasing emphasis on higher efficiency of current facilities through management schemes and less emphasis on additions to the inventory of facilities. Programs are developed to discourage single-occupant automobiles because of their major contribution to pollution and fuel consumption in relation to the transportation service produced. The "carrot" rather than the "stick" is used, by making car pooling and the use of mass transit more attractive. Priority or exclusive lanes for transit vehicles and cars containing at least three persons are one form of persuasion; priority parking areas in the most convenient locations are another. However, such measures must produce substantial changes in travel habits before the effects on air pollution, noise, and energy use become noticeable.

PROGRAM ADOPTION AND IMPLEMENTATION

The transportation planning staff presents the results of the evaluation for each feasible alternative to the regional transportation planning policy committee or commission. It may also recommend a preferred plan with reasons for its choice. A first draft of a Transportation Improvement Program (TIP), listing projects to be undertaken annually for at least three years, is prepared. Public hearings are held throughout the region to obtain comment and reactions before the policy group votes for adoption of a program for the metropolitan area.

Implementation is usually the responsibility of the state and local governments rather than of the group that adopted the plan. This can present some difficulties. However, the major elements in the capital improvement program probably require federal aid, which may not be forthcoming if the project is not on consonance with the approved TIP.

CONTINUING STUDY

After the transportation plan has been completed, the staff becomes a permanent continuing transportation study group. Its function is to maintain current data on land use, travel, and transportation system developments. Trends in each of these areas are checked against the predictions made during the initial study. As deviations appear, adjustments to the horizon year forecasts can be made and the consequences of these on the plan currently in effect can be estimated. As mentioned earlier, the staff is also alert to changing goals of the community and to changing national and state policies. At regular intervals amendments to the plan will be prepared for consideration by the policy committee. At perhaps five-year intervals, the plan will be projected to a new horizon year further in the future.

The continuing study group also monitors and, probably, administers federal aid for transportation projects in the region; it may have the same duties with regard to state financial assistance of local programs. It further provides information and technical assistance to decision makers who implement the transportation plan.

REFERENCES

1. Blunden, W.R., *The Land Use/Transport System: Analysis and Synthesis.* Oxford, England: Pergamon Press, 1971, 318 pp.
2. Dickey, John W., *Metropolitan Transportation Planning.* New York: McGraw-Hill Book Co., 1975, 562 pp.
3. Hutchinson, B.G., *Principles of Urban Transport Systems Planning.* Washington, D.C.: Scripta Books (with McGraw-Hill Book Co.), 1974, 444 pp.
4. Institute of Transportation Engineers, *Trip Generation.* Informational Report. Arlington, Va.: 1976. Looseleaf.
5. Lane, Robert, et al., *Analytic Transport Planning.* London, England: Duckworth, 1971, 283 pp.
6. Owen, Wilfred, *The Accessible City.* Washington, D.C.: Brookings Institution, 1972, 150 pp.
7. Owen, Wilfred, *The Metropolitan Transportation Problem.* Garden City, New York: Doubleday and Company, 1966.
8. Smith, Wilbur, and Associates, *Transportation and Parking for Tomorrow's Cities.* Detroit, Michigan: Automobile Manufacturers' Association, 320 New Center Building, 1966.
9. Stopher, P.R. and A.H. Meyburg, *Urban Transportation Modeling and Planning.* Lexington, Mass.: Lexington Books, 1975, 345 pp.
10. U.S. Department of Transportation, *Environmental Assessment Notebook Series: Highways.* Washington, D.C.: U.S. Government Printing Office, 1975, 7 vols.
11. U.S. Federal Highway Administration, *Policy and Procedure Memorandum 90-1 (Environmental Impact and Related Statements); Policy and Procedure Memorandum 90-4 (Process Guidelines—Economic, Social and Environmental Effects on Highway Projects).* Washington, D.C.: U.S. Government Printing Office, Sept. 1972.

14

SYSTEMS PLANNING ELEMENTS

Transportation systems have grown increasingly more complex in the past 25 years. In urban areas they now include not only street and highway networks but also bus and rail (where applicable) transit networks. The previous chapter discussed transportation planning in urban areas. The basic goal of any transportation system is the development, maintenance, and operation of a system capable of accommodating travel in an efficient, safe, and economical manner.

The same general principles of transportation planning apply whether the system is a statewide network, a regional system, or an urban system. The philosophy of planning at each level is so markedly different, however, that techniques and approaches are also different. This chapter describes the elements of transportation systems planning; then it discusses classification before covering inventory and needs studies. A section on forecasting is followed in turn by a section on economic analyses and on system evaluation. The chapter is concluded by a discussion on programming and implementation. This chapter will be oriented toward statewide transportation systems. When any section is similar to the material in Chapter 13, "Urban Transportation Planning," reference will be made to the appropriate section rather than presenting a separate discussion.

Planning is the first step in preparing programs for transportation systems. It is an organized, rationally conducted process of collection, analysis, and presentation of facts about all facets of the transportation system. Planning studies are made to identify the needs of a system to compare the condition of the existing network with the present requirements and future demands placed on the system, to determine possible sources for additional resources, and to allocate available money and manpower to the improvement of those parts of the system where needs are greatest.

Planning studies furnish information about the system's adequacy, use, needs, cost, and financing. Other studies of administrative practices and regulations are needed to provide essential supplementary information used in the transportation planning process.

The characteristics of transportation planning can be summarized as:

1. Planning *anticipates* those needs that will be created over a planning horizon of 15 to 25 years.
2. Planning is *comprehensive,* requiring [Ref. (1)] :
   "*a.* That economic, population and land use elements be considered fully.
   "*b.* That estimates be made of the future demands for both public and private movement of both people and goods.
   "*c.* That terminal facilities and traffic control systems be included in the planning.
   "*d.* That the entire area within which the forces of development are interrelated should be included— not just as it exists now, but as it is expected to be urbanized within the forecast period."
3. Planning is *cooperative,* implying formal agreements between various agencies.
4. Planning is *continuous,* recognizing that transportation plans need frequent and regular reevaluation and updating.
5. Planning represents a *decision,* not a conclusion, or final answer.

SYSTEMS PLANNING ELEMENTS

Transportation must be viewed as more than movement, because it both serves and shapes future development. Transportation is a function that comprises part of the "quality of living" and must be considered part of and in conjunction with comprehensive state or community development plans. The general urban transportation planning process is depicted in Fig. 13-1. The statewide process can be viewed as composed of six main elements, as shown in Fig. 14-1. The general highway planning process, prior to the formation of State Departments of Transportation (DOT's), has evolved as indicated in Fig. 14-2 [Ref. (2)]. Transportation planning involves more than conceiving long-range goals and objectives. The planning process must outline the mechanism in which to translate the broad plan into short-range programs and schedules with adequate monitoring and updating. Today, transportation planning is beginning to take on this broad viewpoint, through multimodal planning, considering broad social, economic, energy, and environmental consequences.

The above discussion suggests that modern statewide transportation planning follows a systems planning approach, and indeed it does or should. In fact, it is a part of comprehensive planning which includes activities that:

1. Attempt to guide physical change and development while fully considering all consequences and factors.
2. Use the best possible techniques for identifying and measuring consequences.
3. Apply established principles to develop plan alternatives.

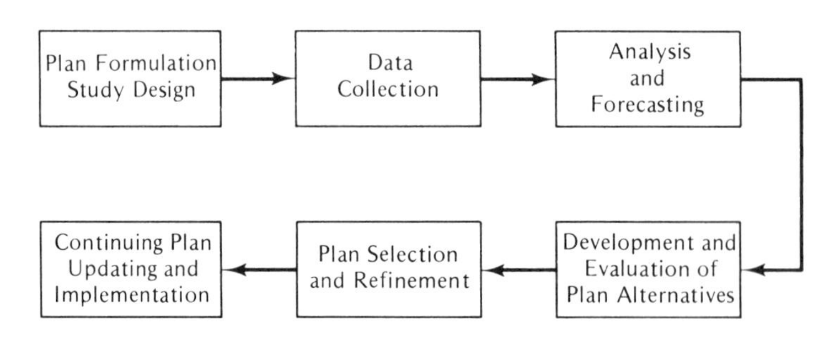

FIGURE 14-1
ELEMENTS IN THE TRANSPORTATION PLANNING PROCESS

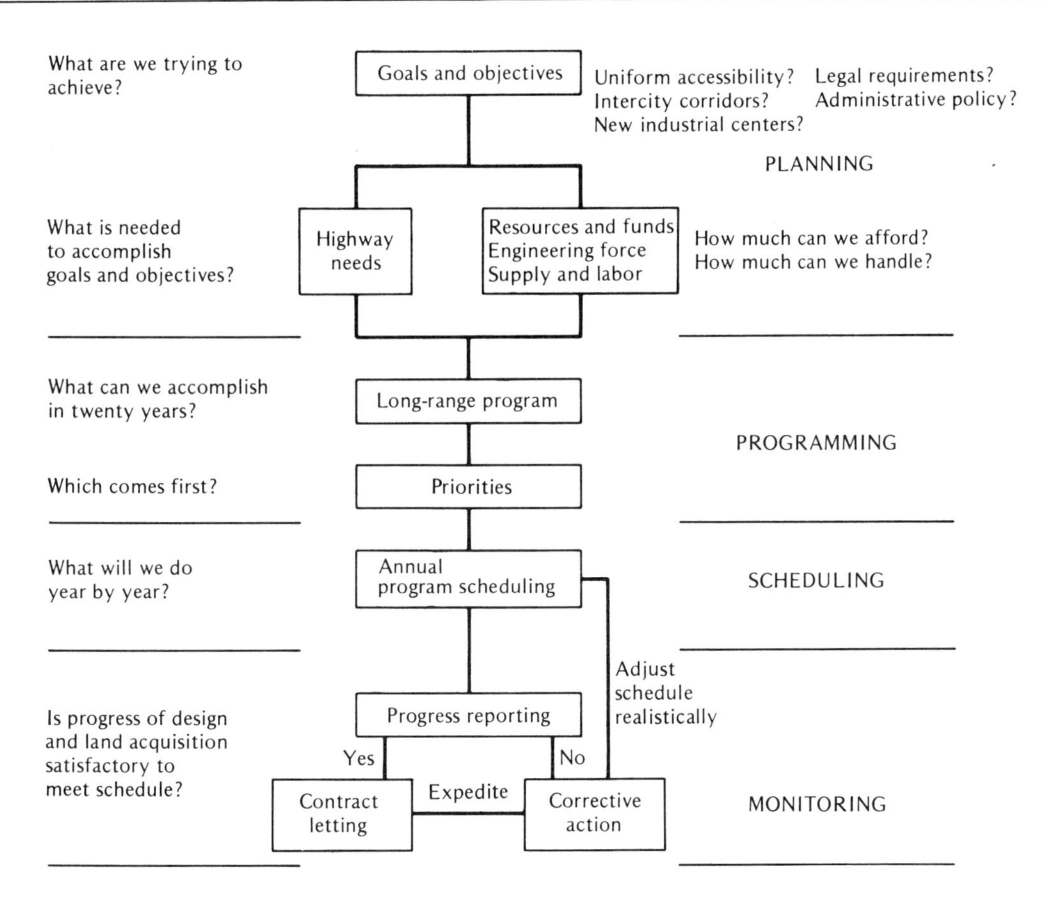

FIGURE 14-2
BASIC COMPONENTS OF THE HIGHWAY DEVELOPMENT AND MANAGEMENT SYSTEM [REF. (2)]

The widespread availability of digital computers combined with improved capabilities to forecast and to perform economic analyses led to a type of systems approach to evaluating alternative urban transportation plans in the 1960's. In order to avoid short-run solutions to long-run problems, a systems approach is essential. Systems analysis can be defined as a way of viewing a group of independent elements working together toward a common goal. Some system terms and characteristics of systems are

1. A *system* may be defined as a set of objects, either fixed or mobile, and all the relationships that may exist between the objects. All systems have subsystems and are members of a higher system (e.g., a through street as a part of the street network).
2. An *environment* may be defined as a set of objects, either fixed or mobile, that is outside the system, but significantly affects the system. The environment, a metropolitan area for example, is the aggregate of all external conditions and influences significantly affecting the system. For example, the street network is influenced by staggered work hours or carpool programs.
3. An *open system* is one in which a change in the environment produces a change in the system and vice versa. Most traffic engineering systems are open.
4. A *closed system* is one in which the environment does not have any influence upon the system.
5. An *objective* is a desired aim of the system—i.e., to reduce travel time and delay.
6. An *alternative* is one of several (two or more) means

of accomplishing an objective—i.e., prohibiting parking vs installing a new signal system.

7. A *criterion* is a standard of judgment, a rule or test, whereby an alternative can be evaluated to see if it succeeds in obtaining a given objective [e.g., to achieve an overall travel speed of 20 mph (32 kmph)].
8. *Resources* are the available materials to be consumed in a given project (man-hours, dollars, etc.).
9. A *constraint* is a limitation imposed (budget, time periods, etc.).
10. *Interfaces* are the places where the systems interact.
11. A *model* is the formulation of a logical structure that describes systems mathematically and applies the criterion to alternatives to determine if the objectives are met with the available resources (e.g., minimum time path algorithm).
12. A *decision maker* is an authority that decides which alternatives should be used to accomplish the objective of a system.

There are three basic reasons why methodologies are required to support systems planning and programming at the state level [Ref. (3)]:

1. To provide the information needed to formulate regional and state policy in those areas that either are currently the responsibility of state transportation agencies or at least should be in the future. There are five broad areas: (a) to determine the level of funding for transportation and the trade-offs between transportation programs and nontransportation programs such as health, education, recreation, and water resource programs at the federal, state, and local levels; (b) to help direct state policy toward issues such as land use development policy, recreational development opportunities and objectives of the state, interagency cooperation with regional and local interests, and water and natural resource conservation; (c) to interface, effectively, transport investment decisions with regulatory decisions by the transportation regulatory agencies on issues such as price regulation and entry and exit to markets; (d) to effectively integrate public policy decisions with decisions being made in the private sectors on locational choices, development schemes, and economic growth, and (e) to effectively integrate decisions affecting the movement of both freight and passengers.
2. To define and effectively allocate resources among and within the various transport modes. Statewide planning is required to predict funding sources, whether federal, state, or local; to predict the degree of uncertainty about those sources; and to provide mechanisms for transferring or generating additional funds and identifying modal budget constraints and area minimums. It is also required to effectively settle priorities of investment programs and determine the appropriate modal trade-offs of alternative programs.
3. To ensure equity in providing transportation services throughout the state. This involves making service-level trade-offs for geographical areas, e.g., rural versus urban, and interpersonal trade-offs for users and nonusers of the transportation system, including the poor, the aged, the handicapped, and those with less than average mobility.

The general elements of a transportation system analysis are

1. Selection of goals and objectives; definition of the problem.
2. Determination of the system elements and variables significant to the system under study.
3. Formulation of a systems model.
4. Generation and specification of alternative designs.
5. Evaluation of alternatives.
6. Selection of a design.
7. Implementation of the chosen design.
8. Review and redefinition of the problem; establishment of a continuous process.

The current practice in statewide transportation planning is such that certain activities are necessary in order to achieve some semblance of systems planning. The next five sections present some of the activities.

CLASSIFICATION

Functional classification is the key to transportation planning, since it groups facilities into classes or systems according to the character of service that they are expected to provide. Classification contributes to an orderly solution of many problems by (1) determining the relative importance of various facilities, (2) establishing the basis for assigning levels of service or design standards, (3) evaluating deficiencies (i.e., the difference between the standards and the present geometrics or level of service), (4) determining the resulting needs, and (5) determining the cost for improvements. Functional classification, then [Ref. (4)],

1. Establishes logical integrated systems that bring together all facilities that should be under the same jurisdiction because of the type of service rendered.
2. Aids in assigning responsibility for each class of facilities to the level of government having the greatest basic interest.
3. Groups facilities that require the same degree of engineering and administrative competence together in order to apply a consistent level of competence to each group, with suitable variation between groups.
4. Relates geometric design standards to the facilities comprising each classification type.
5. Establishes a basis for long-range programming, improvement of priorities, and fiscal planning.

In the case of highways, the major difference between roads can be conceptually displayed in Fig. 14-3. Variations then exist according to the degree of mobility or land access service provided. At one extreme, freeways (i.e. controlled access facilities) provide for mobility and not land access, while a local road in the county provides access to land, but is rarely used by through traffic [Ref. (4)]. (See, also, Fig. 13-2.)

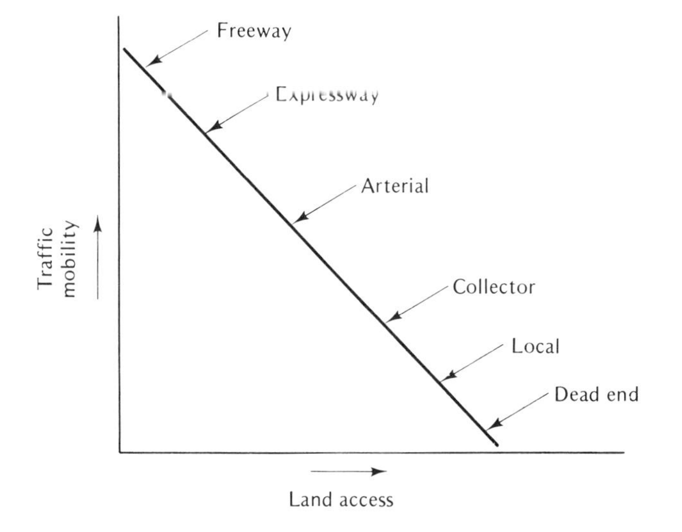

FIGURE 14-3
RELATIONSHIP BETWEEN FUNCTIONALLY CLASSIFIED SYSTEMS: TRAFFIC MOBILITY VS. LAND ACCESS

The different road systems that are generally established as part of a rural functional classification study are as follows [Refs. (5) and (6)].

1. *The land access function.* Land access roads, normally secondary county highways, township roads, and local streets, do not serve through traffic, but provide access to abutting property, and are of local interest primarily. Traffic volumes are usually quite low on land access roads.
2. *The collector function.* These are secondary roads that provide access to higher-type roads and connect small communities and nearby areas, and in addition serve adjacent property. These include primary county highways and secondary urban arterial highways, and other collector roads.
3. *The major arterial function.* Major arterial highways provide primarily for relatively high volumes of traffic between major traffic generators, and should be designed to facilitate traffic movement and discourage land access. This functional classification includes primary state highways, which are of statewide interest, and major urban arterial highways.
4. *The freeway function.* Routes of this type would include the Interstate System and supplemental freeways, connecting large population centers, or carrying heavy volumes of traffic long distances in and around metropolitan areas. Such highways serve through traffic only, and provide no access to abutting property.

Other transportation facilities are functionally classified in a similar manner and are presented in other references. For example, for a discussion of airport classification, see *Planning and Design of Airports* by Robert Horonjeff [Ref. (7).]

INVENTORY AND NEEDS STUDIES

The success of transportation planning relies heavily on having a good source of current and accurate data. This means of data must be readily accessible and periodically updated. Computerized, coordinated information systems have been suggested as an answer to handling the vast quantities of data needed in highway planning. The Federal Highway Administration has recommended a coordinated data processing system, which could help overcome organizational fragmentation and assist in the handling of data. Usually there are several sections in the planning division of a state DOT, each responsible for collecting and processing some class of data and for supplying these data to various users. In many cases, these users are individually coding, storing, and processing the same data element, thus duplicating each others' effort. In a coordinated data system, the section that collects the data is responsible for maintaining the basic computer file of those data [Ref. (8)].

The first group of data suggested includes road inventory, traffic characteristics, condition rating, road life, historical maintenance, and accident information. It is also suggested that a separate file be made for each of these kinds of information. The second group includes information on finances, cost indexes, and highway statistics. Although these data should also be in separate files, the retrieval of data from them is frequently on a different basis from that for the first group.

Many state highway departments have recorded their data by control sections, which serve as the basic unit for measurement of all data and communications. A control section is simply defined as a section of road that has homogeneous characteristics based upon factors such as: (1) soil conditions, (2) traffic conditions, (3) design standards, (4) political jurisdiction, (5) administrative jurisdiction, (6) type of development, and (7) type of construction. It is also necessary to establish the definition of the control section in the field so that reference can be made to a consistently defined road section. Mile posts, coordinates, and other means have all been used to relate field data to the base of reference used in the office.

Inventories include, in addition to the traditional road inventory [Ref. (9)], road life studies, travel inventory, and a socioeconomic inventory. Inventories of all modes of transportation are basically a means of establishing a data base.

Needs Studies

Past practice in statewide transportation planning placed emphasis on producing needs studies. Essentially, needs studies were concerned with only one mode, the highway system. Needs studies were defined as [Ref. (10)]:

A comprehensive highway needs study is a careful examination and evaluation of conditions that affect motor vehicle travel on rural roads and city streets. Its objective is to determine the adequacy of a state highway system, to suggest improvements, and to ascertain the estimated cost of bringing these deficiencies up to a designated standard. When properly completed, the study can serve as a firm base for legislative and administrative action to improve transportation. Likewise, the report provides for a better informed public.

The basic questions being explored in a needs study include [Ref. (10)]:

"*1.* What are the social and economic factors which cause and will continue to cause a demand for the improvement of highway facilities, and how can these factors be measured?

"*2.* What classes and types of highways, and how many miles of each are needed to supply present and expected future needs, and what will it cost to build and maintain them?

"*3.* Can the state and subordinate governmental units finance the needed expenditures under current tax levies and, if not, what are the alternative financial proposals?

"*4.* How can the road and street systems be effectively administered?"

A needs study traditionally consists of the following steps:

1. *An inventory of the existing highway network.* This involves carrying out an inventory of each section of highway and its present level of service.
2. *Traffic volumes and characteristics.* Obtaining data of the present volumes of traffic as carried by the various sections of highway and of the various types of traffic, such as private cars, trucks, and buses.
3. *Existing conditions of roadway.* A study of the structural and surface condition of the various sections of highway, together with an estimate of maintenance costs and remaining service life.
4. *Classification.* This involves a classification of all roads within the network according to a system that may be administrative, financial, or functional.
5. *Analysis and forecasting of future growth of traffic.* This is the most important single element of the whole highway needs study.
6. *Establishing satisfactory and appropriate design standards.* The anticipated traffic demand for the design year and its assignment within the network having been arrived at, sets of standards are formulated. These standards include specifications for desirable lane width, shoulder width, width of median, design speed,

capacity, stopping sight distance, maximum gradient and curvature, type of structure, and surface, as well as intersection and access controls.

7. *Comparison of existing roadway conditions with acceptable standards (i.e., comparison of items 3 and 6 above).* The set of minimum standards having been established, each section of the existing network is compared to these standards so as to identify deficiencies. This includes not only the identification of those sections of the roadway that will be deficient by the design year, but also those sections that will become deficient before the design year.
8. *The identification of necessary improvements.* Based on the above comparison, a list is made of deficient sections of the present roadway and of the improvements required to correct these deficiencies. These improvements will include resurfacing, widening, and reconstruction of existing highways and, where necessary, the construction of new highways. In all cases, the design standards arrived at under item 6 should be used.
9. *Financial.* Estimates are made of the costs of the work shown to be required under the study, the estimates giving a breakdown of each section of improved or new highway for presentation to the administrative agencies in order to allow them to make the necessary decisions about future work. Together with this, an outline of the work program and the respective costs should be submitted, this outline drawing particular attention to the most urgent needs.

Figure 14-4 depicts the various steps in a highway needs study, showing how the various studies interact with a needs study. Beginning in 1971 the U.S. Department of Transportation (DOT), in cooperation with the state DOT's or highway agencies, began a series of transportation needs studies (the first was the 1972 National Transportation Needs Study) to determine the needs of all modes of transportation. The 1974 effort was expanded to the 1974 National Transportation Study [Refs. (11) and (12)].

As shown in Fig. 14-4 and described above, an important element of a needs study is a fiscal study. Until the needs are placed in the proper perspective with the financial ability of the state to meet these identified deficiencies, they remain nothing more than a "wish list." Thus the final output of a comprehensive needs study is a revised listing of needs that can be implemented in some order of priority.

FORECASTING

Forecasting is the art (and/or science) of identifying the potential magnitude of change over some future years. It is the predicting or estimating of future values—vehicle miles of travel on a system, the number of person or vehicle trips between two areas, etc.

"In State or regional transportation planning studies, forecasts of population and economic activity are a basic input to projection of travel demand. It should be recognized there is uncertainity in forecasting" [Ref. (13)]. Accuracy and complexity must be traded off against increased time and cost of forecasting. Good, sound planning requires good forecasts, and long-range transportation

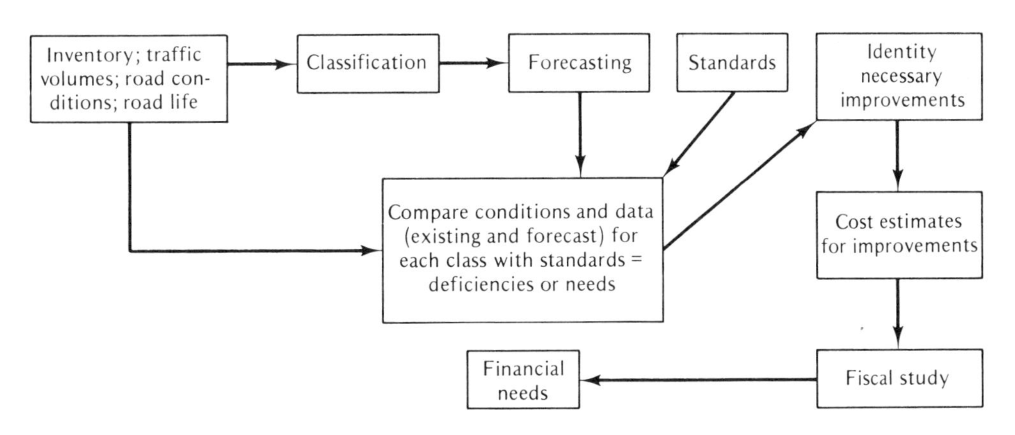

FIGURE 14-4
STUDIES AND ELEMENTS COMPRISING A NEEDS STUDY

planning must be a part of the framework for overall comprehensive planning. This in turn should be a part of statewide or community planning. Good forecasting must consider the underlying factors affecting the estimated change. Thus improvements in forecasting ability and accuracy are dependent upon a better understanding of travel behavior. What are the motivations for travel? To what extent are travel patterns influenced by factors such as auto ownership, income, education, leisure time, etc.? Forecasting depends on data availability and the establishment of data relationships.

Demand forecasts are used to answer the following questions:

1. How rapidly is the use of the link increasing or decreasing?
2. Where is the traffic on this link coming from and where is it going; how does this link fit into the total network?
3. Will improvements on this link divert traffic from other links, and will changes in other transportation links affect the traffic on this link?
4. Will improvements on this link generate wholly new traffic that did not exist before, either by filling a latent need for transportation or by stimulating new socioecónomic development?
5. Will congestion or changes at the nodes or terminals at both ends of the link affect the traffic on the link?
6. Will plans for other public or private investment affect this link, by changing the traffic or by competing for the same resources that could be used to improve this link?

Forecasting actually strives to provide travel demand estimates that serve as the basis for decisions on transportation investments for both new and improved facilities. Forecasting travel also generally involves the forecasting of land use, population, and economic activity (possibly only the consideration of previously made forecasts of these elements).

Forecasting methods range from pure guesses through very sophisticated mathematical model results. Most transportation demand forecasts range from trend analysis (or fitting a line) to multiple regression analysis. Fitting a trend line means to show the pattern of relationship by means of a line that "best fits" the data. This is the trend line and the line that is projected forward to a target year. Trend lines may be obtained by:

1. "Time-series analysis" (which is the analysis of only one statistical series).
2. "Correlation analysis" (which is the process of analyzing more than one statistical series and the relating of each series to the other to establish a pattern of interrelated movements over time).

Forecasting is extremely significant for good transportation planning and decision making. A poor forecast can hardly help but result in poor planning. Therefore, strong emphasis on improving travel forecasts should be obvious. Like planning, transport demand forecasting is based on the basic premise that "there is an *order* to human travel behavior, which can be measured and described" [Ref. (13.] Orderliness in travel provides the basis for intelligent forecasting, which is necessary so that solutions will be directed toward problems of the future—not just those of the present or past. Further discussion on transportation forecasting can be found in the work of Oi and Shuldiner [Ref. (14)], Peskin [Ref. (15)], or Hutchinson [Ref. (16).

ECONOMIC ANALYSES

Transportation decisions, particularly systemwide transportation decisions, require a firm basis. A complete analysis involves more than economics, but a good economic analysis should be the basis for all decisions. In a transportation economic study the following steps should be used:

1. Recognize the problem requiring study.
2. Identify possible alternative solutions to be compared and determine preliminary estimates of the differences between alternatives.
3. Analyze the preliminary estimates and determine which alternatives justify further study.
4. Prepare detailed estimates of these alternatives chosen for final comparison.
5. Determine the interest rate i, or minimum rate of return, and perform the economic analysis.
6. Choose between alternatives.

The six most widely used methods of engineering-economic analysis are

1. *Equivalent Uniform Annual Cost.* This method combines all investment costs and all annual expenses into

a single annual sum that is equivalent to all disbursements during the analysis period if spread uniformly over the period. When more than one alternative is being examined, the one with the lowest equivalent uniform annual cost is the most economical. The total annual transportation costs for alternative j = Annual Construction Costs for alternative j + annual maintenance costs + annual vehicle operating and other user costs.

2. *Present Worth.* The present worth method combines all investment costs and all annual expenses into a single present-worth sum, which represents the sum necessary at time zero to finance the total time stream of disbursements over the analysis period. Of the alternatives compared, the one with the lowest present worth is the most economical.
3. *Equivalent Uniform Annual Net Return.* The equivalent uniform annual net return method is the equivalent uniform annual cost method plus the inclusion of an income factor or benefit factor. The answer indicates the amount by which the equivalent uniform annual income exceeds (or is less than) the equivalent uniform annual cost. The alternative having the greatest equivalent uniform annual net return is the one of greatest economy.
4. *Net Present Value.* The net present value method gives the algebraic difference in the present worths of both outward cash flows and inward flows of incomes or benefits. It is the same in principle as the present worth of costs method, but includes the factor of annual income. The alternative having the greater net present value is the one with greatest economy.
5. *Benefit/Cost Ratio.* The benefit/cost ratio method expresses the ratio of equivalent uniform annual benefit to the equivalent uniform annual cost. Any alternative that has a benefit/cost ratio above 1.0 is economically feasible, and the alternative that has the highest incremental benefit/cost ratio is indicated as the preference.
6. *Rate of Return.* The rate of return method determines that interest rate, or discount rate, which will equalize the negative costs and the positive returns or benefits. As in the benefit/cost ratio method, the rate of return method, in highway proposals, usually compares two alternatives in order to develop a differential benefit. The higher the rate of return, the greater the economy.

When several alternatives are being considered by using the benefit/cost or rate of return methods, an incremental rate of return and incremental benefit-cost analysis should be conducted. The two most critical and sensitive factors in the engineering-economic analysis techniques are the interest rate i and the length of the analysis period n. The first of these ought to be close to the current market cost for borrowing. The length of the analysis period should be the smaller of the life of the facility or the time period of forecasting.

In summary, engineering-economic analyses do not result in a final decision. The results must be presented to the decision makers along with social and environmental factors (that cannot be measured in monetary terms).

For a more detailed discussion of this topic the reader is referred to the work of Winfrey [Ref. (17)] or Grant and Ireson [Ref. (18)].

EVALUATION

The evaluation process should consider those alternatives which are consistent with the local transportation objectives. To accomplish this in a reasonable manner, an evaluation methodology must be utilized. A broad array of factors reflecting both the user's and the community's hierarchy of values must be considered in making an intelligent choice among alternatives. The prominent factors are usually of the following form:

1. Physical
2. Aesthetic
3. Social
4. Economic
5. Fiscal
6. Political
7. Environmental

Alternatives need to be structured so as to allow:

1. Testing and forecasting
2. Determination of consequences
3. Derivation of values of consequences
4. Determination of the total value of all consequences of each alternative

Also important in the evaluation of alternatives are

1. Consideration of uncertainty
2. Time constraints
3. Quality of information
4. Subjective aspects

In evaluating transportation alternatives, many complex factors are involved, and a truly mathematically optimal alternative is not possible. Rather the aim is to choose the alternative that satisfies the constraints while attaining the objectives of the system to the highest degree possible. Evaluation of transportation plans has evolved in the past twenty years from a very straightforward benefit-cost approach to a complex cost-effectiveness analysis. Efficiency evaluation involves engineering-economic techniques, which were discussed above (e.g., how economically efficient is an alternative?). Effectiveness evaluation of transportation alternatives typically have included techniques that may be classified as:

1. Ranking schemes
2. Rating schemes
3. Visual techniques
4. Cost-effectiveness techniques

An evaluation analysis generally might include the following steps:

1. Measure the extent to which a particular alternative achieves the individual parameters.
   a. Rank on a discrete scale, or
   b. Rate on a continuous scale
   c. Use subjective values for nonquantifiable factors
2. Establish the relative importance (weight) of the several factors.
3. Calculate a combined "grade" for each of the alternatives.

For a more detailed discussion of evaluation procedures and techniques, the reader is referred to an FHWA report on the subject [Ref. (19)].

PROGRAMMING/IMPLEMENTATION

The final major element in transportation systems planning and one that is often underemphasized is that of programming the alternatives chosen. In other words, a statewide transportation plan is only a paper plan until it is translated to a series of projects, which may be thought of as programming or as implementation.

Priority programming for transportation development is the rational selection of proposed projects on the basis of relative urgency, systematically scheduled within limits of available resources to carry out legislative and administrative objectives. The principles involved are applicable to the programs of all transportation agencies. Details vary among systems, depending upon the complexities of the problems involved and policy objectives for system development. However, all systems are generally influenced by both legislative and administrative actions.

The legislature shapes transportation programs by defining the responsibilities of the various levels of government, establishing financing plans for each system, and setting long-range objectives for system development.

Within legislative controls, administrative actions affecting priority programming processes include:

1. Establishment of firm programming policies and procedures. These are related to needs and must be within estimated revenues available for various projects.
2. Establishment of systematic engineering analysis procedures for evaluating relative urgency of proposed improvements.
3. Establishment of continuous scheduling. This includes a review and control process to ensure optimum use of available funds and manpower, which permits adjustments needed to meet changing conditions and unforeseen emergencies.

Since funds are insufficient to meet all needs immediately, there must be choices of which improvements should be undertaken. Even with sufficient funds, project selection would still be necessary, since it is physically impossible to undertake all work simultaneously.

All transportation facilities are not of equal functional importance—either between modes or within a given mode. Likewise, not all needs are equally urgent. A policy of selecting improvement projects to provide greatest possible benefits from available funds is of utmost importance to continued economic development.

The prime objective in setting goals for system development is to obtain greatest benefits from available revenues. In general, this means providing for faster rates of construction on facilities of higher functional importance while at the same time taking care of essential requirements on facilities of lesser importance.

For each system there are numerous possible budget allocation criteria. Some of these are relative needs of the various systems, proportion of total travel served by the systems, degree of acceleration deemed desirable for more important facilities, and policies for area distribution of funds. To permit cost-effectiveness comparisons of differing combinations of these variables, a clear-cut definition of importance according to functional characteristics and services provided is essential.

When results of the needs and fiscal studies are available, alternative construction budgets for each system are carefully investigated and evaluated. Budgets should remain firm over a reasonable period of time to insure continuity of operations. Above all, policies and objectives for system development must be clearly stated and understood.

Design standards are engineering yardsticks for determining the character of improvements required to obtain safety and economy in an integrated transportation network. They must be related to functional classes of facilities and not to administrative or legal systems made up of several different functional classes of facilities.

Basic concepts and criteria must be consistent with good engineering practices. All proposed improvements must be reasonable and justified, with provisions to insure adherence to established principles. It must also be recognized that the utility of the formal appraisal as a foundation for priority programming is only as good as the system established for keeping basic data and information updated to reflect changing conditions.

From the transportation needs study or similar sources one should have information on deficiencies, costs of correcting the deficiencies, and information on social, economic, and environmental consequences. In the case of highways, each roadway project should be rated, relative to all other roadway projects, as to:

- The roadway's structural conditions and ability to carry loads imposed upon it.
- The roadway's capacity to move traffic at reasonable speeds without undue congestion.
- The roadway's adequacy of alignment, both horizontal and vertical.
- The adequacy of roadway and pavement widths.
- The roadway's accident experience.

Each bridge project should be rated, relative to all other bridge projects, as to:

- The bridge's load-carrying ability.
- The bridge's horizontal clearance.
- The bridge's vertical clearance.

Each type of deficiency must be rated separately to identify roadway sections or structures critically deficient and urgently in need of improvement. Where projects are rated on an average or combined score for all deficiencies, the identity of a single serious deficiency is frequently lost.

The rating procedures for each type of deficiency should not be so complicated as to be burdensome. However, they must be sufficiently detailed for uniform, consistent, and relative ratings, or the results will be meaningless. Ratings should be based on conditions observed in the field during the needs appraisal and upon all available factual data pertaining to each type of deficiency.

The separate ratings for each type of deficiency should be included on the individual project data processing cards developed for the rating study. They provide the basic engineering data for project comparisons and for development of initial priority arrays of improvements in order of relative urgency.

Firmly established priority programming procedures, which include current factual data on existing conditions and realistic evaluation of needs, will insure greatest benefits from available funds.

When one moves to programming for all modes of transportation on a statewide basis, the tie between planning studies and data and programming has not been completed. The situation is much more complex when one is attempting to determine which of several airport improvements is the most urgent relative to several highway improvements, or port development needs, or other transtation needs. This fact was highlighted in the 1974 Transportation Research Board Conference on Statewide Transportation Planning [Ref. (3)]. As state DOT's continue to wrestle with this problem, new techniques for programming total transportation will, it is hoped, evolve.

REFERENCES

1. State Highway Commission of Wisconsin, *Highways I, the Basis for Planning.* Wisconsin Development Series, Wisconsin Dept. of Resource Development, Madison, 1967.
2. Bureau of Public Roads, "Programming and Scheduling Highway Improvements," *Highway Planning Technical Report No. 4,* April, 1966.
3. Transportation Research Board, *Issues in Statewide Transportation Planning,* Special Report 146. Washington, D.C.: Transportation Research Board, 1974, p. 95.

4. Grunow, R., "Functional System Classification," Panel paper presented at the Region 3 Highway Planning and Research Conference, Chapel Hill, N.C., March 25-27, 1963.

5. Bureau of Public Roads, *1968 National Highway Functional Classification Study Manual,* Washington, D.C.: U.S. Dept. of Transportation, April 1969.

6. Berry, D.S., "Highway Classification and Needs Studies," Paper presented at Illinois Highway Engineering Conference, University of Illinois, Urbana, Illinois, March 3, 1965.

7. Horonjeff, Robert, *Planning and Design of Airports.* New York: McGraw-Hill Book Co., 1975.

8. Blessing, W., "Coordinated Data System for Highway Planning," Bureau of Public Roads (now the Federal Highway Administration), *Highway Planning Technical Report No. 7.* Washington, D.C.: Bureau of Public Roads, May 1968.

9. Berry, D.S., Lectures on Highway Inventories, Road Life Studies, and Highway Planning Studies, Northwestern University and University of California, 1955-1967.

10. McCormack, C., "Creating, Organizing and Reporting Highway Needs Studies," Bureau of Public Roads, *Highway Planning Technical Report,* Number 1. Washington, D.C.: Bureau of Public Roads, September 1963.

11. U.S. Department of Transportation, *1972 National Transportation Report,* July 1972. Also *1972 National Highway Needs Report.* Washington, D.C.: Government Printing Office, May 1972.

12. U.S. Department of Transportation, *1974 National Transportation Report,* 1974. Also *1974 National Highway Needs Report.* Washington, D.C.: Government Printing Office, January 1975.

13. Governor's Committee for Transportation, *Transportation and Pennsylvania's Future.* Harrisburg, Pa.: Commonwealth of Pennsylvania, January 1969.

14. Oi, W. and P. Shuldiner, *An Analysis of Urban Travel Demands.* Evanston, Illinois: Transportation Center, Northwestern University, 1962.

15. Peskin, H., *Some Problems in Forecasting Transportation Demands.* Princeton, N.J.: Mathematica, 1965.

16. Hutchinson, B.G., *Principles of Urban Transport Systems Planning.* New York: McGraw-Hill Book Co., 1974.

17. Winfrey, R., *Economic Analysis for Highways.* Scranton, Pa.: International Textbook Company, 1969.

18. Grant, E.L., and W.G. Ireson, *Principles of Engineering Economy,* 6th ed. New York: Ronald Press, 1975.

19. Carter, E.C., L.E. Haefner, and J.W. Hall, Report to the Federal Highway Administration, Department of Transportation, *Techniques for the Evaluation of Factors Relevant to Decision Making in the Federal-Aid Highway Program.* Washington, D.C.: Federal Highway Administration, January 1972.

BIKE ROUTE

15

TRANSPORTATION ENGINEERING MANAGEMENT AND ADMINISTRATION

Sound administrative practice requires that responsibility for all technical activities necessary to build and operate a transportation system—research, planning, design, construction, maintenance, and operations—be placed within a properly constituted organization. For effective activities in each of these broad areas, this organization should meet the following criteria:

1. The functional activities and/or modal organizational units should have a relatively equal rank.
2. Some effective means must be provided to insure coordination of all functions and of all modal organizations.

TRANSPORTATION ENGINEERING ACTIVITIES

Responsibilities of transportation engineering can be divided or grouped into eight major categories:

1. Planning
   a. Systems
   b. Short-range
2. Design
   a. Systems
   b. Project
3. Studies and surveillance
4. Construction
5. Operations and control
6. Maintenance

7. Research
8. Administrative support (for the above seven functions)

Systems planning was discussed in Chapters 13 and 14, and there was some discussion of short-range planning (transportation systems management, or TSM) in Chapter 11. Geometric design, which pertains to design details or project design was also discussed in Chapters 6, 7, and 8. Systems design was briefly covered in Chapter 14. Traffic studies were covered in Chapter 5, and transportation studies were discussed in Chapters 13 and 14. Operations and controls were also discussed in Chapters 9 and 10. Therefore, only traffic design, operations and control, and research will be discussed below.

Traffic Design

The traffic engineering unit has responsibility for the functional design of highway improvements as they affect operations. Areas of traffic engineering review include the following:

1. The preparation of detailed designs and standards for traffic control equipment and installations.
2. Review of proposed arterial highways as to their ability to carry the expected volumes at desirable speeds. Such design elements as alignment, cross section, access control, sight distance, intersections and interchanges, and street lighting should be reviewed.
3. Redesign of existing highways and intersections to increase capacity and safety.
4. The design of off-street parking and terminal facilities and access connections.
5. The establishment of standards and the review of subdivision plans as to traffic flow patterns, land use, access control, setbacks, and driveways.
6. The development of design criteria to be met for different levels of demand.

System Operations Control

From the information provided by its surveillance activities, the traffic engineering unit can determine where and how traffic may be regulated. This regulation may require the use of traffic control devices and, frequently, enabling laws and ordinances.

The traffic engineering unit should be responsible for the design, installation, operation, and maintenance of all traffic control devices used on the street or highway systems under its jurisdiction. This responsibility can be by direct supervision of operations or through administrative control of work executed by another agency of government or a private contractor.

Recommendations for the laws and ordinances necessary for traffic regulation such as one-way streets, through streets, turn controls, speed zones, curb parking and loading controls, public transit control, and street lighting should be initiated or reviewed by the traffic engineering unit. This permits a program of "reasonable" traffic control based on factual data.

Research

Because of the complex and dynamic nature of transportation, research is a very important function of a good transportation organization. The development of new study techniques, traffic control measures, new materials for construction, better design, etc. are all possible results of basic or applied research. A program of research, including all aspects of all (almost all, since some apsects of some modes are in the private sector) modes of transportation, should be established.

TRAFFIC AND TRANSPORTATION ENGINEERING ORGANIZATIONS

There are major differences in the size, the scope of activities, and the legal authority of city, county, and state transportation operations. The type of administrative organization necessary is strongly influenced by the level of government. This section will present a discussion of municipal transportation organization, followed by state transportation organizations and county organizations, concluding with a brief description of the federal organization.

Municipal Transportation Organizations

In cities, street transportation is a critical element in city operation and in the economic life of the area. Because of the wide variation in population, geographic size, and area, there is a wide variety of transportation organizational patterns. The type of transportation organization also tends to vary somewhat with the type of city government; the three basic forms are

1. The "strong mayor" form, in which the legislative activities are vested in the city council with the mayor serving as the chief administrative officer. Normally, the mayor has the power to veto matters of legislative importance.
2. The council-manager form, in which the mayor usually serves as a member of the city council and the city manager is chief administrator. The city manager is appointed by and serves at the pleasure of the council.
3. The commission form, in which each elected commissioner generally serves as the head of an operating department. The commissioner, who may be elected at large or by districts, constitutes the legislative body and may choose one of its members to serve as chairman.

Traffic or transportation engineering functions have most often been included in city public works departments but may be placed in others such as public safety. Figure 15-1 illustrates a separate traffic engineering department. The traffic engineering department has equal status with other important departments such as public works, fire, and police. It has complete authority and responsibility over all engineering aspects of traffic operations. Departmental status for traffic or transportation engineering assures the necessary authority and responsibility for establishing intergovernmental relationships. The department head has a rank equal to the heads of other city units with whom he must work. However, its relatively small size and the rivalry of the larger departments may inhibit the cooperation and coordination so vital for efficient operation of traffic engineering functions.

A recent development in larger cities or metropolitan areas of over 1,000,000 population has been the grouping

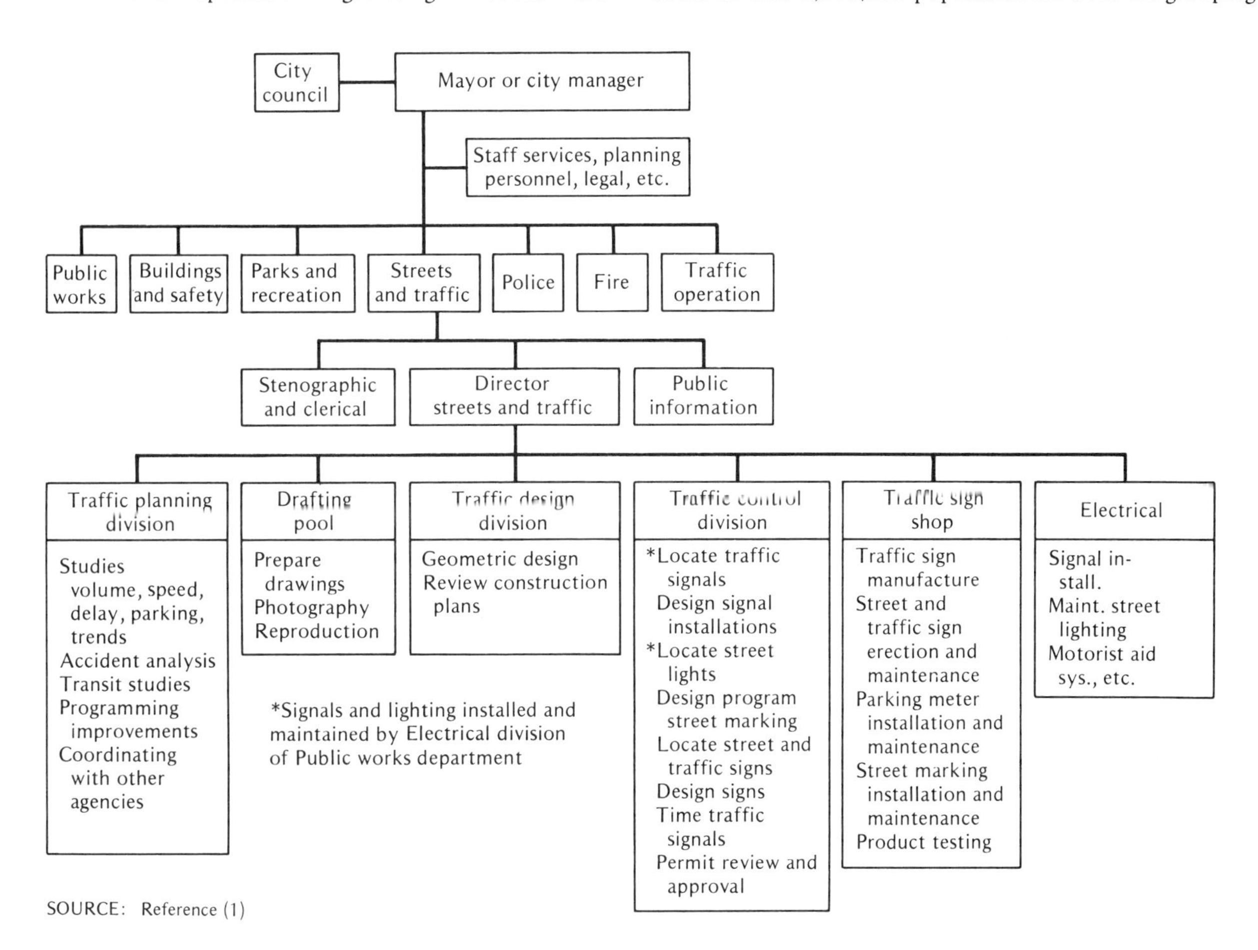

FIGURE 15-1
TYPICAL ORGANIZATION OF A LARGE CITY WITH A SEPARATE TRAFFIC ENGINEERING DEPARTMENT

of traffic engineering and other transportation functions into a single transportation department. The complexities of highway system planning, construction, operation and maintenance have resulted in the consolidation of these functions into a single agency.

Figure 15-2 shows the traffic engineering function as a division of the public works department. In this case, the division is usually responsible for traffic planning, geometric design, and traffic control devices.

A typical arrangement of traffic engineering in a department is shown in Fig. 15-3.

State Organizations

State highway and transportation organizations have undergone major changes in the past decade or so, with most states forming departments of transportation (DOT's). Regardless of the type of organizational arrangement, the effectiveness of the various functions is basically determined by the effectiveness of management coupled with the technical capabilities and dedication and sincerity of personnel. The DOT organizations can be classified as [Ref. (2)] :

1. Modal. Primary divisions are categorized by mode (e.g., highways, aviation, urban transit, railroads, water). Most functions are performed under each division.
2. Functional. Operating divisions are responsibile for a specific function for all modes, such as planning, design, construction, and safety.
3. Mixed Organization. Both modal and functional divisions exist at the operating level. Some functional activities related to all modes occur at the administration division.

Active and explicit expressions of support from the executive and legislative branches of state government help any type of organization to function more effectively. Also important are the functions of citizen participation and coordination with the private transportation sector. Figure 15-4 shows the basic structure for three types of state DOT's. Figure 15-5 is the organization chart for one state DOT.

County Organizations

The principles of state organizations generally apply to county transportation organizations but are somewhat dependent on the extent of urban development. Urban counties are more similar to larger municipalities, whereas rural counties tend to have a traditional roads or highway department. Figure 15-6 shows a typical organization of transportation within a public works department, and Fig. 15-7 shows an urban county organization of a new county DOT.

Federal Organization

Since 1912, when Congress made the first of a continuing series of appropriations to the states for road construction, the federal government has been active in highway transportation. A federal-aid highway system has been established to insure continuity and uniformity on a nationwide basis. The Federal Highway Administrator administers the highway program of the federal government. The federal role has not been as a constructor and maintainer of highways in competition with the states, but as a partner who provides federal matching funds and sets standards for uniformity. The funds have been established for a variety of systems. In a similar manner, the Federal Aviation Administrator is responsible for the federal airport and airway program, the Federal Railroad Administrator for the federal railroad program, the Urban Mass Transportation Administrator for the federal mass transportation program, etc. Figure 15-8 shows the 1976 organization of the Federal Department of Transportation. Figure 15-9 shows the organization of the Federal Highway Administration and Fig. 15-10 the organization of the Urban Mass Transportation Administration, both as of 1976.

Because of the federal funding programs the organizational structure of federal agencies has strongly influenced similar organizational arrangements in the various states.

STAFFING

Regardless of the organizational placement of transportation engineering functions at the various levels of government, the provision of an adequate staff is vital. In addition to personnel with an engineering background, personnel with specialized advanced education in traffic engineering and/or transportation planning are almost essential in the larger agencies. And, of course, support personnel such as draftsmen, clerks, and technicians will be required.

A modern multimodal DOT-type agency requires personnel from a multitude of disciplines, including public relations, economics, geography, environmental sciences, sociology, management, computer science, and, of course, engineering.

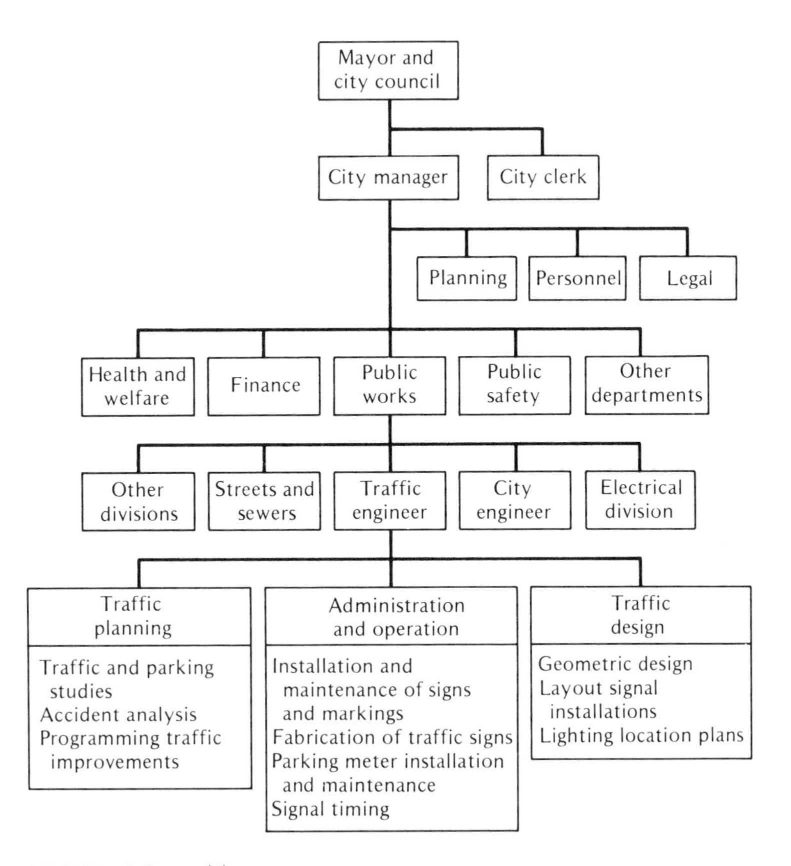

SOURCE: Reference (1)

FIGURE 15-2

TYPICAL ORGANIZATION OF A SMALLER CITY WITH A TRAFFIC ENGINEERING DIVISION WITHIN THE PUBLIC WORKS DEPARTMENT

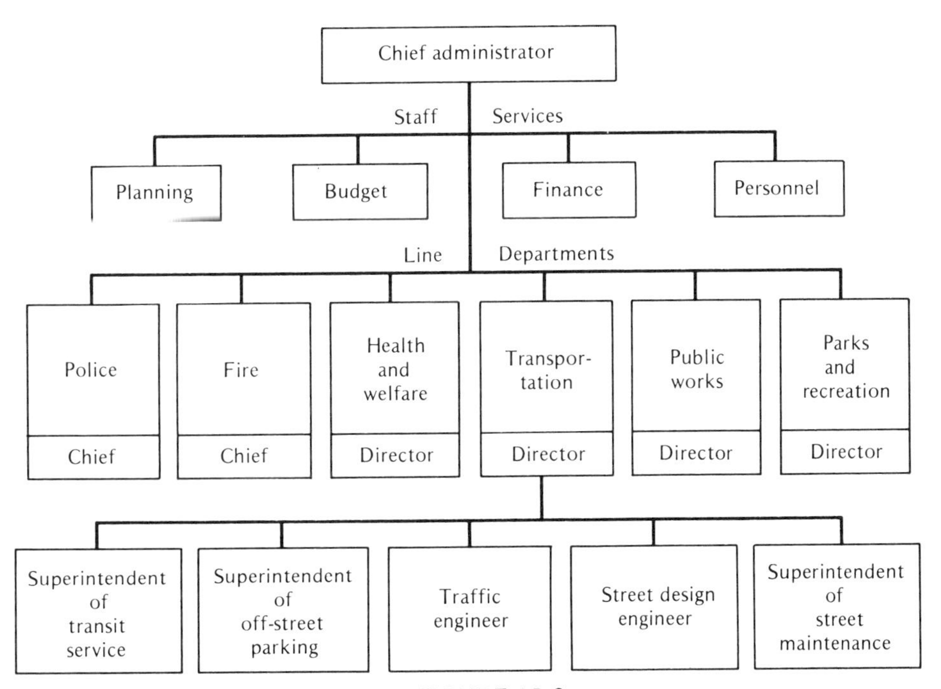

FIGURE 15-3

TRAFFIC ENGINEERING IN A DEPARTMENT OF TRANSPORTATION

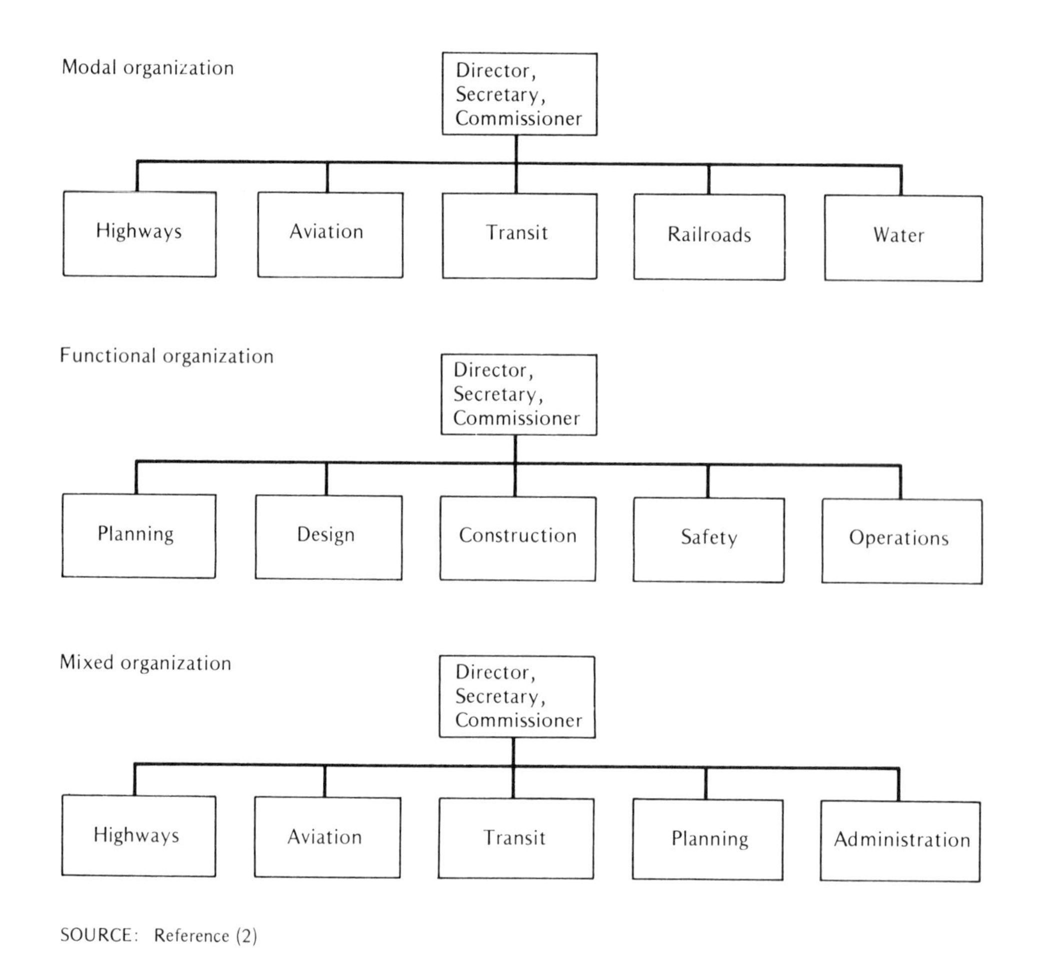

SOURCE: Reference (2)

FIGURE 15-4
BASIC ORGANIZATIONAL STRUCTURES OF STATE TRANSPORTATION DEPARTMENTS

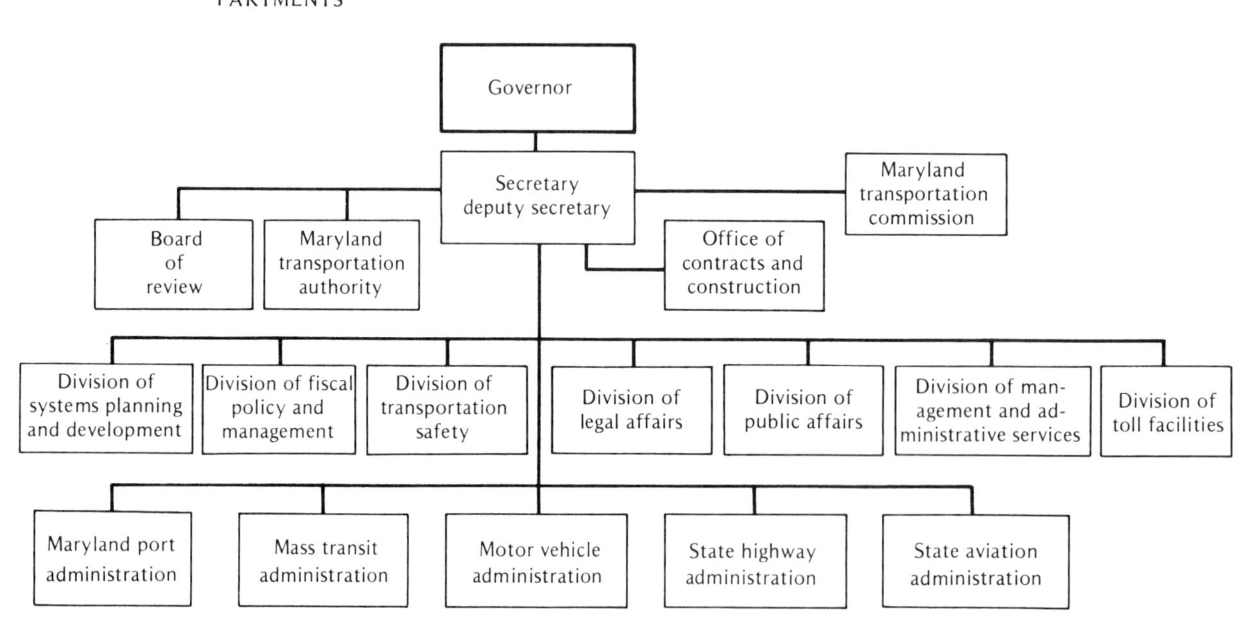

FIGURE 15-5
ORGANIZATION CHART: MARYLAND DEPARTMENT OF TRANSPORTATION

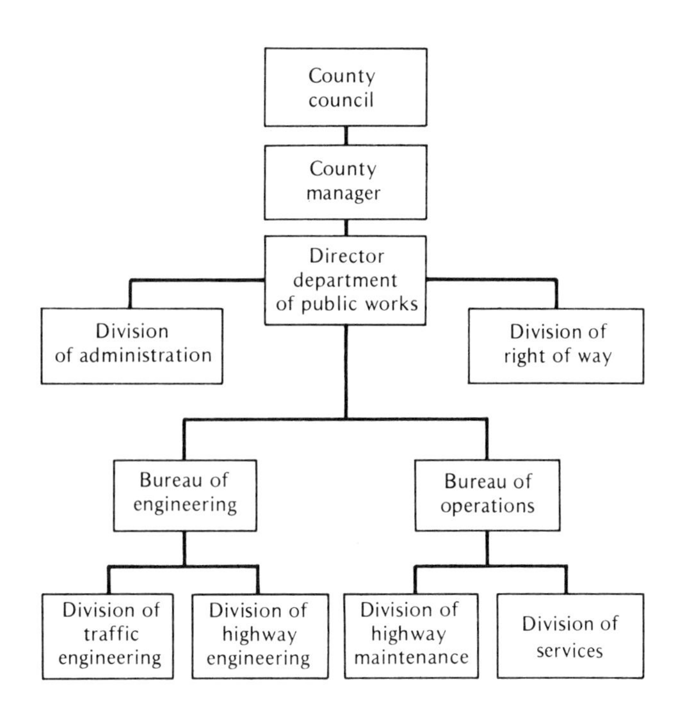

SOURCE: Reference (1)

FIGURE 15-6

TYPICAL ORGANIZATION OF A COUNTY PUBLIC WORKS DEPARTMENT, SHOWING LOCATION OF THE TRAFFIC ENGINEERING FUNCTION

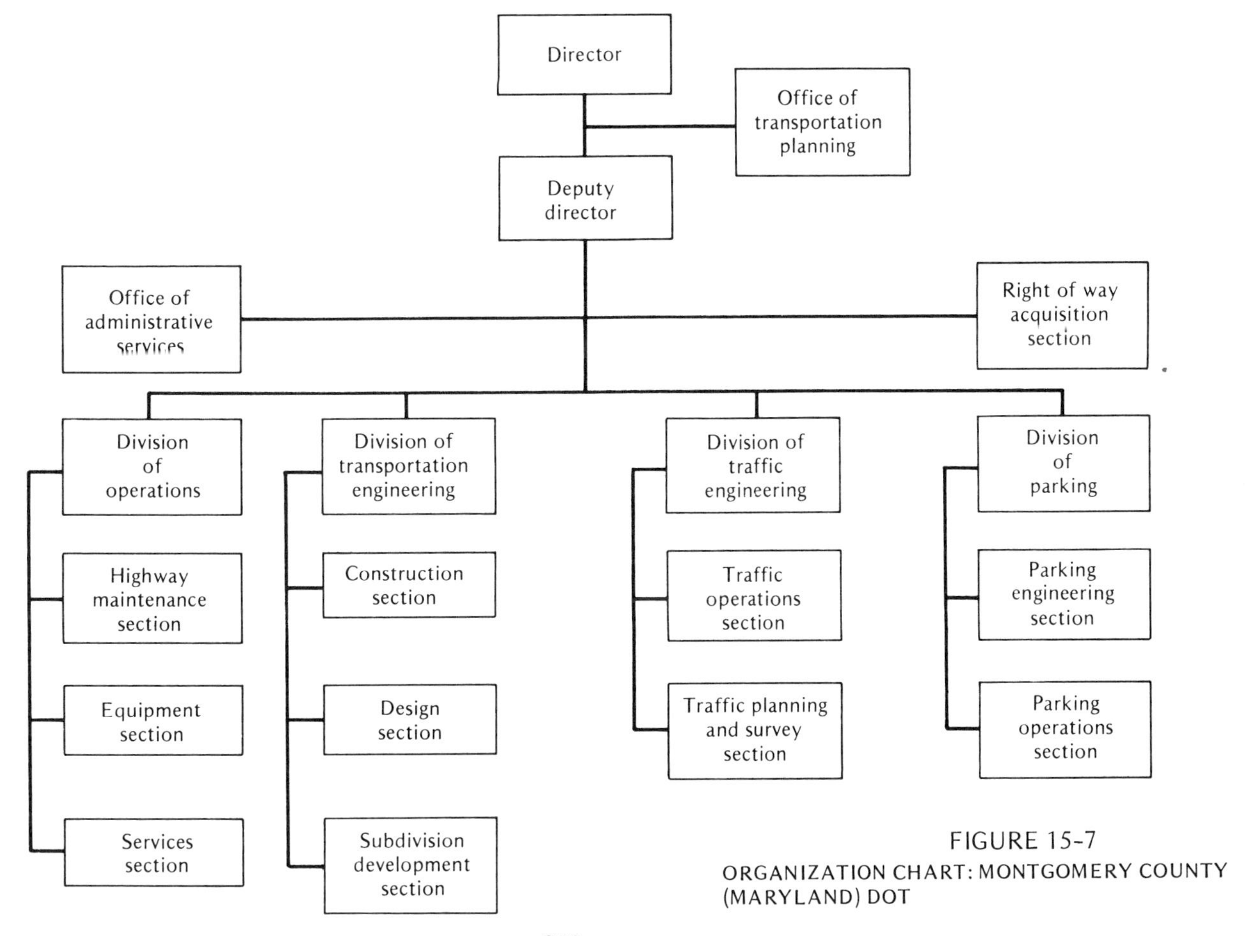

FIGURE 15-7

ORGANIZATION CHART: MONTGOMERY COUNTY (MARYLAND) DOT

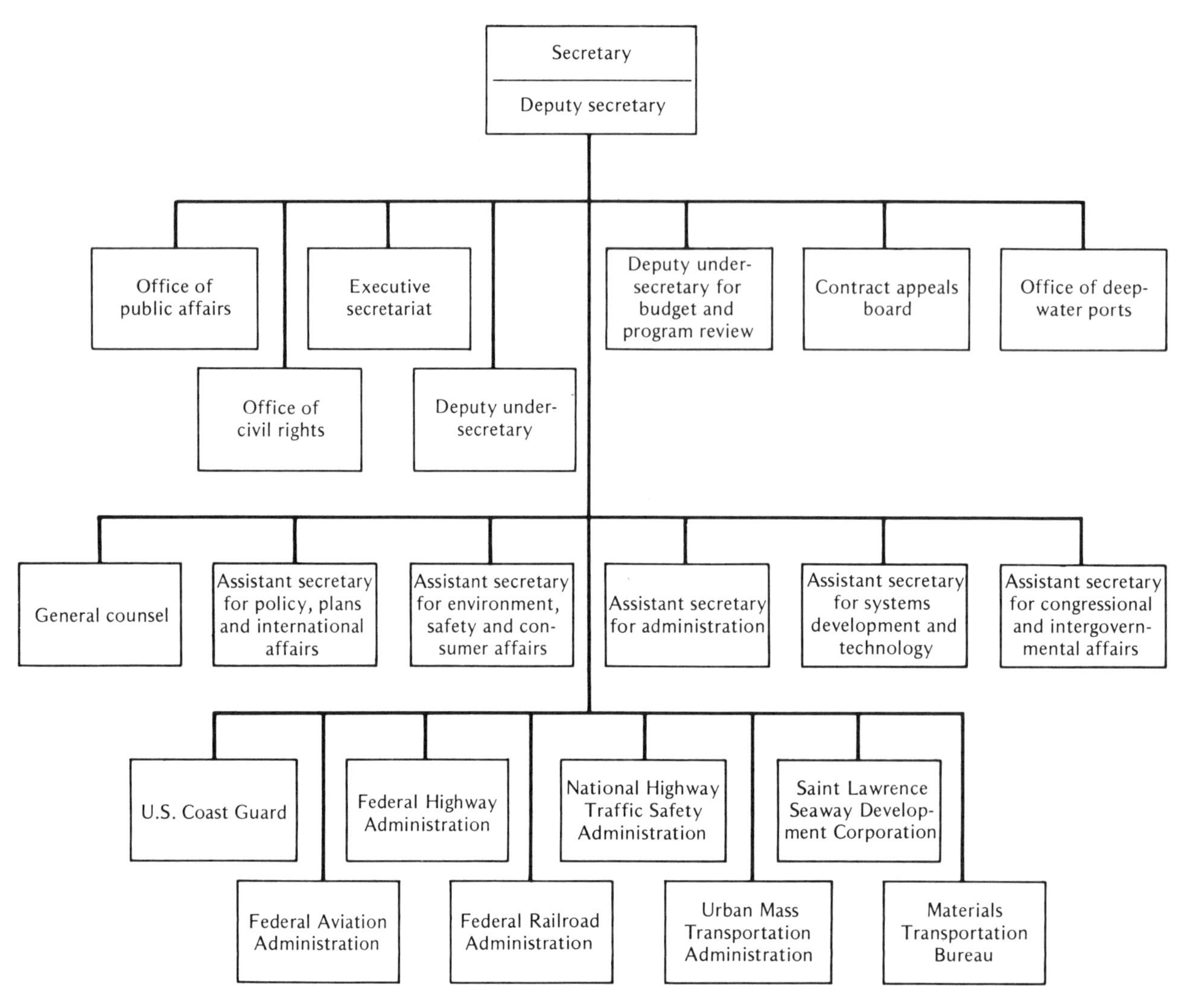

FIGURE 15-8
ORGANIZATION CHART FOR THE U.S. DEPARTMENT OF TRANSPORTATION

PARKING ADMINISTRATION AND FINANCE

The transportation engineer may find himself concerned with the administration and financing as well as operation of parking facilities. Design and operations were discussed in Chapters 8 and 10. There has been a trend toward considering the off-street parking problem as one in which the transportation engineering unit has a responsibility. The combined curb and off-street system raises some unique problems in street management because of the intense feelings generated by parking problems as well as the mixed private-public responsibilities in parking. Some of these problems are considered in this section.

To ensure street space for traffic movement, effectively distributed off-street parking is essential to provide terminal facilities for the automobile. A useful way to classify off-street parking is by type of ownership and operation.

Ownership and Operation

Private interests have been able to "fully" meet parking demands in very few cities. However, in shopping centers,

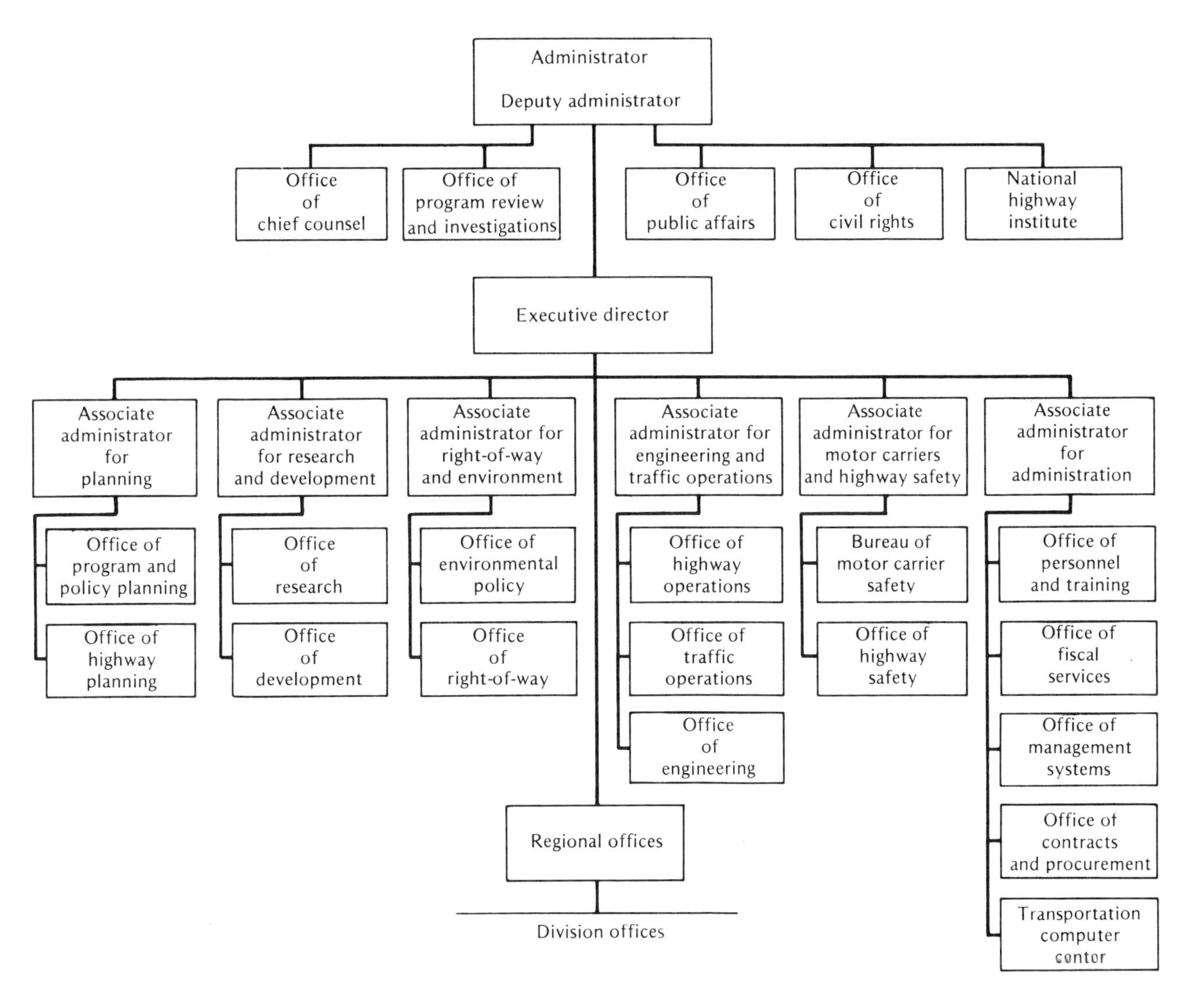

FIGURE 15-9
ORGANIZATION CHART FOR THE FEDERAL HIGHWAY ADMINISTRATION

supermarkets, etc., parking is fully provided by private enterprise. Private facilities are generally located in areas of maximum demand to ensure a profitable venture. Organizations of businessmen have been successful in providing and operating parking facilities through corporations and leasing the facilities to commercial operators. Some businessmen have concluded that, although off-street parking is a costly venture, failure to provide terminal facilities may be more costly in the long run.

Combined ownership-operation is a form of municipal subsidy that furnishes an appreciable part of parking capacity in downtown areas. It ranges from city-owned lots leased to private operators, to city-financed multistory parking structures operated by the highest bidding commercial enterprise on a long-term lease. The advantages of private enterprise are combined to some extent with those peculiar to municipalities, such as eminent domain, financing, absence of real estate taxes, and permanency.

Municipal acquisition and operation of off-street parking facilities have resulted largely from the failure of private enterprise to solve the parking problem satisfactorily. A city can offer minimum-cost parking facilities for the public, since profits are not realized. Also, a parking system planned to meet a community problem of parking

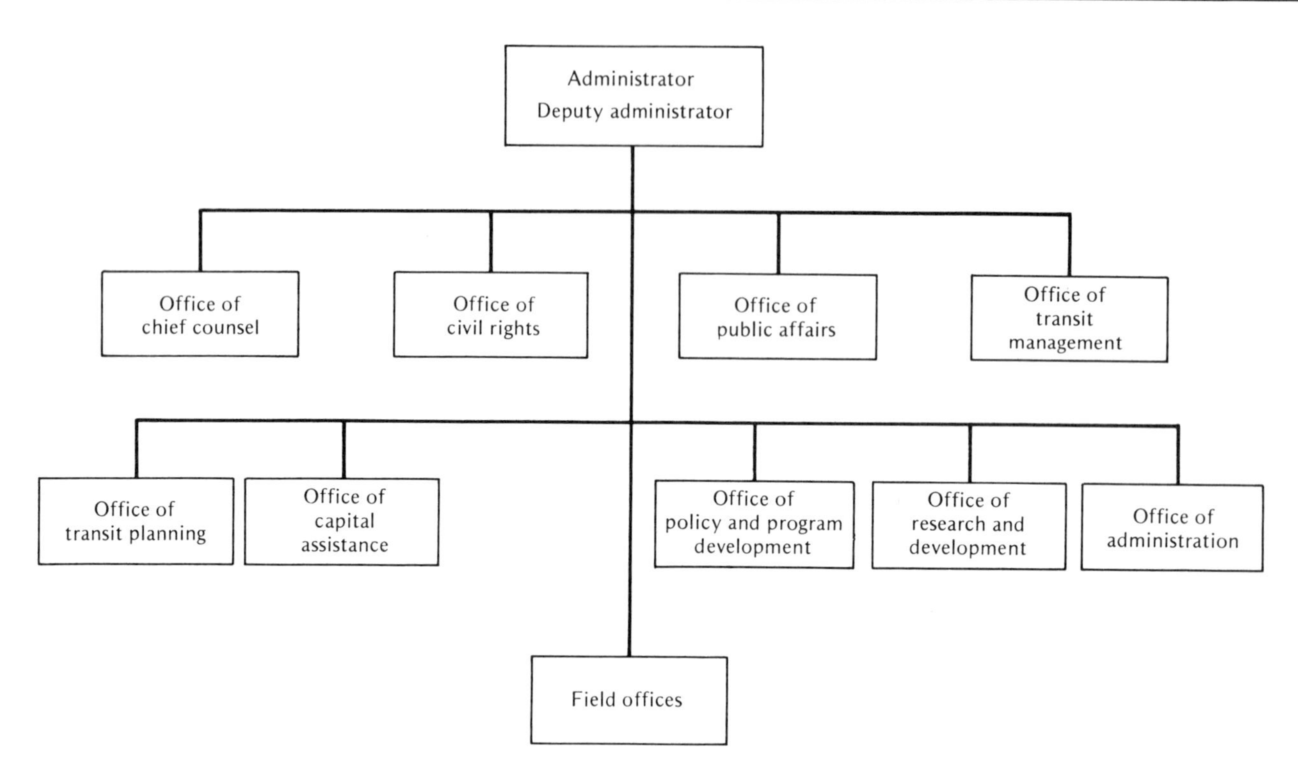

FIGURE 15-10
ORGANIZATION CHART FOR THE URBAN MASS TRANSPORTATION ADMINISTRATION

shortages can be formulated with due consideration given to the distribution of traffic movement. The principal objection made to municipal ownership is that government is in competition with private enterprise.

Off-Street Parking Administration

City governments have assumed the responsibility of administering their off-street parking in one of four ways:

1. Placing the function in an existing department, such as Streets or Traffic Engineering.
2. Setting up a separate municipal department of off-street parking.
3. Appointing a parking board or commission to coordinate the efforts of other city departments.
4. Creation of an autonomous parking authority.

There are, of course, advantages and disadvantages to each of the administrative organizations. The subdepartment is the simplest way to create an organization, but the added function is often subordinated to the department's primary activities.

The separate department organization can attract capable management personnel through its prestige, salary, and authority. However, political pressure and uncertain tenure are not removed by creation of a department.

The parking board or commission is intended to be an advisory-type function. It is composed of civic-minded citizens motivated by the public interest. Although the board may voice public opinion and can act as a buffer between the city administration and the parking function, its powers to act are very limited, and the functions of planning and execution are separated.

A parking authority is a separate, legal entity, completely autonomous in the planning, financing, construction, and operation of a city or district off-street parking program. Created by the city, it combines public responsibility with the business initiative and efficiency of private enterprise. The authority has the power to finance operations by revenue bonds but is controlled by a governing board appointed by the mayor or council. The main disadvantage of an authority is its immunity from city government. Its autonomy may bring coordination difficulties with other city activities related to traffic and parking.

All of the methods mentioned for administering an off-street parking program have been used successfully. Local circumstances will largely determine the best choice.

Methods of Finance

Private financing of parking developments is very little different from the financing of any business. In city programs, a variety of workable financing plans have been used successfully. Parking projects include land acquisition, facility construction, and operation.

A pay-as-you-go financing plan using current budget expenditures has been used by some cities by carefully working the costs into the city budget over a period of years. This plan is most applicable for modest developments, where land can be obtained at reasonable cost.

General obligation bonds have been used for financing parking improvements, pledging the full faith and credit of the municipality to insure the bonds. This method is subject to charter limitations on indebtedness. The debt-service charges and amortization of the bonds are spread equally over the property on the city tax rolls. Opponents of general obligation bonds for parking facilities argue that downtown merchants receive benefits out of proportion to their taxes.

Revenue bond financing is often used for public parking improvements with the income from the facility pledged to retire the bonds. Such bonds do not constitute an obligation on the city's general fund. The interest charges are higher than for general obligation bonds. In some cases, local merchants and property owners buy a bond issue carrying low interest rates. In other cases, the income from on-street parking meters is pledged to increase the security of the revenue bonds.

Where state law permits, many cities and counties have created special assessment districts to finance parking improvements. This system of financing attempts to proportion parking development costs among the nearby properties that benefit from the improvement. The difficulty with such a method is that the degree of benefit varies, and equitable bases for determining each property owner's share of the cost are difficult to establish. Some of the bases that have been used are:

1. Assessed valuation of the property.
2. Front footage of the property.
3. Area of the property.
4. Nearness to the improvement.
5. Anticipated benefits from the improvement.
6. Some combination of part or all of the above measures.

Another method of financing that does not depend on the city general fund is to provide off-street parking facilities with funds from other parking operations, including on-street meters. In this method a portion of the gross revenues is set aside in a fund for acquisition of additional sites and necessary improvements. However, this method suffers from the same weakness of other pay-as-you-go plans in that a long time is required to accumulate funds while the parking problem continues.

Other methods used to finance off-street parking facilities include using income from city-owned public utilities and the leasing of sites for long terms with a percentage of the revenues pledged as rent. Further discussion of parking principles and operation can be found in a 1971 publication of the Transportation Research Board [Ref. (3)]. A 1977 supplement to this publication deals with parking revenue control and points out the problems that occur with different methods of financing of off-street parking [Ref. (4)].

TRANSPORTATION DEPARTMENT ADMINISTRATION

The administration of a traffic (transportation) engineering department or division of an engineering or public works department should be of much interest to the transportation engineer, for some of the transportation engineers will eventually become administrative heads of such departments or divisions. A typical transportation engineering department head will have the responsibility for administering the various functions listed in the first section of this chapter. Since the activities indicated in that section are either briefly described there or have been discussed in previous chapters of this book, some additional functions that are often regarded as nonengineering will be discussed in this section. These functions include: (1) public relations and contacts, (2) intragovernmental and intergovernmental cooperation, (3) records and files, (4) budgets, (5) traffic/transportation committees, and (6) highway safety program elements.

Public Relations and Contacts

One of the most important yet least emphasized aspects of transportation engineering is public relations. Almost everyone in a community is affected by the transportation engineer's decisions, since they have far-reaching effects. Thus, it is necessary that he communicate clearly and effectively with the community. In addition to dealing with the general public, the city transportation engineer must frequently defend his ideas and decisions to elected officials.

There are numerous ways in which communication with the public can be accomplished. One of the most effective ways is by attending meetings (especially to speak) of local organizations and interest groups, such as:

- Parent-teacher associations
- Civic associations
- Neighborhood planning boards or advisory councils
- Various city commissions (planning, senior citizens, etc.)
- Local service groups (Chamber of Commerce, Lions Club, Civitans, Rotary, etc.)

One-way communications to inform the public of a new facility or revised traffic plan, etc. can use the following:

- Press releases
- News media: television, radio, newspapers
- City newsletters
- Local-interest newspapers
- Public meetings

In communicating with local officials or the governing body of the jurisdiction, the transportation engineer must be prepared to present factual information to support his recommendations:

- Written reports
- Charts
- Slides
- Other visual aids

A significant percentage of a traffic engineer's time may be spent in responding to citizen complaints, and in order to be consistent and to document the complaint, a standard form for recording is recommended. Such a form is shown in Fig. 5-11, which allows recording of the date; name, address, and telephone number of the complainant; results of studies and investigations; and recommendations. The most common complaints include speeding, requests for signals or stop signs, and parking. A policy should be adopted to conduct studies and other activities so as to keep ahead of the public, thus minimizing complaints.

Governmental Relations

The traffic or transportation engineer must cooperate with several other agencies within his own jurisdiction (e.g., the city government). Such intragovernmental cooperation includes working with the following agencies:

- Police department
- Legal staff
- Auditor/comptroller
- Planning departments
- Transit manager
- Public works department (possibly other units within the department)

Cooperation can be achieved in several ways, including interagency committees, department head (or division chief) conferences, etc. However, personal contact and persuasion are extremely important both at supervisory and working levels.

The other category of governmental relationships involves cooperation with nearby jurisdictions and with state and federal agencies on such matters as planning, finance, design, construction, operations, and on occasion, maintenance. Cooperative arrangements vary from informal agreements and committees to formal agreements (e.g., UMTA grant agreement) with specific agencies. The formal agreements may be established on a continuing basis, such as the metropolitan planning organization (MPO) agreement with the FHWA, UMTA, state agencies, and local jurisdictions to continue the transportation planning process. Other formal agreements may be for specific projects, such as the expansion of an airport terminal or the purchase of new buses.

Records and Files

In the course of the traffic or transportation engineer's day many pieces of information are generated, and some

Date received ______________

Name ______________________________
Address ______________________________
Phone No. ______________ Home ____________ Work ____________

Nature of complaint:

Necessary studies to be conducted:

Results:

Recommendations Date __________ Forwarded to __________

Final Action Date __________

FIGURE 15-11
STANDARD COMPLAINT FORM

of it needs to be preserved. Recording of complaints was discussed above. Other records are necessary for the following purposes:

- *Operations.* For continuing reference, volume data, speed studies, designs, signal studies and plans, signing studies, etc.
- *Legal purposes.* For attorney requests, agency legal needs, etc.
- *Maintenance.* For replacement of signs and markings, signal timing, lighting, etc.
- *Design.* For checking implementation of a project and for answering complaints.
- *Complaints.* For timely answering of complaints, records of past studies, etc. can be invaluable.

The types of files that should be maintained include:

- Correspondence
- Work orders
- Designs (geometric and construction)
- Signs and markings
- Signals (timing, locations, types of equipment)
- Accident records
- Traffic volumes
- Other traffic study data (speed, delay, etc.)
- Street lighting (location, type, etc.)
- Other

Depending on the size of jurisdiction, many such files will or should be computerized, especially accident records, signs and markings, inventory, and traffic volumes.

Budgets

Every jurisdication has financial constraints, and various functions must be performed within a certain restricted budget. A budget delineates what money can be spent for which function, and places maximum limits on the amount of money available for any particular function. Since the final decision on how much money is allotted to transportation functions is made by the elected officials, it is very important that the transportation engineer present a well-prepared, well-documented, complete, and realistic budget for their consideration. All programs requiring significant expenditures and all essential programs must be identified and documented, as must the exact nature (specific project) of improvements needed. Generally, the transportation engineer must establish priorities within the recommended budget, considering such criteria as cost, safety, and benefits. An economic analysis of alternatives is highly desirable.

Traffic/Transportation Committees

The Model Traffic Ordinance provides for a traffic commission as follows [Ref. (5)]:

1. Membership: traffic engineer, chief of police, chairman of the traffic committee of the city council, representatives from the city engineer's and city attorney's offices, and other city officials or representatives of unofficial bodies.
2. Duties: coordinate traffic activities, supervise preparation and publication of traffic reports, receive complaints, and make recommendations.

State statutes or local ordinances may also provide for the creation of either traffic or transportation committees or commissions. These committees/commissions can assist the traffic/transportation engineer by acting as advisory bodies; by disseminating data, reports, and other information to the public, and, in other ways, serving as a public relations arm of the traffic/transportation engineer. Very often the traffic/transportation committee of the city council or of the Chamber of Commerce can perform many public relation functions, and the transportation engineer should establish good rapport with these two units.

Highway Safety Program

The head of the transportation/traffic engineering unit is concerned with various aspects of the Federal Highway Safety Program, particularly with Standard 13, Traffic Engineering Services. One of the major efforts, in this regard, of the local transportation engineer is to coordinate the activities at the local level with the programs and the activities at the state level (the governor's highway safety representative).

Much of the highway safety activity at the local level involves public relations and a continuing learning process about funding eligibility, amounts, etc. Some of the specific technical activities that are safety-related and in which the transportation engineer should be involved are as follows:

1. Operational surveillance. Traffic reports from a traffic control center, helicopter, or other source.
2. Training. Many safety-related seminars, conferences, short courses, etc. are eligible for federal safety funds, and have greatly enhanced the training available to the transportation engineering unit.

Finally, Standard 13 has given the transportation/traffic engineer a significant increase in political/financial power in dealing with elected officials. Since many of the traffic engineering activities are safety-related, the administration of the highway safety program elements has been a welcome requirement.

REFERENCES

1. Baerwald, John E., editor, *Transportation and Traffic Engineering Handbook*. Englewood Cliffs, N.J.: Prentice-Hall, Inc., 1976.
2. Transportation Research Board, *Issues in Statewide Transportation Planning*, Special Report 146. Washington, D.C.: Transportation Research Board, 1974.
3. Transportation Research Board, *Parking Principles*, Special Report 125. Washington, D.C.: Transportation Research Board, 1971.
4. Transportation Research Board, Circular 184, *Parking Revenue Control*. Washington, D.C.: Transportation Research Board, June 1977.
5. National Committee on Uniform Traffic Laws and Ordinances, *Uniform Vehicle Code and Model Traffic Ordinance*. Washington, D.C.: 1968 (with 1972 and 1976 revisions).

APPENDIX

Selected References in:

Air Transportation
Railroad Transportation
Water Transportation
Pipeline Transportation
Pedestrian Transportation

General References

Harmon, G.M., ed., *Transportation: The Nation's Lifelines.* Washington, D.C., Government Printing Office, 1968, 208 pp. (Industrial College of the Armed Forces, National Security Management, Vol. 7).

Hay, W.W., *An Introduction to Transportation Engineering.* New York: John Wiley & Sons, 1961, 505 pp.

Hennes, R.G., and Martin Ekse, *Fundamentals of Transportation Engineering,* 2nd ed. New York: McGraw-Hill Book Co., 1969, 613 pp.

Paquette, R.J., Norman Ashford, and P.H. Wright, *Transportation Engineering: Planning and Design.* New York: Ronald Press, 1972, 760 pp.

U.S. Department of Transportation, *National Transportation: Trends and Choices (To the Year 2000).* Washington, D.C.: U.S. Government Printing Office, 1977, 412 pp.

Air Transportation

Federal Aviation Administration, *Advisory Circular Series* (on various topics, in several publications per year—some of which are printed by the U.S. Government Printing Office, Washington, D.C.).

Horonjeff, Robert, *Planning and Design of Airports,* 2nd ed. New York: McGraw-Hill Book Co., 1975, 460 pp.

Howard, G.P., ed., *Airport Economic Planning,* Cambridge, Mass.: MIT Press, 1974.

International Civil Aviation Organization, *Aerodromes: International Standards and Recommended Practices,* 7th ed., Montreal: International Civil Aviation Organization, 1976, 125 pp.

Schriever, B.A., and W.W. Seifel, eds., *Air Transportation 1975 and Beyond: A Systems Approach,* Cambridge, Mass.: MIT Press, 1968.

University of Toronto, *Readings in Airport Planning.* Toronto: Center for Urban and Community Studies and the Department of Civil Engineering, 1972, 473 pp.

Rail Transportation

American Railway Engineering Association, *Manual of Recommended Practice,* 1965.

Hay, William W., *Railroad Engineering.* New York: John Wiley & Sons, 1953.

Inglis, R.A., and J.P. Paton, *An Introduction to Railway Engineering.* London: Chapman & Hall, 1953, 200 pp.

Railway Engineering and Maintenance Cyclopedia. New York: Simmons-Boardman Publishing Corp., 1955.

Raymond, William G., Henry E. Riggs, and Walter C. Sadler, *Elements of Railroad Engineering,* 6th ed. New York: John Wiley & Sons, 1947.

U.S. Army Transportation School, *Railway Construction and Revitalization.* Fort Eustis, Va.: 1968, 66 pp. (Reference Text 602)

U.S. Department of Transportation, *Rail Planning Procedures Report.* Washington, D.C.: Federal Railroad Administration, Sept. 1975, 196 pp.

Water Transportation

American Association of Port Authorities, *Port Planning, Design and Construction,* 2nd ed. Washington, D.C.: American Association of Port Authorities, 1973, 514 pp.

American Society of Civil Engineers, *Pile Foundations and Pile Structures,* Manual of Engineering Practice No. 27, 1946.

Big Load Afloat. Washington, D.C.: American Waterways Operators, Inc., 1965.

Bruun, Per. *Port Engineering,* 2nd ed. Houston: Gulf Publishing, 1976.

Bulson, P.S., "The Theory and Design of Bubble Breakwaters," *Proceedings of Eleventh Conference on Coastal Engineering.* American Society of Civil Engineers, September 1968.

Cornick, Henry F., *Dock and Harbour Engineering.* London: Charles Griffin and Co., Ltd., 1959, Vol. 2.

Lee, Theodore T., "Design Criteria Recommended for Marine Fender Systems," *Proceedings of Eleventh Conference on Coastal Engineering.* American Society of Civil Engineers, September, 1968.

Minikin, R.R., *Wind, Waves, and Maritime Structures,* 2nd ed. London: Charles Griffin and Co., Ltd., 1963.

1967 Inland Waterborne Commerce Statistics. Washington, D.C.: American Waterways Operators, Inc., April 1969.

Quinn, Alonzo, *Design and Construction of Ports and Marine Structures,* 2nd ed. New York: McGraw-Hill Book Co., 1972, 611 pp.

U.S. Army Corps of Engineers, *Shore Protection, Planning and Design,* Technical Report No. 4, 3rd ed., 1966.

Pipeline Transportation

American Society of Civil Engineers, *Report on Pipeline Location.* New York: American Society of Civil Engineers, 1965. 74 pp. (Manuals and Reports on Engineering Practice No. 46).

Bell, H.S., ed., *Petroleum Transportation Handbook.* New York: McGraw-Hill Book Co., 1963.

See *Transactions,* ASME, various years.

See *Transactions,* ASCE, Journal of the Pipeline Division, various issues.

See current periodicals such as:
Pipeline Industry
The Oil and Gas Journal

Pedestrian Transportation

Tough, J.M., and C.A. O'Flaherty, *Passenger Conveyors: An Innovatory Form of Communal Transport.* London: Ian Allan, 1971, 176 pp.

Compiled by

Elizabeth R. Carter

Associate Librarian
Institute of Transportation Studies
University of California at Berkeley

and

Everett C. Carter

Professor of Civil Engineering and
Director, Transportation Studies Central
University of Maryland

Index

A

Acceleration:
  motor vehicles, 26–29
  rapid transit trains, 195
Access control, 85
Accidents:
  alcohol as a factor, 40–41
  characteristics, 40–42
  prevention, 42–43
  rates, 42
  record analyses, 65, 76–77
  studies, 76–77
  transit, 41–42
  trends, 39–40
  types, 40–42
  victims, 40–41
Administration:
  transportation engineering, 237–250
  urban transportation planning, 210
Aesthetics (*see* Esthetics)
Air transportation, 8, 251–252
Alcohol as an accident factor, 40–41, 140–141
Alinement (also spelled alignment):
  coordination of horizontal and vertical, 97–98
  horizontal, 94–97
  vertical, 88–94

Automobiles:
acceleration and deceleration, 26–29
dimensions, 25–26
emission standards, 29, 31
energy efficiency, 10, 12
operating cost, 29, 31
Average Daily Traffic (ADT), 82, 83, 85

B

Barricades, 159
barricade warning lights, 159
Barriers:
median, 101
pedestrian, 117
Benefit-cost analysis, 231
Bicycles:
accidents, 41–42
density of traffic stream, 56–57
dimensions, 39
energy efficiency, 12
flow rate, 56–57
role in transportation, 37
speed, 39, 56–57
Bicyclists:
travel patterns, 38–39
trip purposes, 37–38
Border areas, 102
Budgets, 249–250
intersection, 169–172
lanes, 180–181
Buses (*see also* Transit systems):
accidents, 41–42
articulated, 190, 199
capacity, 198–199
dimensions, 26, 190, 199
electric (trolley coaches), 193
energy efficiency, 12
interior, 191
maintenance facilities, 193
role in local transit, 188–189
Bus roadways (busways), 188–191
Bus stops, 187–188, 190–192, 198–199
Bus systems (*see* Transit systems)
Bus terminals, 132–133, 190, 192

C

Capacity:
bicycle traffic stream, 57
Capacity: (*Contd.*)
bus stop, 198–199
highways, 54–58
transit systems, 198–200
Captive transit riders, 186
Car pools, 189, 219
Central Business District (CBD), 16–17, 19–22
Centrifugal force, 91, 94–95
Channelization, 118–119
Charter transit service, 198
Choice transit riders, 186
Cities (*see* Urban areas)
Citizen participation in planning, 210, 221
Climbing lanes, 89, 102
Club bus service, 198
Coefficient of friction, 28–29
Collision diagram, 76–77
Committee, traffic/transportation, 250
Communications, with the public, 248
Commuter railroads (*see* Urban railroads)
Computer application, 66
Concentration (*see* Density)
Condition diagram, 76–77
Conflicts, intersectional, 107–108, 147
Continuing transportation studies, 221
Cordon line counts, 68, 70
Correlation analysis, 230
Cross sections, 98–102
lane widths, 98
pavement slope, 98
side slope, 99
Crosswalks, 117
Curb parking, 17, 21, 122–123
Curbs, 99–101
barrier, 100
design, types, 100
mountable, 100
Curb use control, 180
Curvature, circular, 94–95

D

Data analysis and interpretation, 65
Data collection and sampling, 64–65
Data storage and retrieval, 65–66
Deceleration:
motor vehicles, 26–29
rapid transit trains, 195
Delay:
signalized intersection, 171, 173

Delay: (*Contd.*)
studies, 68, 71, 74
Density:
bicycle traffic stream, 57
contours, 52-53
motor vehicle traffic stream, 45, 52-55, 57
pedestrian traffic stream, 58
studies, 71-72
Design designation, 85
Design speeds, 85
Design standards, 228-229
Design vehicles, 83-84
Dial-A-Bus, 198
Ditches, 99
Drainage, 102
Drivers, 32-33, 139-141
age, 32-33, 139-141
alcohol and drugs, 40-41, 140-141
controls, 139-144
driving strategy, 33-34
education, 142-143
licensing, 139-141
observation of regulations, 74
point system for violations, 142-143
reactions, 32-33
route selection studies, 74
vision, 32-33
Driveways, 110-112
Driving task, 33-34

E

Economic analyses, 230-231
Economic forecasts, 229
Education (*see* Drivers, education)
Efficiency in transportation systems, 7-8
Energy efficiency and use, 10-12, 220-221
Enforcement, rules of the road, 140-143
Engineering economic analyses, 229, 231
Environmental impact studies, 220-221
Esthetics, 102-103
Evaluation of transportation alternatives, 231-232
Experimental studies, 65, 78

F

Fencing, 102
Financial responsibility, motor vehicles, 138
Financing:
studies, 229
Financing: (*Contd.*)
transit systems, 204-206
urban transportation, 219
Flow rate (*see* Volume)
Forecasting, 229-230
Friction coefficient, 28-29, 94-95
Functional classification:
highways, 217-218, 226-227
transit routes, 186-187

G

Gaps in traffic streams, 45, 51-52
Garages (*see* Parking, garages)
Geometric design:
bus terminals, 132
controls, criteria, 82-86
elements, 85-98
highways, 81-105
interchanges, 110-115
intersections, 107-110, 115-118
parking facilities, 117-119, 121-132
pedestrian facilities, 117-119
streets, 107-119
truck terminals, 132-134
Glare, 103
Goals, 224-226
Goods movement studies, 212
Governmental relations, 248
Grades, 88-89
Grade separations, 107-119
Guardrail, 102
Gutter, 101

H

Headways:
motor vehicle traffic stream, 45, 51
transit systems, 199-201
Highways:
classification, 226-227
finance, 229-232
inventories, 66, 212, 227-228
planning studies, 227-228
safety program, 250
statistics (U.S.), 10, 50
transportation, 8, 10-11
Home interview studies, 75, 212
Horizontal alignment, 94-97

I

Interchanges:
  adjacent, 114-115
  cloverleaf, 113
  diamond, 111
  directional, 113-114
  ramps, 110-111, 113, 115
  types, 110-112
Intersection controls, 167-168
Intersections:
  auxiliary lanes, 117
  design elements, 115-117
  design principles, 108-110
  sight distance, 115
  turning radii, 115-116
  types, 108-109
Interview studies, 75, 212
Inventories, 66-67, 212, 227-228

J

Jitneys, 197

L

Landscaping, 102
Land use:
  forecasting models, 215-216
  inventories, 212
  plans, 216-217
Lane controls, 180-181
Levels of service, 54-59, 170-171
Lighting, 103-105
Light rail transit (*see* Semimetros)
Load factor, 172
Luminaire, 103-104

M

Markings (*see* Traffic markings)
Medians, 41, 101
Mobility in transportation systems, 7-8
Modal split analysis, 214-215
Model traffic ordinance, 138-139, 250
Models, systems analyses, 223-226
Motor vehicles:
  acceleration and deceleration, 26-29
  accidents (*see* Accidents)
  dimensions, 25-26
Motor vehicles: (*Contd.*)
  driver eye height, 29
  operating costs, 29, 31
  safety standards, 29
  use (U.S.), 12
  volume (*see* Volume)
  weight, 25-26, 145
  wheel and axle loads, 26, 145

N

National Committee on Uniform Traffic Laws and Ordinances, 138
Needs studies, 228-229

O

Observance-of-regulations studies, 74
Observation studies, 65, 67-74
One-way streets, 181-182
Operating costs:
  automobiles, 29
  bus systems, 205
Organizations, transportation, 238-246
Origin-destination studies, 75, 212

P

Paratransit, 196-198
Parking:
  administration, 244-247
  central business district, 17, 19-22
  curb, 17, 21, 122-123
  demand, 17, 21
  dimensions, 122-126
  duration, 17, 22
  garages, 125-132
  inventory, 66
  off-street facilities, 122, 124-130
  on-street (curb), 17, 21, 122-123
  ownership and operation, 244-246
  restrictions (curb), 180
  studies, 66, 72-73, 75
  supply, 17, 20
  surface lot design, 125-126
  surface lot versus structure, 127
  turnover, 19, 22
  walking distances, 19, 22
Passenger cars (*see* Automobiles)

Passengers:
boarding and alighting at transit stops, 189
miles of travel by mode (U.S.), 9
peaking patterns on transit systems, 203–204
travel on highway system (U.S.), 9–10
Passing sight distance (*see* Sight distances)
Peak hour factor, 172
Peaking patterns, 15–16, 18, 203–204
Pedestrians:
accidents, 40–42
alcohol, 41
crosswalks, 117
density of traffic stream, 34–36, 58
dimensions, 34–35
flow rates, 58
gap acceptance, 34, 37
references, 252
safety islands, 119
space requirements, 36
speeds, 34, 58
trip-making, 34
volume, 56, 58
walking distances, 19, 22, 37–38
Pipelines, 9, 252
Population forecasts and studies, 216, 229
Programming, 232, 233
Public relations and hearings, 221, 248

Q

Queueing, 60, 124, 170

R

Railroad-grade-crossing accidents, 42
Railroads, 8, 196, 252
Rapid transit (*see also* Transit systems), 41–42, 193–201
Rate-of-return analysis, 231
Rating techniques, 232–233
Records and files, 248–249
Regulations:
authority, 138
drivers, 137–144
vehicles, 144–145
Resistances to motion, 27
Rest areas, 102
Reversible lanes and streets, 180, 182
Roadside interview studies, 75
Rules of the road, 137–138, 148–149
Rural accidents, 40–42

S

Safety:
highway, 137–145
motor vehicle standards, 29, 139
program standards, 139
School crossing studies, 74
Semimetros, 193, 199
Shoulders, 98–99
Sight distances:
horizontal, 96–97
intersectional, 115
passing, 85–88
stopping, 85–86
Signals (*see* Traffic signals)
Signs (*see* Traffic signs)
SIGOP, 178
Silhouette, 103
Skidding friction coefficient, 28–29
Spacing of vehicles in traffic stream, 45, 52
Speed:
analysis methods, 46–47
bicycles, 57
motor vehicles, 45–47, 52, 54–58
pedestrians, 58
spot studies, 71
transit vehicles, 200–201
Spiral curves, 95–96
Staffing, 240
Statewide transportation planning, 223–234
Statistical studies, 65, 76
Stop signs, 168
Streetcars, 193, 199
Studies (*see* specific study listings)
Surveillance and control, 178–179, 196, 238
Symbol signs, 152–155

T

Tapers, 117
Taxis, 75, 197–198, 212
Terminals:
bus, 132–133, 190, 192
parking (*see* Parking, garages; Parking, off-street facilities)
rapid transit, 194

Terminals: (*Contd.*)
transportation plans, 219
truck, 132-134
Time-series analysis, 230
Time-space diagram, 175, 177
Traffic assignment, 215
Traffic conflict studies, 72
Traffic control devices, 66-67, 102, 147-165
Traffic delineators, 158
Traffic density (*see* Density)
Traffic flow (*see* Volume)
Traffic markings:
inventories, 66-67
letters, 160
materials, 158-159
reflectorization, 159
types, 157-158
Traffic metering, 178-179
Traffic operations, 147-183
Traffic signals:
computer control, 178
criteria, 169
detectors, 163-164
equipment, 162, 175, 178
history, 161
installation, 168
interval, 161
inventories, 66-67
maintenance, 168
offset, 163
phase, 161
pretimed, 163, 175, 178
pretimed operation, 172
purpose, 168
special signal controls, 179
traffic actuated, 163
traffic actuated operation, 174, 178
warrants, 169
Traffic signal systems:
alternate, 115
linear, 174
network, 175
progressive, 174-175
simultaneous, 175
Traffic signal timing, 169, 174-177
Traffic signs:
design, 149-152
illumination, 154
installation, 156
inventories, 66-67
Traffic signs: (*Contd.*)
lettering, 152-156
maintenance, 156-157
materials, 154
message legibility, 152, 154
overhead, 152, 156
reflectorization, 157
uniformity, 149-155
Traffic stream flow, 45-61
Traffic surveillance and control (*see* Surveillance and control)
Traffic volumes, 82-83, 228
Training for transportation engineers, 250
Transit systems:
accessibility, 187-188
accidents, 41-42
area coverage, 37, 187-188
capacity, 198-200
economic characteristics, 203-206
fares, 201-202
headways, 199-201
inventories, 67, 212
labor, 203-204
local transit, 188-193
maintenance facilities, 193, 195-196
major activity center service, 198
management, 203-206
networks, 185-187
operation, 198-203
paratransit, 196-198
peaking patterns, 203-204
planning, 218-219
rapid transit, 193-196
right-of-way, 186-187, 189-190, 193-194
routes, 185-187, 189-190, 193-194
safety, 41-42, 202-203
speed, 200-201
stops (*see* Bus stops; Rapid transit)
studies, 72, 74-76
terminals (*see* Bus terminals; Rapid transit)
vehicles (*see* Buses; Rapid transit)
Transportation, 3, 7
Transportation engineers, 3-4
Transportation Improvement Program (TIP), 221
Transportation planning (*see* Statewide transportation planning; Urban transportation planning)
Transportation System Management (TSM), 182-183, 219
Transportation systems:
characteristics, 7-10
energy efficiency, 10-12

Transportation systems: (*Contd.*)
planning elements, 224–227
statistics (U.S.), 9
TRANSYT, 178
Travel demand forecasting, 229–230
Travel demand formulation, 213–215
Travel time, 45–47, 55
Travel time and delay studies:
highways, 68, 71
transit systems, 74
Trip distribution, 213–214
Trip generation, 213
Trip lengths, 12, 14
Trip purpose:
classification, 213
relation to walking distance, 19, 22
urban areas (U.S.), 15, 17
Trolley coaches, 193
Trucks:
acceleration and deceleration, 26–29
dimensions, 26, 84
emergency stopping distances, 29–30
routes, 182
terminals, 132–134
turning movements, 168–170
turning paths, 83–84
use surveys, 75, 212
weight, 26

U

Ubiquity of transportation systems, 7–8
Unbalanced flow, 181
Uniform Vehicle Code, 138–143, 148–149, 250
Urban areas:
central business district (*see* Central Business District)
peaking patterns, 15–16, 18
travel characteristics, 11–17
trip making, 12–16
Urban railroads, 196
Urban transportation planning, 209–221
Utilities, 102

V

Vehicle equipment, 144
Vehicle inspection, 144–145
Vehicle registration, 144
Vehicles (*see* Automobiles; Buses; Motor vehicles; Rapid transit; Trucks)
Vertical alignment and curves, 88–94
Volume:
Annual Average Daily Traffic (AADT), 67
bicycles, 57
cyclic patterns, 48–50
design hour, 82–83
directional imbalance, 51
hourly person trips, 16, 18
motor vehicles, 45, 47–58
pedestrians, 56, 58
studies, 67–70
transit passengers (maximum), 200

W

Walking distances, 19, 22, 37–38
Water transportation, 8–9, 252
Weaving on freeways, 113–114

Photo Credits

The following photos were provided as a courtesy of the listed agency:

(opposite title page)	Transportation Research Board
(opposite Foreword)	Oregon Department of Transportation
(opposite Acknowledgments)	Washington Metropolitan Area Transit Authority
(Chapter 1)	San Francisco Bay Area Rapid Transit District
(Chapter 2)	Institute of Transportation Engineers
(Chapter 3)	California Department of Transportation
(Chapter 4)	Arizona Department of Transportation
(Chapter 5)	Highway Users Federation for Safety and Mobility
(Chapter 6)	Virginia Department of Highways and Transportation
(Chapter 7)	Ontario Department of Highways
(Chapter 8)	Highway Users Federation for Safety and Mobility
(Chapter 9)	Arizona Department of Transportation
(Chapter 10)	Virginia Department of Highways and Transportation
(Chapter 11)	Institute of Transportation Engineers
(Chapter 12)	Washington Metropolitan Area Transit Authority
(Chapter 13)	Washington Metropolitan Area Transit Authority
(Chapter 14)	Highway Users Federation for Safety and Mobility
(Chapter 15)	Oregon Department of Transportation